Embedded Cooling of Electronic Devices

Conduction, Evaporation, and Single- and Two-Phase Convection

WSPC Series in Advanced Integration and Packaging

ISSN: 2315-473X

In the 21st century, innovative integration and packaging drives success in the marketplace, establishing the "look and feel" of electronic products and their price point, and enabling differentiation and branding. The rapid development and proliferation of design methodologies, assembly techniques, and thermal solutions for components ranging from integrated circuits to photonics, and from sensors and MEMS to mixed signals and power amplifiers, drive the often frenetic pace of exploration, discovery, and innovation characteristic of the electronic industry.

This recently introduced WSPC book series on "Advanced Integration and Packaging of Electronic Systems" (AIPack), with a particular focus on 2.5D and 3D form factors, will provide a "technology commons" for the packaging community, a place where ideas can be articulated, examined, and disseminated rapidly and efficiently through both print and electronic media. It is our hope that product developers will find AIPack a convenient and reliable source of technical information; that researchers and students will be guided to the challenges and barriers facing key technologies, and that technology managers and investors will find the tools and data needed to make broad technology assessments and comparisons for emerging packaging and integration technologies.

Published

Vol. 8: *Embedded Cooling of Electronic Devices: Conduction, Evaporation, and Single- and Two-Phase Convection*
edited by Madhusudan Iyengar, Justin A. Weibel and Mehdi Asheghi

Vol. 7: *Thermoelectric Energy Conversion Devices and Systems*
by Kazuaki Yazawa, Je-Hyeong Bahk and Ali Shakouri

Vol. 6: *Solder Materials*
by Kwang-Lung Lin

Vol. 5: *MEMS Packaging*
edited by Y. C. Lee, Yu-Ting Cheng and Ramesh Ramadoss

Vol. 4: *Cost Analysis of Electronic Systems (Second Edition)*
by Peter Sandborn

More information on this series can also be found at https://www.worldscientific.com/series/wssaip

WSPC Series in Advanced Integration and Packaging — Vol. 8

Embedded Cooling of Electronic Devices

Conduction, Evaporation, and Single- and Two-Phase Convection

Editors

Madhusudan Iyengar
Google Platforms, USA

Justin A Weibel
Purdue University, USA

Mehdi Asheghi
Stanford University, USA

World Scientific

NEW JERSEY • LONDON • SINGAPORE • BEIJING • SHANGHAI • HONG KONG • TAIPEI • CHENNAI • TOKYO

Published by

World Scientific Publishing Co. Pte. Ltd.
5 Toh Tuck Link, Singapore 596224
USA office: 27 Warren Street, Suite 401-402, Hackensack, NJ 07601
UK office: 57 Shelton Street, Covent Garden, London WC2H 9HE

Library of Congress Control Number: 2023039953

British Library Cataloguing-in-Publication Data
A catalogue record for this book is available from the British Library.

WSPC Series in Advanced Integration and Packaging — Vol. 8
EMBEDDED COOLING OF ELECTRONIC DEVICES
Conduction, Evaporation, and Single- and Two-Phase Convection

ISBN 978-981-127-793-1 (hardcover)
ISBN 978-981-127-937-9 (ebook for institutions)
ISBN 978-981-127-938-6 (ebook for individuals)

For any available supplementary material, please visit
https://www.worldscientific.com/worldscibooks/10.1142/13458#t=suppl

Desk Editors: Logeshwaran Arumugam/Amanda Yun

Typeset by Stallion Press
Email: enquiries@stallionpress.com

https://doi.org/10.1142/9789811279379_fmatter

Tribute

Professor Avram Bar-Cohen was a giant in the fields of electronics packaging and cooling and a pioneer who significantly influenced academia, industry, and governmental research. Beyond his technical accomplishments and leadership, he was a fantastic mentor for individuals and the community. He fostered a culture of innovation, collaboration, and high technical achievements for several decades and founded vibrant and healthy international conferences, such as the *IEEE ITherm* and the *ASME InterPack*. Since his sudden passing in 2021, awards have been renamed in his honor, tributes have poured in, special issues of the *IEEE Transactions on Components, Packaging, and Manufacturing Technology* (TCPMT) and the *ASME Journal of Electronic Packaging* have been published, and articles have been crafted to honor his legacy. Avi's more recent passion was through his leadership of the DARPA IceCool program, which led to breakthrough research on embedded cooling of ultrahigh performance chips. We expect his achievements to be a beacon that shines a light for future progress in the field. This book was originally initiated by Avi to represent this field of embedded cooling, and we dedicate its completion to his memory.

As Avi's former student (Madhusidan Iyengar) and research collaborators (Justin Weibel and Mehdi Asheghi), we would like to offer our heartfelt reflections on Avi below.

Dr. Madhu Iyengar, Google: For me personally, Avi was a teacher, mentor, and friend over the last 25 years and whose passing I mourn and grieve deeply. Avi inspired fearlessness and total commitment toward the progress of the field, with a cheerfulness and exuberance that was infectious. He will be missed, but will continue to guide us through our memories.

Dr. Justin Weibel, Purdue University: I feel extremely fortunate to have received guidance and support from Avi early in my career, inspired by his unmatched passion for research in this area. He and his envisioned programs sparked the curiosity of many engineers and scientists who have been attracted to this field. As we all readjust, the aftershocks of his work will be long-lasting.

Dr. Mehdi Asheghi, Stanford University: Avi was a larger-than-life character and a tremendous *force* and *advocate* for our community. He was kind and supportive but, at the same time, had high expectations of the colleagues around him. For me, he was more than a friend and colleague, an inspiring character and mentor. He will be missed, but we take solace in remembering him anytime we get together with colleagues and friends.

https://doi.org/10.1142/9789811279379_fmatter

About the Editors

Madhu Iyengar is a principal engineer at Google and a lead in innovative product development and path-finding for IT hardware and physical infrastructure, including chip packages, server systems, and data centers. Madhu is also a technical lead manager with technical and managerial responsibilities and with a mission to develop and deliver excellent thermal systems, products, and advanced technologies, focusing on end-to-end, chip-to-chiller, infrastructure stack optimization and innovation opportunities in collaboration with cross-functional partners. Prior to working at Google, Madhu was a hardware architect/engineer at Facebook and a senior engineer at IBM. He has co-authored over 115 technical papers in journals and conference proceedings, holds more than 290 US patents, is an elected fellow of the American Society of Mechanical Engineers (ASME), and is a member of IEEE and ASHRAE. Madhu has served as an associate editor for the IEEE Electronics Packaging Society's *IEEE Transactions on CPMT*, the *ASME Journal of Electronic Packaging*, and the *Electronics Cooling* magazine. He has also served as a voting member of the ASHRAE TC9.9 Mission Critical

(Data Center) Facilities Committee and as the general chair of the 2016 IEEE *ITherm* conference. He is currently chair of the Thermal Technical Working Group for the IEEE Electronics Packaging Society's Heterogeneous Integration Roadmap. Madhu has a PhD in mechanical engineering from the University of Minnesota, USA, and a BE in mechanical engineering from the University of Pune, India.

Justin A. Weibel is an associate professor in the School of Mechanical Engineering at Purdue University and the director of the Cooling Technologies Research Center (CTRC), a graduated NSF I/UCRC that addresses the research and development needs of companies and organizations in the area of high-performance heat removal from compact spaces. He received his PhD in 2012 and his BSME in 2007, both from Purdue University. Dr. Weibel's research group explores methodologies for the prediction and control of heat transport to enhance the performance and efficiency of thermal management technologies and energy transfer processes. He has been a key contributor to the development of transformative cooling technologies supported by the DARPA TGP, DARPA ICECool, NAVSEA NEEC, ONR NEPTUNE, SRC CHIRP, ARPA-E ASCEND, and ARPA-E COOLERCHIPS programs, in addition to numerous sponsored research projects that transition these technologies to the industry. Dr. Weibel has supervised 26 PhD and 14 MS students, current and former, and co-authored over 190 refereed journal and conference papers. He has been recognized as an Outstanding Engineering Teacher and Outstanding Faculty Mentor in the College of Engineering at Purdue University. He recently received the 2020 ASME Electronic & Photonic Packaging Division (EPPD) Young Engineer Award and the 2021 ASME K-16 Outstanding Early Faculty Career in Thermal Management Award.

Mehdi Asheghi received his PhD degree from Stanford University, California, USA, in 1999. He completed his postdoctoral training at Stanford University in 2000. He is currently an adjunct professor with the Department of Mechanical Engineering at Stanford University, where he has co-advised/supervised 25 PhD students and is involved in more than 10 major research project initiatives funded by the NSF, DARPA, SRC, ARPA-E, and the high-tech industry (including Google, Intel, AMD, ADI, Ford, and Toyota). He is the author of more than 200 book chapters, journal publications, and fully reviewed conference papers. His research interests include, but are not limited to, the following topics: the design and optimization of advanced microcoolers for microprocessors and power electronics, the development of technologies at the energy, food, water, and environment nexus, such as water harvesting from the atmosphere and CO_2 capture from power plants, the development of advanced metamaterials for revolutionary thermal management solutions, and the thermal characterization of nanostructures relevant to semiconductor devices. Dr. Asheghi served as the general chair and program chair for ASME 2017, the 2015 *International Technical Conference and Exhibition on Packaging and Integration of Electronic and Photonic Microsystems*, and the *Intersociety Conference on Thermal & Thermomechanical Phenomena in Electronic Systems (ITHERM)* in 2014 and 2012.

https://doi.org/10.1142/9789811279379_fmatter

Contents

https://doi.org/10.1142/9789811279379_0001

Chapter 1

Compact Thermal Modeling of Emerging Cooling Technologies for Processors

Zihao Yuan*, Sherief Reda†, and Ayse K. Coskun‡

**Boston University, Boston, MA, USA*

†Brown University, Providence, RI, USA

‡Department of Electrical and Computer Engineering, Boston University, Boston, MA, USA

Abstract

Emerging cooling solutions, such as single-phase liquid cooling, two-phase cooling, and thermoelectric coolers, have been shown to achieve better cooling performance and efficiency than traditional cooling mechanisms, especially for extreme hotspots and large thermal gradients in chips. Maximizing computing performance under permissible temperature limits in future systems can be achieved with the help of co-design and optimization of processors together with these emerging cooling methods. To facilitate design space exploration and optimization with new cooling technologies, there is a need for fast and accurate thermal models for emerging cooling techniques. This chapter presents steady-state compact thermal modeling methodologies for the aforementioned emerging cooling solutions. The chapter also provides cooling evaluation and validation results for these emerging cooling solutions using their corresponding compact thermal models.

Keywords: Thermal modeling, Emerging cooling technologies, Thermal simulation

1. Introduction

Over the past few decades, processor performance has grown exponentially due to the downscaling of transistor dimensions. However, this performance boost has also translated into higher power densities, higher operating temperatures, and reduced reliability [1]. High power densities that reach 1–2 kW/cm^2 have already become a major challenge for modern processors [2]. Localized hotspots resulting from these high power densities not only decrease the lifetime of processors [1] but also increase transistor delays and leakage power [3]. In addition, the heterogeneity in on-chip heat distribution incurred by these hotspots is expected to become more severe with the integration of heterogeneous architectures on a single die, such as a collection of CPUs, GPUs, accelerators, and FPGAs. Existing cooling solutions, such as the conventional heat sink and forced-air cooling via fans, are not efficient enough to mitigate such high-power-density hotspots and may lead to undercooling. Emerging cooling techniques, including thermoelectric coolers (TECs) [4], liquid cooling via microchannels [5], microchannel-based two-phase cooling [6], two-phase vapor chambers (VCs) with micropillar wick evaporators [7], and hybrid cooling (e.g., liquid cooling via microchannels and TEC) [8], have shown better cooling performance[1] against existing cooling solutions.

Commercial multiphysics simulators, such as COMSOL Multiphysics [9] and ANSYS [10], are typically used to design and simulate the thermal models of emerging cooling solutions. However, these tools require significant effort to construct system-specific models. Such tools also incur long simulation times and large memory requirements (e.g., simulating a millimeter-scale chip takes from hours to

[1]Higher/better cooling performance refers to achieving lower maximum temperatures and thermal gradients.

multiple days and easily requires tens of gigabytes of memory). An additional challenge is that some of the existing thermal models are based on in-house prototypes; hence, the simulation results cannot be easily reproduced. Thermal models that are designed for targeting specific platforms and scenarios are also not easily applicable to arbitrary systems. To facilitate design automation and optimization, a thermal model itself should be modular and applicable to various chip floorplans and power profiles. Therefore, there is a need for fast, accurate, and modular thermal models for the above emerging cooling technologies.

Compact thermal modeling has been designed to overcome the thermal simulation challenges of long simulation times and large memory requirements. A compact model leverages the duality between electrical and thermal properties to model temperature [11]. In this approach, the chip is represented as a network of thermal nodes, and the chip temperature is modeled based on an equivalent resistor–capacitor (RC) network of these thermal nodes. This RC representation is then converted into a set of matrix equations, as shown in Eq. (1) [12]:

$$GT(t) + C\dot{T}(t) = U(t) \tag{1}$$

where G, C, and U are the equivalent thermal conductance, thermal capacitance, and power dissipation matrices, respectively. T is the node temperature matrix that needs to be solved. The equivalent RC network is solved using differential solvers to get the temperature of each node. A number of researchers have developed compact thermal models (CTMs) to enable fast thermal modeling of the emerging cooling techniques in target chips with a reasonable trade-off in accuracy [13–17].

This chapter elaborates on cutting-edge compact thermal modeling methodologies for a set of promising emerging cooling technologies. The organization of the chapter is as follows:

- We first discuss how to model the steady-state temperature behavior of a processor using the compact thermal modeling approach.

- We then provide a detailed overview of several emerging cooling solutions and their corresponding CTMs, including TEC, liquid cooling via microchannels, hybrid cooling (of liquid cooling via microchannels and TEC), microchannel-based two-phase cooling, and two-phase VCs with micropillar wick evaporators.
- The chapter also provides cooling performance evaluation results for the aforementioned emerging cooling solutions by leveraging their corresponding CTMs. In our analysis of the cooling performance results, we show the potential of co-designing computing systems with cooling systems.

2. Compact Thermal Modeling

The heat diffusion equation is generally used for describing the heat conduction and calculating the temperature distribution of a chip. For homogeneous materials, the corresponding heat equation is shown as follows:

$$\rho c_p \frac{\partial u}{\partial t} = k\left(\frac{\partial^2 u}{\partial x^2} + \frac{\partial^2 u}{\partial y^2} + \frac{\partial^2 u}{\partial z^2}\right) + q \tag{2}$$

In Eq. (2), k is the thermal conductivity (W/mK), ρ is the density of the material (kg/m^3), c_p is the specific heat (J/kg K), q is the power density (W/cm^2), and u is the temperature (°C) of a location (x, y, z) at time t. There is a well-known duality between heat flow and electric current. The heat flow (W) passing through a thermal resistor (°C/W) can be represented as an electric current (A) flowing through an electrical resistance (Ω). The corresponding temperature difference (°C) is equivalent to the voltage drop (V). In addition, there is a thermal capacitance (J/°C) that describes how much heat can be absorbed, which is represented as an electric capacitance (F). Node temperature can then be modeled as the node voltage of an electric RC circuit, as shown in Figure 1(a). To model a chip with multiple heat sources, heat conduction from each vertical and horizontal node is modeled as a thermal resistance. Node n_k represents the temperature of the circuit block, and the current source, i_k, represents the power consumption of the corresponding node. v_0 is the ambient temperature, and C_{k0} represents the thermal capacitance of

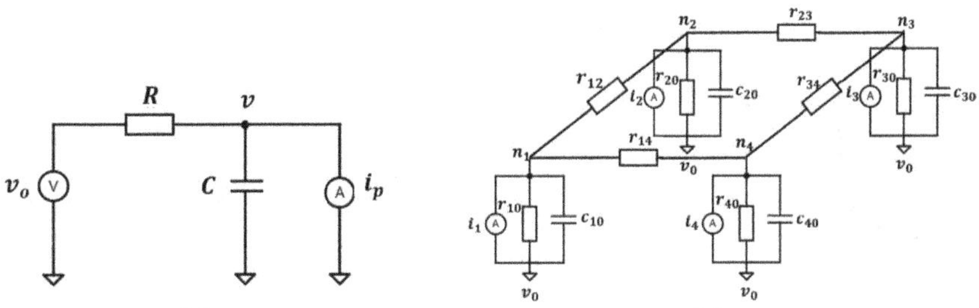

(a) (b)

Figure 1. (a) A simple thermal RC circuit [11]: R is the thermal resistor, C is the thermal capacitor, v_0 is the ambient temperature, and v is the temperature of the node. (b) An equivalent RC network to model temperature distribution [11].

the node. A thermal RC network can be built based on the above parameters, as shown in Figure 1(b).

Various CTM-based thermal simulators have been designed to model the full-chip temperature behavior and emerging cooling solutions [11, 13, 17, 18]. HotSpot, an architecture-level thermal simulator that adopts the compact thermal modeling methodology to estimate the temperature distribution of a processor, is particularly attractive because it is substantially faster than computational fluid dynamics (CFD)-based models, and it is straightforward to integrate HotSpot with power and performance simulators to enable co-design and optimization of computing and cooling systems [12, 18]. HotSpot allows users to define the number of layers, the physical geometry of the chip stack, the power dissipation of each block, and the thermal properties of each layer as inputs. The *grid* mode in HotSpot provides finer granularity simulation by dividing the chip into a number of grid cells and computing the temperature for each cell. The default grid cell is shown in Figure 2(a). HotSpot constructs a thermal RC network based on these grids and uses a differential equation solver to get the temperature distribution of the chip stack. In addition, Meng *et al.* introduced an additional functionality that supports the modeling of heterogeneous materials within a single layer by allowing floorplan units to have different heat capacitance and thermal resistivity values [19]. This functionality allows users to assign different

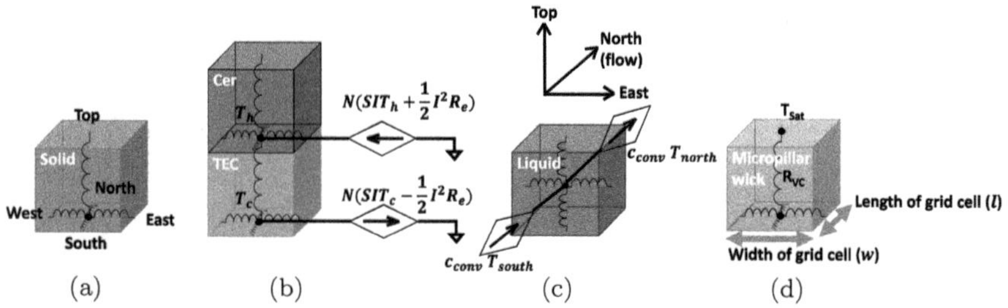

Figure 2. (a) Default grid cell, (b) TEC grid cell [14], (c) liquid cooling via microchannels grid cell [14, 20], and (d) two-phase VCs with micropillar wick evaporators grid cell [16].

thermal properties to individual blocks residing on the same layer (e.g., a layer with logic units and through-silicon vias (TSVs)) [12].

3. CTMs for Emerging Cooling Solutions and Cooling Performance Evaluation

To overcome the challenges of high power density in process cooling, researchers have developed several emerging cooling solutions, such as the following, targeted at various architectures and cooling scenarios: (i) TECs are attractive due to their hotspot mitigation ability; (ii) liquid cooling via microchannels can be used as an interlayer cooling method to solve the thermal-coupling problems for 3D-stacked chips; (iii) compared to TEC and liquid cooling via microchannels, hybrid cooling (of liquid cooling via microchannels and TEC) can enhance cooling performance while minimizing the cooling power; (iv) microchannel-based two-phase cooling uses latent heat of evaporation to remove heat and is expected to achieve higher cooling performance than single-phase cooling (e.g., liquid cooling via microchannels); and (v) unlike active cooling methods (e.g., TEC and liquid cooling via microchannels), two-phase cooling using VCs does not require any additional cooling power on the evaporator side.

Next, we discuss cooling mechanisms and compact thermal modeling methodologies for the aforementioned emerging cooling solutions. In addition, we evaluate the cooling performance of individual emerging cooling solutions using their corresponding CTM [14, 16, 21, 22]. The test chip has dimensions of 20 mm × 20 mm, with a

500 μm × 500 μm hotspot placed at the center. The chip thickness is 750 μm. The background power density is 50 W/cm^2, and the hotspot power densities range from 100 to 2000 W/cm^2. The saturated temperatures for two-phase cooling techniques (two-phase VCs with micropillar wick evaporators and microchannel-based two-phase cooling) are set at 50°C. Water is selected as the coolant.

3.1. *Thermoelectric coolers (TECs)*

TEC units have gained popularity due to their ability to effectively remove heat from high-power-density hotspots [4]. A TEC unit operates based on the Peltier effect: When electric current passes through a TEC unit, heat is absorbed from one side (cold side) and rejected from the other side (hot side) [4, 8]. TEC units are typically placed directly above hotspots. The amount of heat removed by a TEC unit depends on the Seebeck coefficient (S), applied current (I), electric resistance (ρ_{TEC}), and the temperature difference between the cold and hot sides ($T_{\mathrm{h}} - T_{\mathrm{c}}$). Superlattice-based thin-film TECs made of Bi_2Te_3 have high figures of merit (ZT) and can be directly fabricated on the back of a silicon chip [4, 23]. Existing on-chip TEC devices are composed of ultrathin (5–10 μm) Bi_2Te_3-based p–n thermocouples sandwiched between copper mini-headers and are covered with ceramic plate insulation at their outermost surfaces [4]. The terms that contribute to the heat flow in a TEC unit are as follows:

$$Q_{\mathrm{c}} = N\left(\mathrm{SIT}_{\mathrm{c}} - \frac{T_{\mathrm{h}} - T_{\mathrm{c}}}{R_{\mathrm{t}}} - \frac{1}{2}I^2 R_{\mathrm{e}}\right) \tag{3}$$

$$Q_{\mathrm{h}} = N\left(\mathrm{SIT}_{\mathrm{h}} - \frac{T_{\mathrm{h}} - T_{\mathrm{c}}}{R_{\mathrm{t}}} + \frac{1}{2}I^2 R_{\mathrm{e}}\right) \tag{4}$$

where $\mathrm{SIT}_{\mathrm{c}}$ is the Peltier effect term, $\frac{T_{\mathrm{h}} - T_{\mathrm{c}}}{R_{\mathrm{t}}}$ is the thermal conduction term, and $\frac{1}{2}I^2 R_{\mathrm{e}}$ is the Joule heating term, which represents the heat generated by the current passing through the TEC. T_{c} and T_{h} represent the temperatures of the cold and hot sides of the TEC, respectively. R_{t} is the thermal resistance, and R_{e} is the electrical resistance of a TEC unit [22].

To model the thermal gradients caused by the Peltier effect, CTMs of TEC typically use a voltage-controlled current source to represent the heat entering and leaving a TEC grid cell [4, 8]. The chip is divided into grid cells, and the default grid cell arrangement is used to represent the processing layer (Figure 2(a)). A typical TEC grid cell is shown in Figure 2(b). In this figure, the bottom surface of the TEC grid cell represents the cold side of the TEC unit, whereas the bottom surface of the ceramic grid cell corresponds to the hot side.

There have been several prior approaches to creating similar CTMs for TECs and analyzing their cooling performance, for example, a numerical CTM of a TEC that includes both thermal and electrical contact resistances predicts the cooling performance of TECs [4]. Both the temperature measurements from real prototypes and the thermal simulation results from TEC CTMs show a maximum temperature reduction of 15°C owing to the cooling ability of TECs. The thermal resistance of TECs plays a significant role in accurately determining their cooling performance [4]. In addition, TECs' cooling performance is highly correlated with the bias current value.

A recent work has introduced another CTM of TECs to estimate the impact of the bias current on the cooling performance of the TEC units [14]. The authors validated the accuracy of their proposed CTM by comparing its temperature results against those of the TEC model in COMSOL. The TEC device and the chip layers modeled by them in COMSOL are illustrated in Figure 3. This TEC is a superlattice-based thin-film TEC made of Bi_2Te_3 as the bulk material and has a high ZT. The TEC is composed of an array of 7×7 p–n thermocouples and has a total size of 3.5 mm $\times$ 3.5 mm. The legs of the p–n thermocouples are ultrathin (8 μm), and the total thickness of the TEC device, including the ceramic plates, is 100 μm. The width and length of each leg are 400 and 150 μm, respectively. This corresponds to 0.833 p–n thermocouples per mm^2 area, and this value is used when calculating the parameter N (i.e., the number of p–n thermocouples per TEC area) in this model. The detailed parameters of the TEC units are shown in Table 1. They modeled

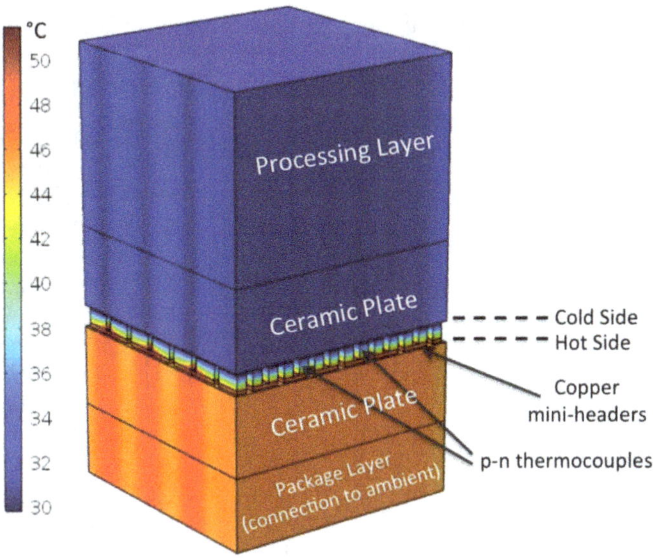

Figure 3. COMSOL model of the TEC device [14].

the processing layer using a 100 μm thick silicon layer at the cold side of the TEC unit and assigned a heat flux value to represent the heat generated by the processor. They placed a cooling package layer with a thickness of 40 μm on top of the chip stack and assigned a heat transfer coefficient (HTC) at the surface open to the ambient conditions to represent the additional cooling package. In COMSOL, they modeled the TECs in detail by defining the individual p–n leg, the copper mini-headers connecting the thermocouples in series, the VDD, and the ground nodes. They defined a multiphysics problem in COMSOL, which combines heat transfer in solids with the thermoelectric effect, an electromagnetic heat source, and thermal-coupling elements. They used the segregated solver in COMSOL to solve the multiphysics problem iteratively for both sub-parts of the problem. Figure 4 shows the thermal map comparison obtained from the two simulations for $q = 20$ W/cm^2 and a TEC current of 4 A. The plots on the left correspond to the temperatures of the processing layer, while the plots on the right show the temperatures on the hot side. They carried out a similar analysis for other q values ranging from

Table 1. The validation setups for TECs, liquid cooling via microchannels, and hybrid cooling CTMs [22].

TEC width and length	$w_{\text{tec}} = l_{\text{tec}}$	3.5 mm
Seebeck coefficient	S	301 μV/K
Thermocouple thickness	h_{thick}	8 μm
Copper mini-header thickness	h_{Cu}	2 μm
Ceramic plate thickness	h_{Cer}	44 μm
TEC electrical resistivity	ρ_{tec}	1.08×10^{-5} Ohm.m
TEC thermal conductivity	k_{tec}	1.2 W/mK
Copper thermal conductivity	k_{Cu}	400 W/mK
Ceramic thermal conductivity	k_{Cer}	175 W/mK
Silicon thermal conductivity	k_{Si}	130 W/mK
Microchannel height	h	100 μm
Microchannel width	w	50 μm
Grid cell width and length	$w = l$	50 μm
Microchannel length	L	10 mm
Coolant thermal conductivity	k_{coolant}	0.6069 W/mK
Coolant specific heat	C_{v}	4,181 J/kg K
Coolant inlet temperature	T_{inlet}	27°C
Coolant density	ρ_{coolant}	998 kg/m^3
Coolant viscosity	μ	8.89×10^{-4} Pa.s
Average coolant velocity	u_{avg}	$\leq$3 m/s

20 to 50 W/cm^2. The reported maximum, average, and root mean square (RMS) errors of their proposed CTM of TECs against those of the model in COMSOL are 3.57°C, 2.07°C, and 2.25°C, respectively. Compared to the COMSOL model, the reported speedup of the CTM of TEC is 16,900×, while the specific speedup number can depend on the number of grids used in CTM simulations and the number of nodes used in COMSOL simulations.

To evaluate the cooling performance of TECs, we build the same CTM of TEC as the previous work [14] with the same TEC unit parameters, as described in the previous paragraph. We vary the TEC current from 1 to 7 A [14], and the temperature results are shown in Table A.1. To remove the heat from the system, we add an additional air convection HTC of 2×10^4 W/m^{2}K to the chip stack to model the heat sink. The results indicate that as the TEC current increases, the hotspot temperature reductions increase. When the

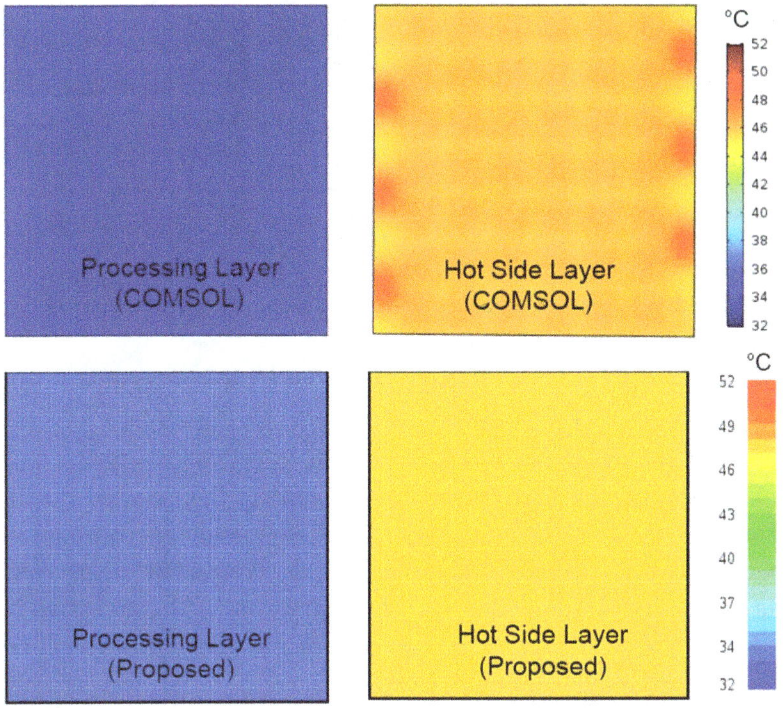

Figure 4. Comparison of thermal maps corresponding to the processing layer and the TEC's hot side between COMSOL and the CTM of TEC. HTC = 10^6 W/m^{2}K, $q = 20$ W/cm^2, and $I = 4$ A [14].

hotspot power density is low (100 W/cm^2), the hotspot temperature reduction is minimal. The reason is that Joule heating is the dominant heat source in this case and can degrade the TEC's cooling performance.

3.2. *Liquid cooling via microchannels*

Liquid cooling via microchannels is an attractive cooling solution that uses the liquid convection effect to remove heat from processors [5, 13]. It is especially useful as an interlayer cooling method to solve the strong thermal-coupling issues in 3D-stacked architectures [13]. Adding an additional liquid microchannel layer between processing layers can efficiently solve the thermal problems due to the vertical

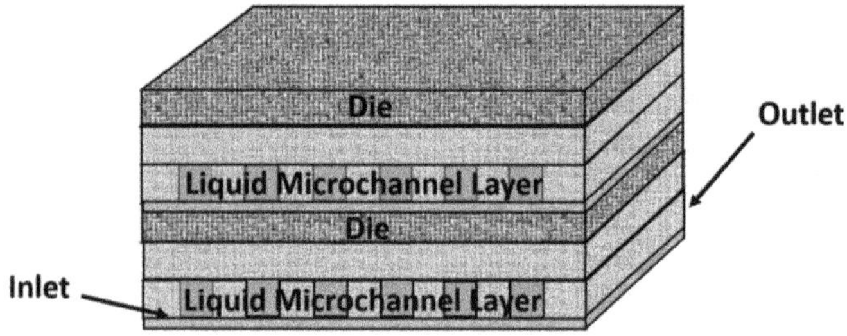

Figure 5. Liquid cooling via microchannels 3D IC architecture [13].

layer stacking. A typical liquid cooling via microchannels in a 3D integrated circuit (IC) is shown in Figure 5. As the coolant is pumped from inlet to outlet, it absorbs heat due to convective heat transfer (liquid convection), and the heated coolant is then cooled down by the chiller outside the system. There are two main contributors to the convective heat transfer of the coolant: (1) convective heat transfer from the walls of the channel to the liquid; and (2) convective heat transfer in the direction of the liquid flow into and out of the current liquid cell.

To build a CTM for liquid cooling via microchannels, one approach (four-resistor model, or 4RM) is to use thermal resistors to represent the convective heat transfer from the sidewalls of the microchannels and a voltage-controlled current source to represent the convective heat transfer along the liquid flow direction [14, 20]. Figure 2(c) shows the liquid grid cell, and the thermal resistances of the sidewalls are defined as follows:

$$R_{\text{top,bottom}} = \frac{1}{h_{\text{f,vertical}} \cdot w \cdot l} \tag{5}$$

$$R_{\text{east,west}} = \frac{1}{h_{\text{f,side}} \cdot h \cdot l} \tag{6}$$

In Eqs. (5) and (6), h, w, and l stand for the height, width, and length of the grid cell, respectively. $h_{\text{f,vertical}}$ and $h_{\text{f,side}}$ are the HTCs for the microchannel forced convection. $h_{\text{f,vertical}}$ and $h_{\text{f,side}}$ can be obtained from empirical correlations or numerical pre-simulations. Prior work provides the following formulas to calculate $h_{\text{f,vertical}}$ and

$h_{\mathrm{f,side}}$, assuming axial heat flux and radial isothermal conditions:

$$h_{\mathrm{f,vertical}} = h_{\mathrm{f,side}} = \frac{k_{\mathrm{coolant}} \cdot Nu}{d_{\mathrm{h}}} \tag{7}$$

where k_{coolant} is the thermal conductivity of the coolant and d_{h} is the hydraulic diameter of the channel defined as $\frac{2hw}{h+w}$. Nu is the Nusselt number, which is a function of the channel's aspect ratio [14, 20]. $h_{\mathrm{f,vertical}}$, $h_{\mathrm{f,side}}$, and Nu may differ under different system assumptions.

The convective heat transfer along the microchannels (i.e., current) is expressed as follows [20]:

$$J_{\mathrm{conv}} = c_{\mathrm{conv}}(T_{\mathrm{south}} - T_{\mathrm{north}}) \tag{8}$$

where T_{south} and T_{north} are the temperatures at the south and north surfaces of the grid cell, respectively. c_{conv} is the liquid convection coefficient, which is defined as $C_{\mathrm{v}} u_{\mathrm{avg,y}} \Delta A_{\mathrm{y}}$. C_{v} is the specific heat capacity, $u_{\mathrm{avg,y}}$ is the average liquid flow velocity, and $\Delta A_{\mathrm{y}} = wh$. The inlet temperature of the coolant is the boundary condition for this model.

Kaplan *et al.* validated the 4RM-based CTM against a COMSOL model of liquid cooling via microchannels. Figure 6(a) illustrates the cross-section of the chip stack, where the liquid microchannel layer is placed on top of the processor layer and an additional 40 μm thick cooling package layer (bulk silicon) is placed on top to provide closure

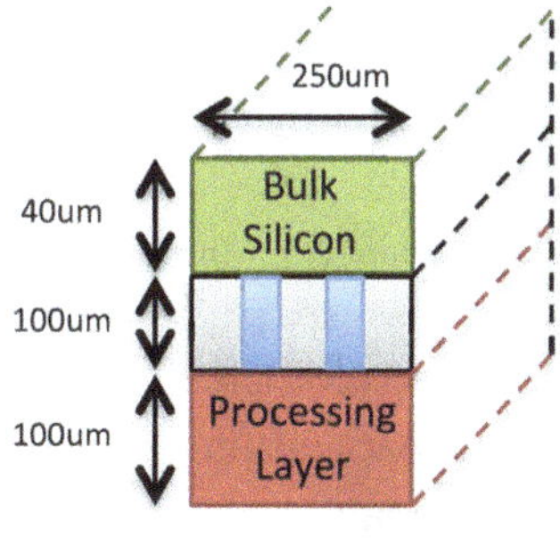

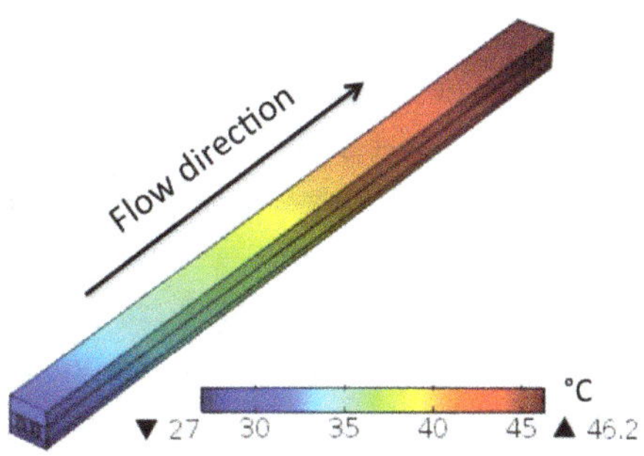

Figure 6. (a) Front view of the thin slice of chip stack for liquid cooling. (b) Side view of the chip stack in COMSOL [22].

to the microchannels with a very small HTC. They simulated a thin slice of this chip stack to reduce the problem complexity in COMSOL, as shown in Figure 6(b). The width and length of the chip stack slice are 250 μm and 5 mm, respectively. The detailed parameters of the model are shown in Table 1. The problem defined by them in COMSOL is a multiphysics problem which combines heat transfer in solids, heat transfer in liquids, and laminar flow elements. They used the segregated solver in COMSOL to solve the multiphysics problem, where the segregated step 1 (corresponding to the laminar flow) is an iterative solver using the GEMRES method, and the segregated step 2 (corresponding to the heat flow) is a direct solver using the PARDISO method. They ran steady-state simulations for different heat flux values of 12.5, 25, 50, and 100 W/cm^2 and different flow velocities of 0.5, 1.0, 1.5, and 2.0 m/s and recorded the maximum temperature of the processing layer obtained from the 4RM-based CTM and COMSOL. Compared to the COMSOL simulations, the 4RM-based CTM has a maximum error of 2.8%, with a speedup of 43,300×.

To further speed up the simulation of the liquid cooling via microchannels, a two-resistor model (2RM)-based CTM can be designed [24]. In the 4RM-based CTM, the boundaries of the thermal cells that conform the solid–liquid interfaces need to be defined, which results in very fine discretization and, consequently, prohibitively large simulation times. The 2RM-based CTM solves this problem by homogenizing the whole microchannel layer into a single porous medium. In this way, the thermal cells are no longer constrained by microchannel dimensions, and even multiple microchannels can be covered by a single thermal cell in this homogeneous medium. A small section of the microchannel layer modeled using 4RM- and 2RM-based CTMs is shown in Figure 7. R_{cond} represents vertical heat conduction, $R_{\text{downstream}}$ stands for the downstream convection heat transfer, C_{Si} is the thermal capacitance of the silicon wall, and C_{coolant} is the thermal capacitance of the coolant. The detailed derivation of this 2RM-based CTM can be found in a recent work [24]. Sridhar *et al.* [13] compared the simulation speed and accuracy of 4RM- and 2RM-based CTMs of liquid cooling via

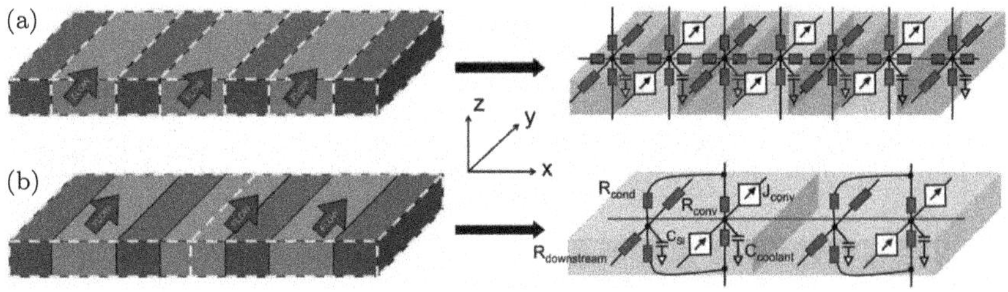

Figure 7. (a) 4RM-based CTM and (b) 2RM-based CTM of a small section of liquid microchannel layer [13].

microchannels on test 3D multiprocessor system-on-chips. The simulation results show that the 2RM-based CTM achieves a 375× speedup when compared to the 4RM-based CTM, with an accuracy loss of 7%.

To evaluate the cooling performance of liquid cooling via microchannels, we adopt the 4RM-based CTM from a recent work [14] and use a 100 μm thick liquid microchannel layer sandwiched between the processing layer and a 400 μm bulk silicon layer as our test chip. The microchannel width and the wall width are selected to be 50 μm. We set the liquid flow velocity at 1.0, 1.5, or 2.6 m/s [14], and the simulation results are shown in Table A.2. We assume that there is no air convection HTC on the top of the bulk silicon layer and that all the heat is removed by the coolant. Compared to TEC cooling, liquid cooling via microchannels cannot effectively reduce the hotspot temperature, which means TECs have a better hotspot mitigation ability.

3.3. *Hybrid cooling*

Hybrid cooling refers to incorporating two or more cooling solutions on the same platform. For example, a hybrid cooling system can be designed using liquid cooling via microchannels and TECs. Liquid cooling via microchannels can effectively remove the background heat on large chips and 3D-stacked architectures. However, as the liquid flows through the microchannels, it heats up, and its ability to further remove heat decreases. Although TECs are favorable for handling

Figure 8. Front view of a hybrid cooling design [22].

high power densities in a small area, the additional cooling power cost becomes substantially higher when cooling a large area. In this case, a hybrid cooling strategy that combines the strengths of liquid cooling via microchannels and TECs can provide a much better cooling efficiency. Several studies have focused on using TECs to target hotspots and liquid cooling via microchannels to remove the background heat [8, 23, 25]. A typical hybrid cooling method using TECs and liquid cooling via microchannels is shown in Figure 8.

There are several ways to create CTMs for hybrid cooling (of liquid cooling via microchannels and TECs). One approach is to use a high HTC value to represent liquid cooling via microchannels in addition to a TEC CTM to model hybrid cooling [8]. This method shows a 10× cooling power reduction when compared to using only liquid cooling via microchannels [8]. Another method involves building a CTM from the hybrid cooling using liquid cooling via microchannels CTM and the TEC CTM [22]. This method incorporates the CTMs for liquid cooling via microchannels and TECs from a previous work [14]. The chip stack is shown in Figure 8. The following are its layers from top to bottom: (1) processing layer; (2) thermal interface material (TIM) layer, where TEC is placed at the center; (3) liquid cooling via microchannel layer; and (4) bulk silicon layer. Kaplan *et al.* validated the accuracy of this hybrid cooling CTM against COMSOL simulations. They created a chip stack with a size of 4.5 mm × 5 m and placed a TEC at the center. A chip width of 4.5 mm corresponds to 44 microchannels and 46 walls. They defined two types of heat fluxes, a uniform heat flux of 20 W/cm^2 and a non-uniform heat flux of 20 W/cm^2 as background heat flux, as well as 1,100 and 1,300 W/cm^2 as hotspot heat fluxes. The hotspot is placed at the center of the floorplan with a size of 500 μm × 500 μm.

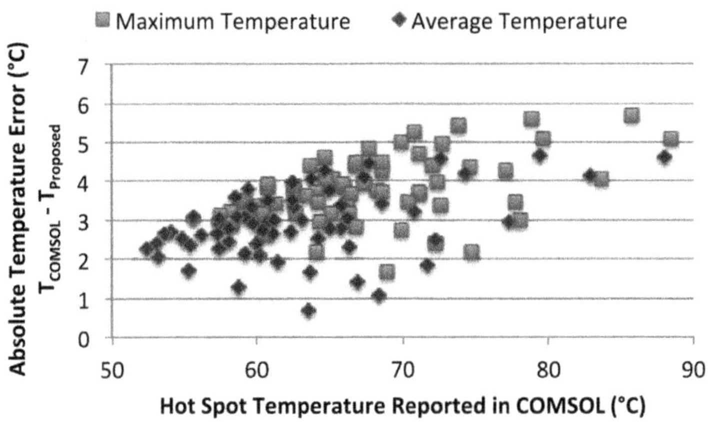

Figure 9. Comparison of maximum and average hotspot temperature results from COMSOL and the proposed hybrid model for all settings [22].

For each heat flux, they varied the liquid flow velocity and the TEC current. The properties of the TECs and liquid microchannels are listed in Table 1. Figure 9 shows the validation results. This hybrid cooling CTM achieves a 1,607× speedup, with a maximum error of less than 5.7°C.

To evaluate the cooling performance of hybrid cooling, we use the compact modeling method from a recent work [22] to model a chip stack, as shown in Figure 8, and show the cooling performance simulation results in Table A.3 [22]. Compared to using only TECs or liquid cooling via microchannels, hybrid cooling achieves better cooling performance in terms of higher hotspot temperature reductions. Most importantly, since the liquid flow velocity and TEC current can be tuned at design time and runtime, these tunable cooling parameters can facilitate researchers' efforts to optimize the hybrid cooling solution with respect to cooling power and cooling performance [21, 22].

3.4. *Microchannel-based two-phase cooling*

Microchannel-based two-phase cooling has a similar microchannel cooling structure as liquid cooling via microchannels, as shown in Figure 5. The difference is that liquid cooling via microchannels

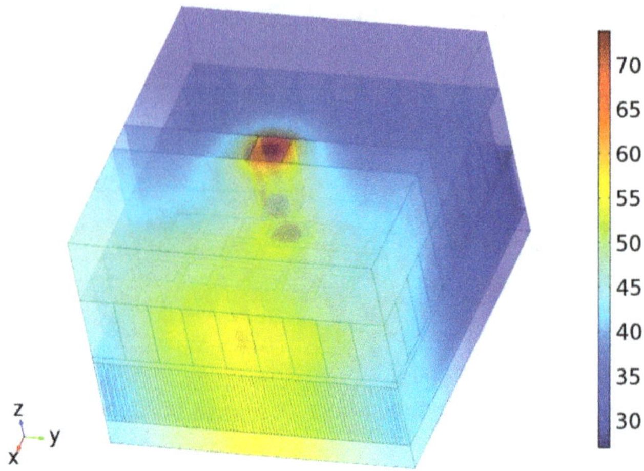

Figure 10. Contour plot for hybrid cooling in COMSOL [22].

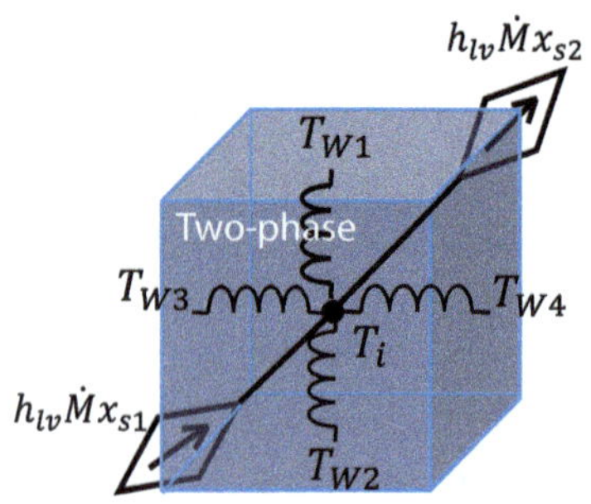

Figure 11. Microchannel-based two-phase cooling grid cell [17].

uses liquid convection to remove heat from the processor, whereas microchannel-based two-phase cooling removes heat through the evaporation of coolant inside microchannels. The heated surface is cooled via latent heat absorption. Generally, latent heat absorption of fluids results in better heat removal than sensible heat absorption [26].

Recent works have built prototypes for the microchannel-based two-phase cooling technique to study the relationship between heat transfer and cooling parameters, such as microchannel geometry (circular/rectangular), fluid (e.g., R-141b/R-134a/water), heat flux, mass flow velocity, and saturation pressure [27–30]. There are also some existing thermal models for microchannel-based two-phase cooling [31–33]. Most of these models rely on uniform heat flux distribution, and the HTC correlations are derived from experimental data from particular studies. Different models show strong disagreements with respect to a particular set of measurements and lead to high simulation error rates [17]. Since the local heat removal capability inside a microchannel is a complex function of flow regime, thermal conductivity of the liquid and vapor phases, local pressure, local temperature, etc. [26], the development of a microchannel-based two-phase cooling thermal model is complicated and still in its infancy.

STEAM, a CTM simulation framework, enables early-stage design and optimization of the microchannel-based two-phase cooling technique [17]. The STEAM simulation flow, as shown in Figure 12, is as follows

1. First, based on an initial physical state of a coolant in the microchannel, STEAM calculates the HTC using the following equation:

$$h_{\mathrm{lv}}\dot{M}x_{s1} + \sum_{k=1} 4\alpha_{\mathrm{W}}(T_{\mathrm{Wk}} - T_i) = h_{\mathrm{lv}}\dot{M}x_{s2} \tag{9}$$

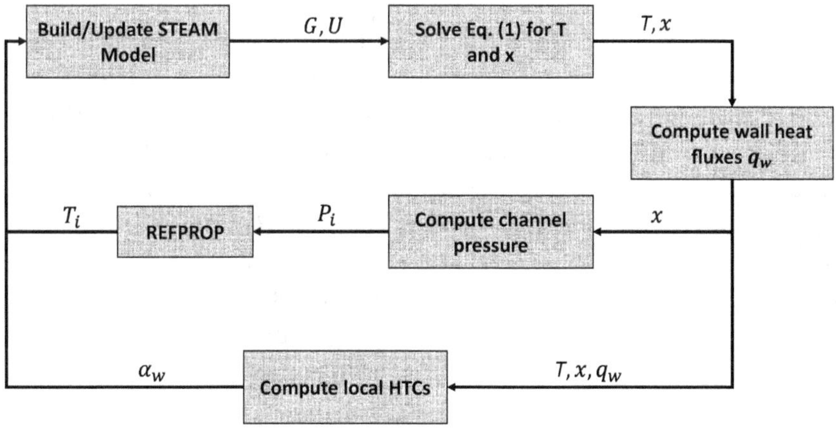

Figure 12. STEAM simulation flow [17].

Equation (9) is the steady-state energy balance equation of the microchannel cell i. This equation describes that the total heat entering the microchannel cell i (from the microchannel cell $i-1$ and local side walls) equals the heat exiting from the microchannel cell i to the microchannel cell $i+1$. The microchannel cell is shown in Figure 11. h_{lv}, M, x, and α_{W} stand for the latent heat of evaporation, mass flow rate, vapor quality, and HTC, respectively. T_i represents the temperature of the microchannel cell i, and T_{W} stands for the local wall temperature.

2. Next, STEAM calculates the thermal conductance and generates the thermal RC matrices, as shown in Eq. (1). The vapor quality x and temperature T are updated by solving the RC matrices.
3. New temperature values are used to calculate the updated HTC as well as the coolant's local pressure P_i. Using the REFPROP software [34], the local temperature of the coolant, T_i, is computed.
4. Equation (1) is updated, and the simulation returns to step 1. This process is repeated until convergence.

To validate the accuracy of STEAM, Sridhar *et al.* compared the results from STEAM simulations with experimental measurements from actual two-phase-cooled IC test vehicles [17]. The test vehicle consists of a single (12.7 mm $\times$ 17.78 mm) silicon die, and 35 resistive heaters in a 5×7 grid are placed on the surface of this die to mimic the different heat-dissipating logic blocks of a large microprocessor. Microchannels were etched on a silicon substrate and attached to the silicon die via a high-conductance thermal interface layer. Finally, the top surface of the structure (with the channels) is hermetically sealed using an insulating pyrex glass. The coolants that flow from top to bottom are selected to be R-236fa or R-245fa, with an inlet temperature of 31.1°C. Each heater block has a temperature sensor placed at the center of the block. The structural properties of the microchannel-based two-phase cooling validation setup are listed in Table 2. The authors tested both uniform and hotspot heat flux cases. In the uniform heat flux case, all the heaters are powered to a uniform heat flux density of 96.4 W/cm^2. The refrigerant used

Table 2. The structural parameters for microchannel-based two-phase cooling CTM.

Individual heater size	2540 μm
Channel width	85 μm
Channel height	560 μm
Number of channels	135
Hydraulic diameter D_h	148 μm
Silicon die thickness	350 μm
TIM thickness	50 μm
Microchannel base thickness	440 μm

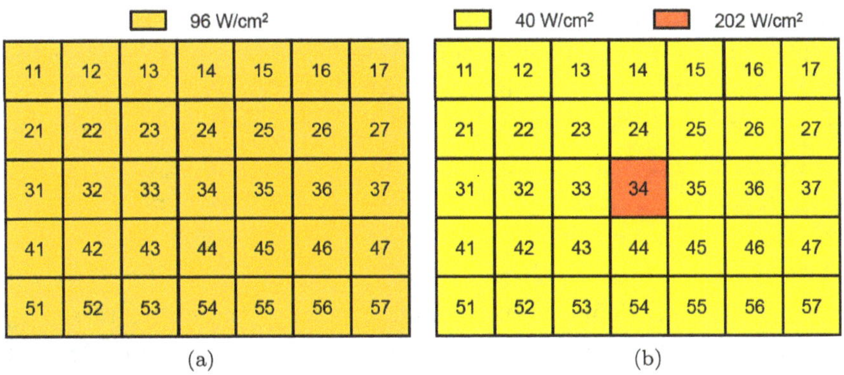

Figure 13. Heat flux patterns in the test structure for (a) uniform heat flux and (b) hotspot heat flux [17].

in this case is R-236fa in a purely liquid state (saturation pressure = 3.32 bar), with a mass flow velocity of 933 kg/m^{2}s. In the hotspot heat flux case, a single heater (at the location of row #3, column #4) is powered to a very high heat flux density (200 W/cm^2), while the rest of the heaters are at a relatively low heat flux density (40 W/cm^2), as shown in Figure 13(b). The refrigerant used in this experiment is R-245fa in a purely liquid state (saturation pressure = 1.84 bar), with a mass flow velocity of 713 kg/m^{2}s. In the STEAM model, the chip is divided into 5×7 grids, and the pressure drops inside the channels are calculated using the homogeneous model. In addition, they adopted three HTC calculation methods proposed by

Bertsch *et al.* [35], Kandlikar *et al.* [36], and Tran *et al.* [37]. The simulations in all the cases converge in 3–4 iterations, taking a total time of about 35 s. The results from the STEAM model are then compared with the temperatures obtained from the various sensors. Sample temperature results for the sensors in column #4 are plotted against simulation results as a function of the distance from the inlet shown in Figure 14. HTCs computed using Bertsch *et al.*'s [35] method result in the most accurate temperature predictions with an average error of 13.8%.

Another microchannel-based two-phase cooling CTM is based on the compact modeling methodology of STEAM [21]. This CTM includes the most accurate HTC correlation reported by STEAM and a linear pressure drop correlation [35]. The flow regime is assumed to be homogeneous. To validate the accuracy of this microchannel-based two-phase cooling CTM, a 0.5×0.25 mm^2 chip is selected as the test chip, with a 0.1 mm thick microchannel layer placed on top of it. An additional silicon bulk layer is placed on top of the microchannel layer to seal the channels and minimize air convection. Both the channel and wall widths are set at 0.05 mm. The coolant is selected to be R-245fa, with an input mass flux of 700 kg/m^{2}s. T_{Sat} of the coolant is set at 31.1°C. Steady-state thermal simulations are carried out using uniform power profiles of 25, 100, and 200 W/cm^2. Among all the simulations, when compared to the STEAM model, this microchannel-based two-phase CTM achieves a maximum error of less than 0.5°C.

To evaluate the cooling performance of microchannel-based two-phase cooling, we build the same microchannel-based two-phase CTM as previous work [21] and use the same structural parameters as those used in liquid cooling via microchannels. We vary the mass flow velocities from 100 to 560 kg/m^{2}s, and the results are shown in Table A.4. We use the correlation embedded in STEAM to estimate the dry-out. The dry-out correlation compares the cooling ability of microchannel-based two-phase cooling with the total heat generated by the chip to determine whether dry-out happens or not. The cooling ability of microchannel-based two-phase cooling is estimated by timing the total mass flow rate of the coolant with the

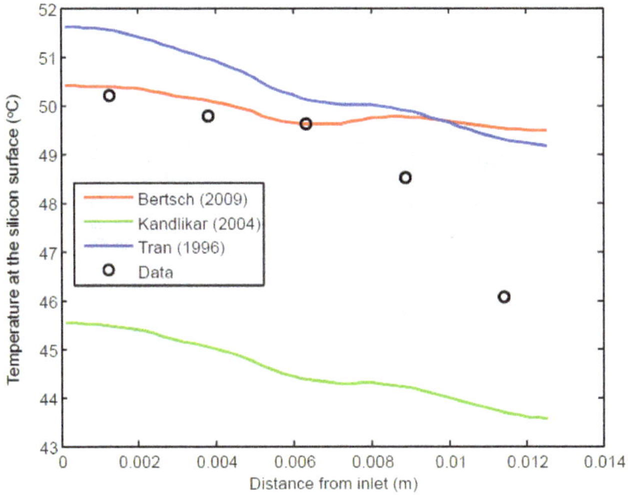

(a)

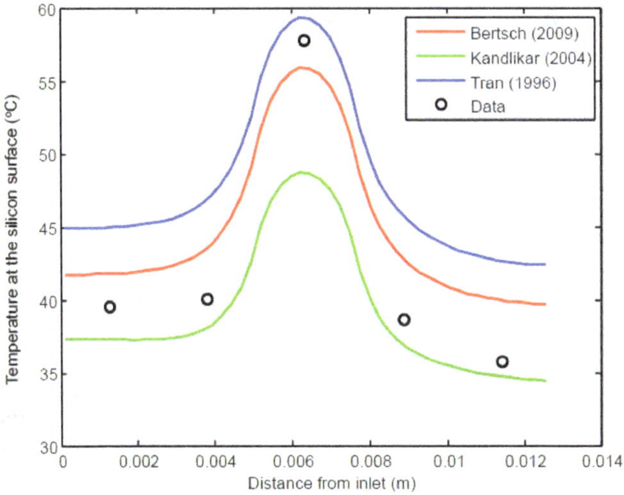

(b)

Figure 14. Comparison of temperatures from column #4 between the STEAM model and the test vehicle for (a) uniform heat flux and (b) hotspot heat flux [17].

specific latent heat of vaporization. Compared to liquid cooling via microchannels, microchannel-based two-phase cooling shows higher hotspot temperature reductions because the latent heat absorption of fluids results in better heat removal ability. This advantage can be

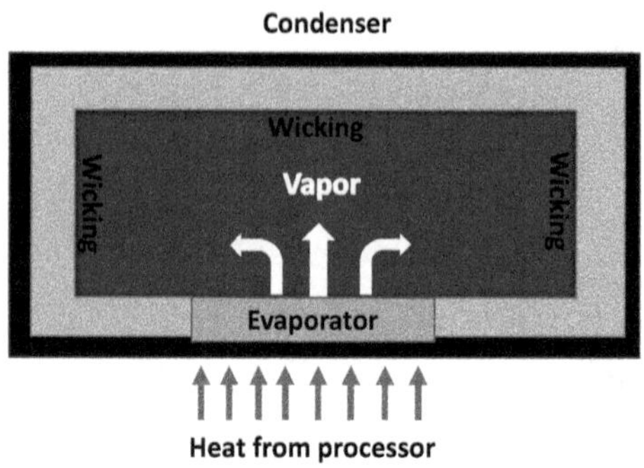

Figure 15. Vapor chamber structure view [50].

enhanced by tuning the pressure to decrease the coolant saturation temperature [16, 17].

3.5. *Two-phase VCs with micropillar wick evaporator*

Two-phase cooling using VCs with micropillar wick evaporators is a passive cooling method that uses a capillary-driven flow conducive to thin-film evaporation through a porous wick placed on the bottom surface of the VC to remove heat from processors [7]. The schematic diagram of a VC is shown in Figure 15. The advantages of two-phase VCs with micropillar wick evaporators are better cooling performance and no pumping power requirements[2] (in contrast to liquid cooling via microchannels and microchannel-based two-phase cooling) [7]. There are two metrics that impact the cooling performance of the VCs: HTC and dry-out heat flux. HTC is a parameter that determines the rate of heat transfer per unit temperature difference in the evaporator. Dry-out heat flux is the thermal limit of a two-phase device. If the hotspot power density of the chip is higher than the dry-out heat flux, the coolant will no longer remain in the two-phase state and possibly cause overheating and damage to the chip. Therefore, a higher dry-out heat flux means more heat can be removed

[2]While there is no pumping power required on the evaporator side, VCs may still need cooling power to reject the heat by means of a fan or other cooling methods.

before the coolant dries out. A number of studies have focused on the cooling performance of VCs with respect to various design parameters, such as wick structure [7, 38–40], working fluid [41–43], charge ratio [39, 44, 45], and VC geometry [7, 46, 47]. Micropillar wick evaporators have been shown to improve the heat transfer rate and enhance the dry-out heat flux owing to their high capillary pumping budget and large evaporation area [48, 49]. The selection of the micropillar array structure has a significant impact on the overall cooling performance of the evaporators as well as the VCs.

To build a CTM for two-phase VCs based on a porous wick that consists of a micropillar array with a periodic and precisely defined geometry, the entire system, including the VC, needs to be divided into small grid cells. The default grid cell arrangement is shown in Figure 2(a). The micropillar wick evaporator is modeled as a separate layer and is placed directly above the processing layer. To model the saturated vapor conditions, an additional virtual temperature node is placed on top of each grid cell of the micropillar wick layer, as shown in Figure 2(d). The temperature of this node is set as the saturated temperature of the coolant, T_{Sat}, and therefore depends on the coolant properties and the pressure inside the VC. One assumption to make when building thermal models for VCs is that T_{Sat} is maintained at a constant temperature. In this way, there is no need to model the condenser [7, 16]. The micropillar wick HTC is highly dependent on the coolant, VC pressure, and micropillar wick geometry (micropillar height, diameter, and pitch). To model the vertical heat transfer of the micropillar wick, a previously established relationship between the HTC and thermal resistance of a grid cell is used to represent the vertical thermal resistance of a micropillar grid cell, R_{VC} [13, 16].

Yuan *et al.* adopted a COMSOL model to extract the HTCs of a wide range of micropillar geometries, coolants, and VC pressures [16, 51]. This COMSOL model is detailed in a prior work [51]. In this COMSOL model, the authors separated the CFD simulation into fluid and heat transfer domains. By coupling these two domains, they iteratively obtained the temperature distribution. They used the

Young–Laplace equation to relate the curvature of the liquid–vapor interface with local pressure. In addition, they used Darcy's law and a volumetric loss function to model fluid flow in a uniform porous medium and the evaporative flux, respectively. Furthermore, they parametrically derived the permeability and HTC for each micropillar wick geometry. The extracted HTCs are stored in HTC lookup tables for various coolants and geometries.

To validate the accuracy of the two-phase VCs CTM, the authors first adopted a dry-out heat flux analytical model from a recent work to ensure the selected geometries and heat flux satisfied the dry-out constraint [49]. They then modeled a 2×2 mm^2 chip with a thickness of 100 μm in COMSOL [51] and used the CTM method [16, 21] to validate the accuracy of the proposed two-phase VC CTM. They ran two sets of simulations in COMSOL: (i) processing layer with a uniform power density; and (ii) processing layer with a non-uniform power density, with a 500×500 μm^2 hotspot placed at the center. Each simulation set has three different micropillar wick geometries and two coolants: water and R-134a. In these validation experiments, since water has a better HTC compared to R-134a, to ensure that the maximum temperatures are less than 85°C, they selected higher power densities for water and lower power densities for R-134a. In the experiment with uniform power density, they set the power density at 100 W/cm^2 for water and 20 W/cm^2 for R-134a. In the non-uniform power density simulations, they set the background power density at 50 W/cm^2 for water and 20 W/cm^2 for R-134a. The hotspot power density was set at 100, 200, and 300 W/cm^2 for water and 25, 50, and 75 W/cm^2 for R-134a. To prevent extremely high chip temperatures, they reduced the saturated temperature of water to 50°C by setting the pressure inside VC to 0.124 bar. As for R-134a, they set its saturated temperature at 50°C under a 13.2 bar pressure. For each COMSOL simulation, they used 592 nodes to simulate the fluid domain and 1,106 nodes to simulate the heat transfer domain. They modeled the same chip using the compact thermal modeling method and use 64×64 grids to compute the steady-state temperatures. The simulation time of the COMSOL CFD model is 45 s, while it only takes 0.21 s to simulate the two-phase VC CTM. Table 3 compares

Table 3. Validation of the proposed two-phase VC CTM.

Simulation	Coolant	P_{bg}	P_{hs}	$\{h, d, i\}$	Avg. error (°C)	Max. error (°C)
Uniform power	Water	100	100	{30,12,36}	0.20	0.19
		100	100	{40,16,48}	0.22	0.21
		100	100	{50,20,60}	0.25	0.25
	R-134a	20	20	{30,12,36}	0.65	0.71
		20	20	{40,16,48}	0.75	0.77
		20	20	{50,20,60}	0.80	0.82
Non-uniform power	Water	50	100	{30,12,36}	0.23	0.23
		50	200	{40,16,48}	0.31	0.26
		50	300	{50,20,60}	0.49	0.45
	R-134a	20	25	{30,12,36}	0.99	1.1
		20	50	{40,16,48}	1.02	1.12
		20	75	{50,20,60}	1.04	1.24

Note: h, d, i stands for height, diameter, and pitch of the micropillar wick evaporator [21].

the temperatures obtained from the above simulations. For both uniform and non-uniform simulations, the proposed model achieves high accuracy, with both the maximum and average errors being less than 0.5°C for water and 1.25°C for R-134a, while achieving a speedup of 214× when compared to the COMSOL CFD simulations. Their two-phase VC CTM provides an accuracy of 98.5%.

To carry out two-phase VCs cooling performance evaluation, we build the same CTM of two-phase VCs with micropillar wick evaporators as in previous work [16, 21] and use the micropillar wick optimization flow proposed by a recent work [16] to select the optimal micropillar wick geometry. We use the dry-out correlation adopted from previous work [16, 21, 49] to calculate the dry-out constraint of the selected micropillar wick geometry. All the optimal micropillar wick geometries selected satisfied the dry-out constraint. The optimal micropillar wick geometries and temperature results are shown in Table A.5. Compared to other cooling techniques, VCs can effectively remove the heat from the system with no additional cooling power costs on the evaporator side. Because of this advantage, researchers have been focusing on integrating VCs with the heat spreader or heat sink of the chip [7, 21].

4. Final Remarks

4.1. *Conclusion*

This chapter discussed various emerging cooling techniques for processors, such as TECs, liquid cooling via microchannels, hybrid cooling (of liquid cooling via microchannels and TECs), microchannel-based two-phase cooling, and two-phase VCs with micropillar wick evaporators. The chapter also elaborated on the compact thermal modeling methodology for full-chip thermal modeling as well as various CTMs of the aforementioned cooling techniques. Using these CTMs, we performed a cooling performance evaluation of these emerging cooling techniques and demonstrated the potential of co-designing computing systems with emerging cooling systems.

4.2. *Limitations of compact thermal modeling*

We identified several challenges with using the compact thermal modeling methodology. First, existing compact thermal modeling methods only support the cuboid grid shape. Grid shapes, such as circular (which is useful for simulating round heat pipes), can only be approximated using several cuboid grids. Such an approximation may lead to accuracy losses compared to FEM models. Second, when considering complex CFD simulations (e.g., turbulent flow), because compact thermal modeling is based on the duality between electrical and thermal properties, it's less accurate to model such pressure-dependent flows. Last but not least, the compact thermal modeling method does not envision the quantum effects in the nanometer scale (40–300 nm). For example, to guarantee the simulation accuracy of CTMs, the minimum grid size has to be higher than 300×300 nm^2.

Acknowledgments

The authors were partially funded by the NSF CRI (CI-NEW) grant #1730316/1730003/1730389 and NSF CCF 1910075/1909027.

The discussions on the thermal models of two-phase VCs with micropillar wick evaporators in this chapter are based on our joint work with Dr. Evelyn Wang (Massachusetts Institute of Technology), Geoffrey Vaartstra (Massachusetts Institute of Technology), and Prachi Shukla (Boston University). We also thank Dr. Fulya Kaplan (Apple Inc.), Dr. Mostafa Said (Brown University), Dr. Arvind Sridhar (IBM), and Dr. David Atienza (Swiss Federal Institute of Technology Lausanne) for their works on the thermal models of liquid cooling via microchannels, TECs, hybrid cooling, and microchannel-based two-phase cooling, all of whom contributed to this chapter.

Appendix A

Table A.1. Hotspot temperatures (°C) of the chip with TECs.

	Hotspot power density (W/cm^2)						
I (A)	100	300	700	1,000	1,300	1,500	2,000
1	71.84	73.87	77.28	80.32	83.36	85.39	90.46
3	71.20	73.23	75.33	78.91	81.94	83.63	89.91
7	68.58	70.71	74.73	77.74	80.75	82.76	87.78

Note: I stands for TEC current.

Table A.2. Hotspot temperatures (°C) of the chip with liquid cooling via microchannels.

	Hotspot power density (W/cm^2)						
u (m/s)	100	300	700	1,000	1,300	1,500	2,000
1.0	87.03	89.18	93.48	96.70	99.92	102.07	107.44
1.5	78.46	80.57	84.79	87.96	91.12	93.23	98.51
2.6	71.39	73.47	77.63	80.74	83.86	85.94	91.13

Note: u stands for liquid flow velocity.

Table A.3. Hotspot temperatures (°C) of the chip with hybrid cooling.

		Hotspot power density (W/cm^2)						
u (m/s)	I (A)	100	300	700	1,000	1,300	1,500	2,000
1.0	1	72.36	74.25	78.03	80.87	83.70	85.60	90.32
	3	71.32	73.21	76.98	79.80	82.63	84.52	89.23
	7	71.09	72.97	76.71	79.53	82.34	84.21	88.90
1.5	1	68.50	70.37	74.12	76.94	79.75	81.63	86.31
	3	67.45	69.32	73.06	75.86	78.67	80.54	85.21
	7	67.14	69.0	72.72	75.51	78.30	80.16	84.81
2.6	1	64.88	66.64	70.46	72.88	76.05	77.91	82.57
	3	63.83	65.68	69.40	72.18	74.96	76.82	81.46
	7	63.46	65.30	68.99	71.76	74.53	76.37	80.99

Table A.4. Hotspot temperatures (°C) of the chip with microchannel-based two-phase cooling.

	Hotspot power density (W/cm^2)						
G (kg/m^{2}s)	100	300	700	1,000	1,300	1,500	2,000
100	76.29	78.01	81.44	Dry-out	Dry-out	Dry-out	Dry-out
300	71.39	73.05	76.37	Dry-out	Dry-out	Dry-out	Dry-out
560	68.68	70.30	73.55	dry-out	Dry-out	Dry-out	Dry-out

Note: G stands for mass flow velocity.

Table A.5. Optimal geometries (height, diameter, and pitch) of two-phase VCs with micropillar wick evaporators and hotspot temperatures (°C) with two-phase VCs.

P_{hs} (W/cm^2)	100	300	700	1,000	1,300	1,500	2,000
MG$_{\text{opt}}$ (μm)	20, 10, 5	30, 10, 5	30, 10, 5	35, 10, 5	40, 10, 5	40, 10, 5	45, 10, 5
T_{max} (°C)	70.69	72.51	76.76	78.76	81.46	83.19	87.63

Note: P_{hs} stands for hotspot power density, MG$_{\text{opt}}$ stands for optimal micropillar wick geometry, and T_{max} stands for hotspot temperature.

References

[1] J. Srinivasan, S.V. Adve, P. Bose, J. Rivers, C.-K. Hu, Ramp: A model for reliability aware microprocessor design, IBM Research Report, 2003.

[2] M. Schultz, F. Yang, E. Colgan, R. Polastre, B. Dang, C. Tsang, M. Gaynes, P. Parida, J. Knickerbocker, T. Chainer, Embedded two-phase cooling of large three-dimensional compatible chips with radial channels, *J. Electron. Packag.*, 138 (2), 2016, 021005.

[3] N.S. Kim, T. Austin, D. Blaauw, T. Mudge, J.S. Hu, M.J. Irwin, M. Kandemir, V. Narayanan, Leakage current: Moore, *Computer*, 12, 2003, 68–75.

[4] I. Chowdhury, R. Prasher, K. Lofgreen, G. Chrysler, S. Narasimhan, R. Mahajan, D. Koester, R. Alley, R. Venkatasubramanian, On-chip cooling by superlattice-based thin-film thermoelectrics, *Nat. Nanotechnol.*, 4 (4), 2009, 235.

[5] B. Dang, M.S. Bakir, D.C. Sekar, C.R. King Jr., J.D. Meindl, Integrated microfluidic cooling and interconnects for 2D and 3D chips, *IEEE Trans. Adv. Packag.*, 33 (1), 2010, 79–87.

[6] S.G. Kandlikar, Fundamental issues related to flow boiling in minichannels and microchannels, *Exp. Therm. Fluid Sci.*, 26 (2–4), 2002, 389–407.

[7] M. Bulut, S.G. Kandlikar, N. Sozbir, A review of vapor chambers, *Heat Transfer Eng.*, 40 (19), 2019, 1551–1573.

[8] K. Yazawa, A. Ziabari, Y.R. Koh, A. Shakouri, V. Sahu, A.G. Fedorov, Y. Joshi, Cooling power optimization for hybrid solid-state and liquid cooling in integrated circuit chips with hotspots, in: *IEEE Proc. of 13th Intersociety Conference on Thermal and Thermomechanical Phenomena in Electronic Systems (ITherm)*, 2012, pp. 99–106.

[9] COMSOL Multiphysics Software, http://www.comsol.com.

[10] ANSYS, http://www.ansys.com.

[11] M. Pedram, S. Nazarian, Thermal modeling, analysis, and management in VLSI circuits: Principles and methods, *Proc. IEEE*, 94 (8), August 2006, 1487–1501.

[12] K. Skadron, M.R. Stan, W. Huang, S. Velusamy, K. Sankaranarayanan, D. Tarjan. Temperature-aware microarchitecture, in: *IEEE Proc. of 30th Annual International Symposium on Computer Architecture (ISCA)*, 2003, pp. 2–13.

[13] A. Sridhar, A. Vincenzi, D. Atienza, T. Brunschwiler, 3D-ICE: A compact thermal model for early-stage design of liquid-cooled ICs, *IEEE Trans. Comput.*, 63 (10), October, 2014, 2576–2589.

[14] F. Kaplan, S. Reda, A.K. Coskun, Fast thermal modeling of liquid, thermoelectric, and hybrid cooling, in: *Proc. of Intersociety Conf. on Thermal and Thermomechanical Phenomena in Electronic Systems (ITherm)*, 2017, pp. 726–735.

[15] A.K. Coskun, J.L. Ayala, D. Atienza, T.S. Rosing, Modeling and dynamic management of 3d multicore systems with liquid cooling, in: *17th IFIP International Conference on Very Large Scale Integration (VLSI-SoC)*, 2009, pp. 35–40, IEEE.

[16] Z. Yuan, G. Vaartstra, P. Shukla, M. Said, S. Reda, E. Wang, A.K. Coskun, Two-phase vapor chambers with micropillar evaporators: A new approach to remove heat from future high-performance chips, in: *Proc. of Intersociety Conf. on Thermal and Thermomechanical Phenomena in Electronic Systems (ITherm)*, 2019.

[17] A. Sridhar, Y. Madhour, D. Atienza, T. Brunschwiler, J. Thome. STEAM: A fast compact thermal model for two-phase cooling of integrated circuits, in: *IEEE Proc. of International Conference on Computer-Aided Design (ICCAD)*, 2013, pp. 256–263.

[18] K. Skadron, M.R. Stan, W. Huang, S. Velusamy, K. Sankaranarayanan, D. Tarjan. Temperature-aware microarchitecture, in: *30th Annual International Symposium on Computer Architecture, 2003. Proceedings,* June 2003, pp. 2–13.

[19] J. Meng, K. Kawakami, A.K. Coskun, Optimizing energy efficiency of 3-D multicore systems with stacked dram under power and thermal constraints, in: *Proceedings of the 49th Annual Design Automation Conference*, 2012, pp. 648–655, ACM.

[20] A. Sridhar, A. Vincenzi, M. Ruggiero, T. Brunschwiler, D. Atienza. 3D-ICE: Fast compact transient thermal modeling for 3D ICs with inter-tier liquid cooling, in: *IEEE Proc. of the International Conference on Computer-Aided Design (ICCAD)*, 2013, pp. 463–470.

[21] Z. Yuan, G. Vaartstra, P. Shukla, S. Reda, E. Wang, A.K. Coskun, Modeling and optimization of chip cooling with two-phase vapor chambers, in: *Proc. of IEEE/ACM International Symposium on Low Power Electronics and Design (ISLPED)*, 2019.

[22] F. Kaplan, M. Said, S. Reda, A.K. Coskun, Locool: Fighting hot spots locally for improving system energy efficiency, *IEEE Trans. Comput. Aided Des. Integrated Circ. Syst.*, 39, 2019, pp. 895–908.

[23] V. Sahu, Y. K. Joshi, A. G. Fedorov, J.-H. Bahk, X. Wang, A. Shakouri, Experimental characterization of hybrid solid-state and fluidic cooling for thermal management of localized hotspots, *IEEE Trans. Compon. Packag. Manuf. Technol.*, 5 (1), 2014, 57–64.

[24] A. Sridhar, A. Vincenzi, M. Ruggiero, T. Brunschwiler, D. Atienza, Compact transient thermal model for 3D ICS with liquid cooling via enhanced heat transfer cavity geometries, in: *Thermal Investigations of ICs and Systems (THERMINIC), 2010 16th International Workshop on*, 2010, pp. 1–6, IEEE.

[25] Y. Hu, S. Chen, L. Peng, E. Song, J.-W. Choi, Effective thermal control techniques for liquid-cooled 3D multi-core processors, in: *International Symposium on Quality Electronic Design (ISQED)*, 2013, pp. 8–15, IEEE.

[26] J. Thome, *Heat Transfer Engineering Data Book III*, Wolverine Tube Inc., 2010.

[27] M. Wambsganss, D. France, J. Jendrzejczyk, T. Tran, Boiling heat transfer in a horizontal small-diameter tube, *J. Heat Transfer*, 115 (4), 1993, 963–972.

[28] S.S. Mehendale, A.M. Jacobi, Evaporative heat transfer in mesoscale heat exchangers, Technical report, Univ. of Illinois, Urbana, IL, 2000.

[29] P.A. Kew, K. Cornwell, Correlations for the prediction of boiling heat transfer in small-diameter channels, *Appl. Thermal Eng.*, 17 (8–10), 1997, 705–715.

[30] D. Wen, Y. Yan, D. Kenning, Saturated flow boiling of water in a narrow channel: Time-averaged heat transfer coefficients and correlations, *Appl. Thermal Eng.*, 24 (8–9), 2004, 1207–1223.

[31] L. Zhang, J.-M. Koo, L. Jiang, M. Asheghi, K.E. Goodson, J.G. Santiago, T.W. Kenny, Measurements and modeling of two-phase flow in microchannels with nearly constant heat flux boundary conditions, *J. Microelectromech. Syst.*, 11 (1), 2002, 12–19.

[32] W. Qu, I. Mudawar, Flow boiling heat transfer in two-phase micro-channel heat sinks — II. Annular two-phase flow model, *Int. J. Heat Mass Transfer*, 46 (15), 2003, 2773–2784.

[33] J. Lee, I. Mudawar, Two-phase flow in high-heat-flux micro-channel heat sink for refrigeration cooling applications: Part I — Pressure drop characteristics, *Int. J. Heat Mass Transfer*, 48 (5), 2005, 928–940.

[34] E.W. Lemmon, M.L. Huber, M.O. McLinden, NIST reference fluid thermodynamic and transport properties–refprop, *NIST Standard Ref. Database*, 23, 2002, v7.

[35] S.S. Bertsch, E.A. Groll, S.V. Garimella, A composite heat transfer correlation for saturated flow boiling in small channels, *Int. J. Heat Mass Transfer*, 52 (7–8), 2009, 2110–2118.

[36] S.G. Kandlikar, P. Balasubramanian, An extension of the flow boiling correlation to transition, laminar, and deep laminar flows in minichannels and microchannels, *Heat Transfer Eng.*, 25 (3), 2004, 86–93.

[37] T. Tran, M. Wambsganss, D. France, Small circular-and rectangular-channel boiling with two refrigerants, *Int. J. Multiphase Flow*, 22 (3), 1996, 485–498.

[38] S. Ravi, R. Dharmarajan, S. Moghaddam, Physics of fluid transport in hybrid biporous capillary wicking microstructures, *Langmuir*, 32 (33), 2016, 8289–8297.

[39] Y.-T. Chen, S.-W. Kang, Y.-H. Hung, C.-H. Huang, K.-C. Chien, Feasibility study of an aluminum vapor chamber with radial grooved and sintered powders wick structures, *Appl. Thermal Eng.*, 51 (1–2), 2013, 864–870.

[40] J. Hsieh, H. Huang, S. Shen, Experimental study of microrectangular groove structure covered with multi mesh layers on performance of flat plate heat pipe for LED lighting module, *Microelectron. Reliab.*, 52 (6), 2012, 1071–1079.

[41] Z. Sun, H. Qiu, An asymmetrical vapor chamber with multiscale micro/nanostructured surfaces, *Int. Commun. Heat Mass Transfer*, 58, 2014, 40–44.

[42] A.A. Attia, B.T. El-Assal, Experimental investigation of vapor chamber with different working fluids at different charge ratios, *Heat Pipe Sci. Technol.*, 3 (1), 2012, 132–143.

[43] T. Shimura, H. Sho, Y. Nakamura, The aluminum flat heat pipe using cyclopentane as working fluid, in: *ITherm 2002. Eighth Intersociety Conference on Thermal and Thermomechanical Phenomena in Electronic Systems* (Cat. No. 02CH37258), 2002, 224–229, IEEE.

[44] T.-E. Tsai, H.-H. Wu, C.-C. Chang, S.-L. Chen, Two-phase closed thermosyphon vapor-chamber system for electronic cooling, *Int. Commun. Heat Mass Transfer*, 37 (5), 2010, 484–489.

[45] S. Lips, F. Lefèvre, J. Bonjour, Combined effects of the filling ratio and the vapour space thickness on the performance of a flat plate heat pipe, *Int. J. Heat Mass Transfer*, 53 (4), 2010, 694–702.

[46] H. Peng, J. Li, X. Ling, Study on heat transfer performance of an aluminum flat plate heat pipe with fins in vapor chamber, *Energy Convers. Manage.*, 74, 2013, 44–50.

[47] H.-Y. Li, M.-H. Chiang, C.-I. Lee, W.-J. Yang, Thermal performance of plate-fin vapor chamber heat sinks, *Int. Commun. Heat Mass Transfer*, 37 (7), 2010, 731–738.

[48] M. Wei *et al.*, Optimization and thermal characterization of uniform silicon micropillar based evaporators, *Int. J. Heat Mass Transfer.*, 127, 2018, 51–60.

[49] S. Adera, D. Antao, R. Raj, E.N. Wang, Design of micropillar wicks for thin-film evaporation, *Int. J. Heat Mass Transfer*, 101, 2016, 280–294.

[50] Z. Yuan, G. Vaartstra, P. Shukla, Z. Lu, E. Wang, S. Reda, A.K. Coskun, A learning-based thermal simulation framework for emerging two-phase cooling technologies, in: *2020 Design, Automation Test in Europe Conference Exhibition (DATE)*, 2020, 400–405, doi: 10.23919/DATE48585.2020.9116480.

[51] G. Vaartstra, Z. Lu, E.N. Wang, Simultaneous prediction of dryout heat flux and local temperature for thin film evaporation in micropillar wicks, *Int. J. Heat Mass Transfer*, 136, 2019, 170–177.

https://doi.org/10.1142/9789811279379_0002

Chapter 2

Microscale Evaporation for High Heat Flux Applications

Damena D. Agonafer*, Mun Mun Nahar*, Binjian Ma†, Zhikai Yang*, Quan H. Chau*, Erdong Song*, Jorge Padilla‡, and Madhusudan Iyengar‡

**Mechanical Engineering and Materials Science, University of Maryland, College Park MD, Washington University, Saint Louis, MO, USA*
†Mechanical Engineering and Automation, Harbin Institute of Technology (Shenzhen), Shenzhen, China
‡Google LLC, Mountain View, CA, USA

Abstract

To improve the capacities and capabilities of microelectronic devices, 3D packaging is an attractive solution which can provide increased functional density at a decreased cost per function. However, classical cooling approaches, such as the use of a heat sink or cold plate on the topmost component, are gradually becoming incapable of maintaining a temperature below the manufacturer-prescribed maximum junction temperature in a 3D-packaged system since the heat is compounded between different die layers. Using micro- and nanofluidic systems to expose the die directly to a dielectric coolant, known as embedded cooling, can provide far lower junction-to-ambient thermal resistances. Further, employing two-phase heat transfer methods can facilitate ultrahigh heat removal to tackle die-level hotspots reaching 1 kW/cm^2 on each high-power tier in a 3D microelectronic device. This chapter presents a brief overview of microscale evaporative cooling technologies that can be implemented in embedded cooling methods capable of removing ultrahigh heat fluxes from 3D chips. The following two major topics are covered in this chapter: (1) device fabrication and integration, and (2) the characterization of thermal and operational performances through temperature measurement and flow visualization. Additionally, we emphasize research findings related to the use of hollow micropillars to create asymmetric evaporating droplet arrays. These droplet arrays can provide an ultrahigh cooling performance that exceeds the maximum heat flux reported earlier in the literature and reaches up to 14.70 kW/cm^2, with a heat transfer coefficient of 7 × 106 W/m^{2}K.

Keywords: Evaporative cooling technologies, 3D microelectronics packaging, evaporating droplet arrays, high heat flux dissipation, fabrication challenges

Nomenclature

R	Droplet radius (m)
Θ	Contact angle (°)
c_{sat}	Saturation concentration
c_∞	Concentration at ambient conditions
J	Diffusive flux (kg/m^{2}s)
D	Diffusion coefficient (m^2/s)
h	Heat transfer coefficient (W/m^{2}K)
$g(\theta)$	Function of the contact angle
ρ	Density (kg/m^3)
ρ_l	Density at the droplet surface (kg/m^3)
ρ_m	Counterion concentration at the mid-plane of a charged surface (kg/m^3)
ρ_∞	Counterion concentration at far end of the bulk (kg/m^3)
p_l	Pressure at the liquid–gas interface (Pa)
p_∞	Pressure far away from the interface (Pa)
p_v	Vapor pressure at the interface (Pa)
T_∞	Ambient temperature (°C)
T_l	Temperature (°C) at the droplet surface
T_v	Vapor temperature (°C) at the droplet surface

m_v	Total evaporation rate (kg/s)
M	Molecular weight (g/mol)
σ_e	Evaporation coefficient (no unit)
σ_c	Condensation coefficient (no unit)
σ	Accommodation coefficient (no unit)
γ_{H20}	Surface tension coefficient of water (N/m)
h_{fg}	Heat of vaporization (kJ/kg)

1. Introduction

IC technology has advanced rapidly in the past several decades to provide enhanced performance and functionality along with reduced power and cost [1]. According to Moore's Law, the number of transistors on a microchip doubles every 18–24 months, while the cost of computers is halved [2]. To achieve better performance, transistor scaling has been pursued by implementing new materials and processing innovations. In the past 15 years, as the minimum transistor dimension has shrunk from 65 to 5 nm, design and fabrication processes have become more complex [3–5], with corresponding increases in manufacturing costs. Therefore, to enable further improvements in system performance without substantial cost increases, recent technological innovation has been directed toward microarchitectural design, die stacking technology, and the development of heterogenous platforms. In this regard, 2.5D and 3D chip-stacking technologies have been introduced as more efficient alternatives to traditional 2D stacking technology [1, 6–13]. In 2.5D ICs, multiple chips are placed side by side on a silicon interposer [3, 14]. In 3D ICs, silicon wafers and/or dies are stacked vertically, and communication between layers is provided by through-silicon vias (TSVs) or wire bonding. In this way, all the layers behave as a single device [3, 14]. One key advantage of 2.5D and 3D technologies is the prospect of heterogeneous integration [15]. In 2.5D, different dies are mounted on an interposer and can accommodate disparate processes or technology nodes. In 3D technology, dies can be stacked on top of each other and electrically connected by TSVs. Utilizing heterogeneous integration, each die can be implemented at the most appropriate technology node. In particular, a high-performance digital logic chip can be created at the latest and greatest (and more expensive) technology node, while lower-performance digital logic devices can be developed using an earlier (and more affordable) technology node. For example, a HBM 3D stacked memory die can be mounted on the interposer, while an accompanying processor die is manufactured via 7 nm process technology [16–18]. Other benefits of using 2.5D silicon packages and 3D die stacking include increased functionality by implementing heterogenous integration, product miniaturization, lower latency, and higher bandwidth [7]. 3D chip stacking can provide additional advantages, including a smaller device footprint, lower power consumption achieved by on-chip signaling, and shorter interconnect lengths. Fabrication cost is also reduced by the increased yield since multiple chips can be accommodated on a single device [7, 12, 19–21]. One example of a commercially available 3D IC is an image sensor manufactured by Monolithic 3D Inc., in which CMOS signal processing arrays are integrated with multiple layers of visible and infrared image sensors. 3D stacking endows this integrated image sensor with multiple capabilities, such as day and night view modes, high imaging resolutions (>20 A/W) [22], high capture speeds, a large dynamic range, a dramatically reduced size, low power consumption [23] (10% of the power consumed by traditional sensors), and lower costs [59]. Although earlier applications of 2.5D and 3D IC technologies are most commonly found in consumer electronics and portable devices, these technologies are expected to be adopted in high-power applications in the near future, such as high-performance computing, the Internet of Things (IOT), graphic processors, and artificial intelligence (AI) processors utilizing high-bandwidth memories (HBMs) [58].

As high-performance applications used more 3D ICs and packed them more closely, their thermal management started to become a challenging issue. Very high temperatures can be formed in a 3D IC due to the following three factors: (i) the overlapping of multiple high-power-density blocks creates a high effective power density; (ii) multiple vertical chip layers increase the total thermal resistance in the path of heat transfer from the chip to the heat sink [24–26]; and (iii) temperature non-uniformity is created by temperature gradients across different dies and by localized hotspots [24, 27]. These large temperature gradients often cause thermal failure, as reported by Li *et al.* [28]. Although they only powered the microbumps under one chip in their tested setup, the unpowered microbumps under another chip failed, with big holes formed in the solder layer due to Joule heating. The heat dissipated from the powered microbumps was transferred horizontally to the bottom of the neighboring unpowered microbumps and created a large temperature gradient of the order of 1,000°C/cm in the lateral direction.

Conventional cooling techniques used in 2D chip packages rely on a remote cooling paradigm, where heat is rejected to a remote single-phase fluid that has been conducted through multiple material interfaces [29, 30]. However, for 3D ICs, such "remote cooling" is not a sufficient thermal management solution because only heat from the topmost die is effectively removed by the heat sink [26, 31–33]. To address the thermal issues associated with 3D IC technology, direct liquid cooling techniques, including immersion cooling and embedded cooling, have been introduced [29, 34, 35]. Immersion cooling involves submerging an electronic component in a dielectric coolant liquid [34–37]. Dielectric liquids are especially attractive for direct liquid cooling since they are electrically non-conductive and chemically compatible with most heat spreading structures and materials. Additionally, they are typically environmentally friendly. Sinha *et al.* immersed 3D IC stacks in a dielectric fluid contained in an enclosure, where heat is removed from the 3D stack to the walls of the enclosure through single-phase convection and conduction. From there, a "synthetic jet/fan air-cooled heat sink" was further employed to ultimately dissipate the heat into the ambient [35]. Two-phase heat transfer has also been utilized in immersion cooling, where electronic components are immersed in a dielectric liquid that undergoes a phase change from liquid to vapor [34]. In embedded cooling [38–40], a liquid coolant is pumped through microchannels [41], in microchannel heat sinks [42], or in manifold microchannel heat sinks [27, 43–46]. The coolants can also be supplied as microdroplets, such as in spray and jet cooling [47–49]. But in all these schemes, the choice of different heat transfer modes will yield tremendous variation in thermal performance.

Although single-phase cooling can dissipate high heat fluxes (1 kW/cm^2) while maintaining a small thermal resistance (0.1°C/W), its efficiency is fundamentally bound by the inverse relationship between the pressure head and the hydraulic diameter [45, 50–54]. Compared to single-phase cooling, two-phase cooling, utilizing either boiling or evaporation, is more desirable, since the large latent heat of vaporization of the fluid can dissipate large amounts of heat. In addition, the nearly constant temperature of the liquid–vapor mixture during the phase-change process can greatly reduce the thermal resistance between the heat source and the ambient environment [55]. While flow boiling in microchannels has shown great potential for dissipating high heat fluxes [56], it presents additional challenges, including very large pressure drops and significant flow instabilities [57–59], thereby limiting the heat removal capacity at the critical heat flux (CHF) [60–62]. In addition, the growth and advection of vapor bubbles in microchannels increases the pressure drop due to the added friction of the two-phase flow and the acceleration of the fluid during vaporization. This increased system pressure during two-phase flow increases the required pumping power and manifests itself as an additional thermal resistance by shifting saturation conditions so that the working fluid vaporizes at a higher temperature. Static Ledinegg instabilities created by the pressure rise during two-phase flow in parallel channel systems lead to flow maldistribution, starving boiling channels of fluid and leading to earlier dry-out [63–65]. Another major challenge for flow boiling is maintaining the optimum exit quality of vapor. The vapor quality, which mainly depends on the inlet coolant temperature and the mass flow rate, determines the different flow boiling regimes [66, 67]. The simultaneous flow of vapor and liquid in a confined channel can take various forms, depending on the distribution and extent of aggregation of the two phases, with each distinct vapor/liquid distribution referred to as a "flow regime".

The literature has identified four primary two-phase regimes: bubble, intermittent, annular, and stratified, as well as numerous sub-regimes [68]. Bubble flow is observed at low gas qualities and consists of dispersed small, spherical bubbles that are transported by the continuous liquid phase. With increasing vapor quality, the bubbles start to grow and coalesce. As the bubbles become larger, they will eventually be confined by the channel walls and continue to elongate axially. In this case, an intermittent flow regime is generated, which is characterized by large vapor "slugs" separated by liquid plugs or bridges. With a further increase in the vapor fraction, the large vapor bubbles merge to form a continuous annular stream consisting of a high-velocity vapor core surrounded by a thin, shear-driven liquid film around the perimeter of the channel. Many researchers have experimentally investigated phase-change heat transfer in microchannels [69–71] and found that very high heat transfer coefficients can be achieved in the annular flow regime at vapor qualities between 15% and 40%. Under these conditions, a large amount of heat is dissipated through the evaporation of the thin liquid film in addition to boiling. When the vapor volume fraction increases to more than 95% (and hence the liquid fraction is less than 5%), the liquid films become thinner, eventually attaining micron and sub-micron thicknesses. These thin liquid films can achieve extremely high heat transfer coefficients due to their decreased conduction resistances. However, once the vapor volume fraction increases further, the stability of these films is shattered, and the films are ruptured [67, 72, 73], which causes local dry-out and decreases the heat transfer coefficient. High vapor quality poses additional challenges, including flow fluctuations and flow bypass, both of which will reduce the local heat transfer coefficient [74]. Researchers have proposed various methods to overcome the challenges associated with boiling in microchannels. For example, since the heat transfer coefficient is independent of vapor quality in the nucleate boiling regime, micro–nano porous coatings have been used to introduce nucleation sites that maintain boiling in the nucleate boiling regime [75]. Another method that has been employed in past studies involves increasing the degree of coolant sub-cooling so that the exit quality is limited [72, 73]. Zhou *et al.* maintained a certain vapor quality by removing excess vapor through a porous hydrophobic membrane [76]. In their heat exchanger, the hydrophobic membrane was able to provide sufficient resistance to retain the liquid and prevent it from leaking out of the device. However, the vapor faces very low resistance to flowing through the membrane into a separate set of parallel channels on the other side of the membrane. The collected vapor is then reintroduced into the coolant flow downstream of the microchannels, where the impact of two-phase flow is less significant on the system. In addition to lowering the mass flux and maintaining a specific vapor quality, this vapor-venting technology can reduce the pressure drop by 60% [64]. Other techniques to mitigate the challenges posed by boiling include using micronozzles and microscale cavities instead of microchannels [77] (achieving a 55% reduction in pressure drop) and creating a flow constriction at the microchannel [78, 79] that prevents reverse flow caused by the Ledinegg instability.

The aforementioned challenges associated with flow boiling are inherently absent in evaporative heat transfer, where an exit vapor quality of 1 is naturally maintained. Compared to boiling, evaporation of liquid at the microscale is a simple process with promising heat transfer performance [80–92]. Evaporation can improve heat transfer by mitigating the resistance associated with vapor bubble formation and growth during boiling heat transfer. The potential of using thin liquid films for enhanced evaporation was first realized by Deryagin *et al.* [93], who demonstrated that the rate of evaporation from capillary tube walls was enhanced by a thin liquid film formed on a wall that was continuously replenished by capillary flow driven by a gradient in the long-range intermolecular force field [94]. This work led to extensive research on surface modifications to control mass and heat transport processes in thin liquid films. Apart from enhancing liquid feeding, a thin liquid layer also minimizes the thermal resistance across the liquid domain. For example, the conduction resistance of water at ambient conditions reaches about 1.67×10^{-6} m^2K/W for a film thickness of 1 μm, which approaches the evaporation resistance of 1.74×10^{-6} m^2K/W [95, 96]. Therefore, an ultrahigh local heat transfer coefficient, exceeding 10^6 W/m^2K, can be achieved from a microns-thick evaporating liquid body [41, 97, 98].

Inspired by naturally occurring biological surfaces that possess favorable chemical and topographical surface features, researchers have developed various forms of superhydrophobic [99–101], superhydrophilic [101–103], and omniphobic [100, 104] surfaces to realize controllable fluid flow. Chemically modified or topographically structured surfaces can affect phase change by modulating several interrelated phenomena [105]. Primarily, the size and morphological features of the surface determine the relative dominance of different thermal resistances, such as Kapitza or conduction resistance, which dictate the nature of solid–liquid–vapor interactions. Secondarily, microscale surface structures not only increase the solid–liquid interfacial area but also provide a path for capillary wicking [106–108] and additional nucleation sites that facilitate phase change [109–111]. Additionally, chemical surface modification or sub-micron/nanoscale surface structures [112–114] can change the surface energy, thus modifying both the surface wettability and the characteristics of fluid flow on the surface. Finally, micro- and nanoscale surface features can increase hydrodynamic slip at the solid–liquid interface by increasing the thickness of the adsorbed film [115–117].

Based on the aforementioned principles, different techniques have been developed to enhance microscale heat transfer, such as surface coatings, the use of wicking structures [94, 108, 114, 118–127], and micro- and nanomembrane devices [128, 129]. Additionally, to utilize enhanced evaporation due to smaller conduction resistances in thin liquid films, the thickness of the evaporating liquid film has been controlled in microchannels by using a shear flow of gas/vapor [130–133] and in microgrooves and microcavities by maintaining a finite thickness in the reservoir [45, 134]. The evaporation of sessile microdroplets (1–100 μm in diameter) has been proposed as another technique for enhancing microscale heat transfer. These droplets can be dispersed on heated substrates [82–84, 92] either by spray and jet impingement techniques [48, 126, 135] or by pumping through a hollow micropillar structure [136, 137]. All these methods seek to realize the same goal of maximizing the region with the maximum evaporation rate, i.e., the contact line region of the evaporating meniscus. The ultrathin liquid layer near the contact line possesses a minimal conduction resistance across the liquid, which yields an ultrahigh local heat transfer coefficient that exceeds 10^6 W/m^{2}K [109, 125, 138, 139]. However, the overall heat transfer rate is restricted by the limited fraction of the total evaporating area occupied by the thin-film region.

One critical aspect of implementing micro- and nanostructures for enhancing microscale evaporation is that their geometry and dimensions can significantly affect heat dissipation through the thin liquid layer. For example, Ranjan *et al.* [109, 125] investigated the interfacial evaporative heat flux for four different microstructures: horizontal wires, rectangular ribs, vertical wires, and packed spheres. Other than the horizontal wires, which performed the worst, the other three structures dissipated almost the same amount of heat. Through experimental and numerical analyses, Farokhnia *et al.* [140] proposed a rational strategy to develop micro-/nanostructures for thin-film evaporation. Their study revealed opposite effects of the non-uniform heat flux along the liquid–vapor meniscus and further demonstrated that the total evaporating area should be used as an optimal criterion for designing the dimensions of the micro-/nanostructures. For vertical rectangular pillars (30 μm high), the optimal ratio of the pillars' height to their pitch was 1.27; and for vertical wires (30 μm high), the optimal ratio was 1.5. At these optimal ratios, a smaller pillar dimension (i.e., a higher micropillar density) will result in a higher interfacial heat flux, irrespective of the pillar height, contact angle, or superheating. For microscale droplets, the evaporation rate depends on the droplet shape [139, 141, 142]. For example, compared to capped spherical droplets, asymmetric droplets exhibit larger average curvatures, k, and a larger curvature induces a higher vapor concentration gradient in the vicinity of the liquid–vapor interface. These droplets also have a larger ratio of their perimeter and interfacial solid–liquid area, (A_{sl}), which drives a greater fraction of heat transport at the contact line region. Both these factors can enhance the evaporation rate, as demonstrated by Saenz *et al.*, who performed evaporation experiments on non-heated substrates [141]. Similarly, Li *et al.* [137] and Ma *et al.* [139] investigated the evaporation of asymmetric droplets confined in hollow micropillar structures. They found that a non-spherical droplet with a triangular footprint, compared to a spherical droplet, can have a 71% greater heat transfer coefficient due to its increased perimeter-to-wetted area ratio, increased radius of curvature, and reduced confinement effects.

Although evaporative cooling using asymmetric droplets is very promising for the thermal management of high-performance computing applications, certain fabrication challenges must be overcome before it can be integrated into 3D stacked chip devices. Various micro-/nanoengineered surfaces have been utilized in evaporative cooling devices, such as porous membrane structures fabricated by etching silicon and copper substrates [129, 143–145]; wicking hierarchical micro/nanostructures, including copper microparticles, carbon nanotubes, and titania nanostructures [102, 107, 108, 127, 146, 147]; and microgrooves and microgap coolers [45, 55, 148]. All different components of these cooling devices, such as micro-/nanostructures, porous membrane structures, microposts/micropillars, and heating and temperature-sensing elements, are often fabricated using conventional micro-/nanofabrication processes in a clean room.

For example, thin conductive and passivation layers are deposited using either chemical- and plasma-enhanced chemical deposition (CVD and PECVD) or physical vapor deposition (PVD), such as sputtering and e-beam evaporation. Surface patterns are created using photolithography, and then, the undesired materials are removed using wet and dry etching processes, including reactive-ion etching (RIE), deep reactive-ion etching (DRIE), and lift-off. The greatest challenge in implementing these cooling schemes is the difficulty of integrating them into the device package [45, 149]. Contamination and clogging of porous microstructures can degrade thermal performance, leading to reliability problems with two-phase embedded cooling technologies [150, 151]. Another potential difficulty is the integration of micro- and nanostructures and TSVs in embedded-cooling systems. Due to the small length scale of the micro- and nanostructures and TSVs, the alignment accuracy of the bonding process is very important for the successful integration of different device components. In addition, the peak temperature of the bonding process normally exceeds 200°C [152], which can cause thermo-mechanical failure during the bonding process, damaging the sample.

This chapter first discusses the different microscale evaporative cooling methods and then highlights the heat transfer characteristics of an evaporative cooling technology developed at the Nanoscale Energy and Interfacial Transport (NEIT) Laboratory at Washington University in Saint Louis. With a pressure-controlled microfluidic system, this technology utilizes stable evaporating microdroplets formed by feeding the working fluid to the top of a micropillar through a central hole. As shown in Figure 1, heat can be removed from a heated substrate using an array of asymmetric droplets. The temperature and velocity fields are measured using planar laser-induced fluorescence (PLIF) and micro-particle image velocimetry (μPIV) techniques. The effects of micropillar size and shape, micropillar aspect ratio, confinement, and micro-convection inside the droplet and the surrounding air domain are investigated using experimental and numerical methods. The results of these studies are briefly presented in Section 3. The dynamic behavior of the solid–liquid–vapor interface is characterized with an optical microscope. The fabrication process for this device

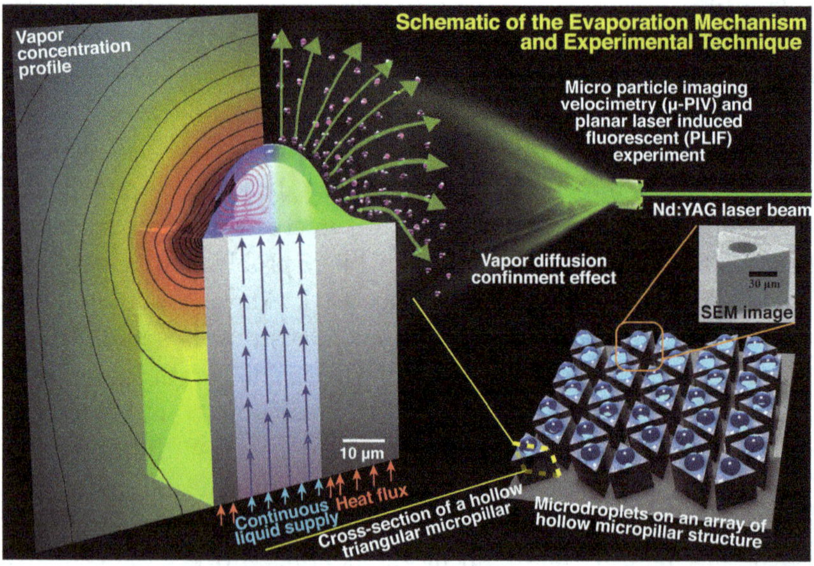

Figure 1. Schematic diagram of evaporation of a non-axisymmetric droplet suspended on a micropillar. The evaporation transport is affected by the conduction resistance across the pillar and droplet, the micro-convection inside the droplet, and the vapor diffusion transport in the ambient gas domain.

with the micropillar array and the associated challenges are also discussed.

2. Background of Evaporative Cooling

2.1. *Evaporation models*

Evaporation is a process in which liquid is changed into vapor at any temperature. Evaporation always occurs at the liquid–vapor interface, when some of the molecules at the interface gain enough kinetic energy to overcome the cohesive forces imposed by the neighboring molecules and escape into the surrounding medium. At a high temperature, a greater number of molecules can gain a higher kinetic energy, which leads to a greater evaporation rate. Evaporation can also be enhanced if there exists a surrounding gas/vapor flow since the evaporated molecules with high kinetic energy are carried away by the flowing gas before they are slowed down by molecular collisions and subsequently readsorbed into the liquid. In a humid environment, evaporation is suppressed since vapor molecules are more likely to condense back into liquid. A higher atmospheric pressure will also reduce the evaporation rate.

Most evaporation events are considered to be limited by the diffusion of vapor molecules in gas medium. One example of diffusion-limited evaporation is gas-assisted thin-film evaporation, which has been demonstrated as an efficient cooling solution to cool the computer modules of a supercomputer named Cray-3. In this method, a very thin liquid film is driven by shear created by a flowing gas in a microchannel. The evaporation of the thin liquid film is the dominant heat transfer mode in such a system [153, 154]. Another practical application of evaporation in the presence of non-condensable gases is the electrospray cooling technique, which can be used for the thermal management of high-density electronic modules. In the electrospray technique, the working fluid is atomized by applying a voltage to a collecting electrode opposite the fluid nozzle. This voltage forces the charges within the liquid to gather at the fluid meniscus. As the applied voltage increases, the electric field strength and the charge density at the surface increase as well, which deforms the meniscus into a conical shape, thus creating the Taylor cone. At the critical electric field intensity, the forces on the charged fluid in the Taylor cone overcome the intramolecular forces of the fluid, such as viscosity, surface tension, and liquid momentum, and a jet of charged liquid is then sprayed from the tip of the cone. The charged fluid particles expelled from the tip of the fluid cone repel each other, generating fine aerosol droplets. Those charged droplets are then accelerated by the electrostatic force toward the collecting electrode surface. The collecting electrode also acts as a thermal exchange surface. As a result, the spray droplets evaporate, thus dissipating a large amount of heat [155].

In these cases, the simple vapor diffusion model, proposed by Popov [156], can be used to model droplet evaporation for estimating the evaporation rate and evaporative flux. According to Popov's model, the evaporative flux is given by

$$J = \frac{D(c_{\mathrm{sat}} - c_{\infty})}{R} \times \left[\frac{1}{2}\sin\theta + \sqrt{2}\,(\cosh\alpha + \cos\theta)^{\frac{3}{2}} \right.$$
$$\left. \times \int_0^{\infty} \frac{\tau\cosh\theta\tau}{\cosh\pi\tau}\tanh\tau\,(\pi - \theta)\,P_{-\frac{1}{2}+i\tau}\,(\cosh\alpha)\,d\tau\right] \quad (1)$$

where D is the diffusion coefficient, c_{sat} is the concentration of saturated vapor at the liquid–vapor interface, c_{∞} is the concentration of vapor at the ambient away from the interface, θ is the apparent contact angle of the evaporating meniscus, and R is the radius of the evaporating droplet. $P_{-1/2+i\tau}$ is the Legendre function of the first kind with the argument

$$\cosh(\alpha) = \frac{r^2\cos(\theta) \pm \sqrt{R^2 - r^2\sin^2(\theta)}}{R^2 - r^2} \quad (2)$$

The rate of change of droplet volume can be expressed as

$$\frac{\mathrm{d}V}{\mathrm{d}t} = -\frac{\pi D\,(c_{\mathrm{sat}} - c_{\infty})}{\rho}\frac{Rg(\theta)}{(1 + \cos(\theta))^2} \quad (3)$$

where the function $g(\theta)$ is defined by

$$g(\theta) = (1 + \cos(\theta))^2\left\{\tan\left(\frac{\theta}{2}\right) + 8\int_0^{\infty}\frac{\cosh^2(\theta\tau)}{\sinh(2\pi\tau)}\tanh[\tau(\pi - \theta)]\mathrm{d}\tau\right\} \quad (4)$$

Equation (4) determines the lifetime of the evaporating liquid body, defined as the time it takes R and/or θ, as well as V, to reach zero [157]. Hu and Larson [84] adopted this model and proposed correlations for the evaluation of evaporation rates. Researchers have found that, in addition to vapor diffusion, droplet evaporation involves several other transport mechanisms. For example, droplet evaporation can be accompanied by self-cooling or interface cooling, where the gas–liquid interface is cooled due to latent heat supplied for phase change by the bulk liquid. As a result, temperature gradients arise on the gas–liquid interface, which can drive heat conduction in the solid, liquid, and gas phases and simultaneously create internal convection events. Furthermore, sessile droplet evaporation is also affected by the substrate's characteristics, specifically its thermal properties and surface wettability. Therefore, researchers have modified and extended the diffusion model. For example, Dunn *et al.* developed a mathematical model incorporating both interface cooling and heat conduction in all three phases [158]. The predicted evaporation rates were in very good agreement with experiments. Saada *et al.* improved the model developed by Dunn *et al.* by adding thermal diffusion effects in the gas phase [159]. Popov improved his earlier model to consider surface wettability effects: This new model can predict the evaporative flux and evaporation rate for a droplet over a range of contact angles ($0° < \theta < 180°$) [156]. Singh *et al.* built a numerical model to predict the evaporation rate, interfacial evaporative flux, and temperature distribution around a sessile droplet [89]. This model considers interface cooling and substrate properties, such as surface wettability and thermal conduction. Saada *et al.* developed a convection–diffusion model to analyze the effect of buoyant convection in the surrounding air on heat and mass transfer during evaporation [160]. These studies have all found that the diffusion model always underpredicts the overall evaporation rate and that the deviation is larger at higher substrate temperatures.

In many evaporative systems, non-condensable gases are removed to maintain a pure vapor environment, which eliminates diffusion resistance and facilitates evaporation [161, 162]. Diffusion resistance can also be negligible in systems where the evaporating surface has very small dimensions, such as nanoporous structures. For example, it can be applied to nanoporous membrane devices, which can be used for thermal management in GaN power amplifiers [163]. Evaporation in a pure vapor environment can also be employed in a vapor chamber used in the cooling unit for microprocessors and chipsets in a computer. Due to increased heat dissipation in these components, improvements are needed in all aspects of the design of the cooling solutions. This mainly includes increasing the heat transfer coefficient of the working medium and reducing the thermal spreading resistance of the heat sink base. Using a vapor chamber instead of a solid metal heat sink can serve those two purposes. In the vapor chamber, vacuum pressure is created and become saturated with the working liquid. As heat is supplied, the liquid immediately evaporates, and the vapor rushes to fill the vacuum. Whenever the vapor touches the colder wall surface, it condenses, thus releasing the latent heat of vaporization. The condensed liquid then returns to its reservoir via capillary force and/or gravity force, ready to be vaporized again, and repeat the cycle [162]. Another example of evaporation in the absence of non-condensable gases is mechanical vapor compression desalination systems. In the vapor compression process, water vapor from seawater is collected and compressed, which condenses the vapor. The saline water evaporates using the heat from vapor compression rather than the direct exchange of heat from a boiler [164]. The fundamental transport mechanism of evaporation in these systems is governed by the gas kinetics across a very thin non-equilibrium vapor region, known as the Knudsen layer [165]. This layer originates when the net flux across the interface disrupts the thermodynamic equilibrium. Unlike evaporation in a medium with non-condensable gases, for evaporation in a pure vapor environment, there is a pressure difference between the interface and the far field that enhances evaporation [166]. Classical kinetic theory has been applied to the kinetically limited Hertz–Knudsen (HK) evaporation model. In this model, the vapor is approximated as an ideal gas in equilibrium with the liquid–vapor interface and the liquid phase. The velocity distribution of the vapor molecules is approximated using the Maxwell–Boltzmann (M–B) velocity distribution function. The MB velocity distribution is strictly applicable only in equilibrium conditions, where the evaporation flux is zero. Therefore, unless the evaporation is happening under near-equilibrium conditions, the M–B velocity distribution is not valid. The following is the derivation of the HK formula [167]. Equation (5) gives the HK formula, giving the mass flux:

$$J = \sqrt{\frac{M}{2\pi k}}\left(\sigma_e \frac{p_l}{\sqrt{T_l}} - \sigma_c \frac{p_\infty}{\sqrt{T_\infty}}\right) \tag{5}$$

The HK formula assumes that the flux of the molecules moving from the liquid surface of the Knudsen layer depends on the liquid temperature at the interface, T_l, and the vapor pressure, p_l, at a temperature, T_l. On the other side, the flux of impinging vapor molecules from the vapor phase to the liquid surface depends on the vapor temperature, T_∞, and the vapor pressure, p_∞, at the boundary of the Knudsen layer.

The HK formula is only valid when the condensing molecules do not destroy the equilibrium created between colliding vapor molecules. This can happen in the case of condensation on a small liquid body (e.g., an aerosol droplet) surrounded by vapor. Later, Schrage assumed that if there is a net vapor flow toward the interface, the distribution function far away from the interface would be a Maxwellian flowing away from the liquid–vapor interface, and the evaporative mass flux across the interface according to Schrage's model (HKS) [168] is

$$J(0) = J(\infty) = \frac{2}{2-\sigma_c}\sqrt{\frac{M}{2\pi k}}\left(\sigma_e \frac{p_l}{\sqrt{T_l}} - \sigma_c \frac{p_\infty}{\sqrt{T_\infty}}\right) \tag{6}$$

Here, T_l is the temperature of the liquid–vapor meniscus, p_l is the vapor pressure at temperature T_l, and T_∞ and p_∞ are, respectively, the temperature and pressure of vapor at the boundary of the Knudsen layer. σ_e is the mean evaporation coefficient, and σ_c is the mean condensation coefficient [169–171].

Although the HK and HKS equations have been used for a long time, they have repeatedly shown inconsistent results. One reason is the lack of consistency in the values of the accommodation coefficients [169]. Many numerical, theoretical, and experimental studies have been performed to determine the coefficient values, where the values were very diverse; for water alone, the coefficient values ranged from the order of 10^{-3} to 1. Attempts to explain such a wide range of coefficient values have produced several classes of experiments [170]. Many researchers assume the correct value to be unity, even though a large number of experimental studies have reported otherwise. Several factors are assumed to cause discrepancies in the coefficient values from experiment to experiment, including problems with the HK relation itself, the configurations of the experiments, the presence of contaminants, and improper surface temperature and pressure measurements [171]. Although it is usually assumed that the evaporation and condensation coefficients are equal, Young concluded that it is not necessarily so [172]. According to Barrett and Clement [167], the evaporation and condensation coefficients are not equivalent; the evaporation coefficient depends only on the liquid properties, whereas the condensation coefficient is a function of the properties of both liquid and vapor. Julin *et al.* [173] and Barclay *et al.* investigated [174] the relation between the accommodation coefficient, σ (but in most cases, σ_c), and the droplet size or curvature. Julin *et al.* reported [173] that the accommodation coefficient slightly increases with droplet radius. For droplets with radii of 2 nm or larger, the curvature did not affect the value of the accommodation coefficient. Barclay *et al.* experimentally [174] measured the accommodation coefficient during condensation. Their findings agreed with those of Julin *et al.* For droplets with radii of 5 nm or smaller, σ decreased significantly with decreasing radius, but remained unchanged for droplets with radii larger than 5 nm.

Another limitation of the HK and HKS theories is that they do not satisfy the conservation of momentum and energy. Additionally, they are valid strictly at equilibrium and low vapor pressures [171]. Therefore, for non-equilibrium conditions, the HK and HKS models are not applicable. In those cases, statistical rate theory is a suitable model [175–179]. Recently, Lu *et al.* developed [166] a model to estimate the heat flux from an evaporating meniscus using the Boltzmann transport equation (BTE) and considering both mass and momentum conservation. Their non-dimensional heat flux has $\frac{\Delta p}{p_o}$ as the driving potential, where Δp is the pressure jump across the Knudsen layer and p_o is the saturation pressure at the liquid–vapor interface temperature. Although the driving potential for the heat flux derived from the HK and HKS models is Δp at low Mach number conditions, the results from these three models match closely.

2.2. *Transport processes and associated thermal resistances during microscale evaporation*

During evaporation from a liquid film, phase change occurs along the contact line region and liquid–vapor interface. Several heat transport processes are involved in this event, with their associated thermal resistances. As shown in Figure 2, the meniscus of a perfectly wetted or partially wetted liquid body can be modeled as four distinct regions with different film thicknesses: (1) the adsorbed film (<10 nm), (2) the transition region (10–1,000 nm), (3) the intrinsic meniscus region (1–10 μm) [134, 138, 180–182], and (4) the bulk region. The adsorbed film and transition region together constitute the thin-film region, which is dominated by long-range intermolecular forces. The thickness of the adsorbed film region can be as low as ~10 nm [183], and in this ultrathin layer, strong attractive forces

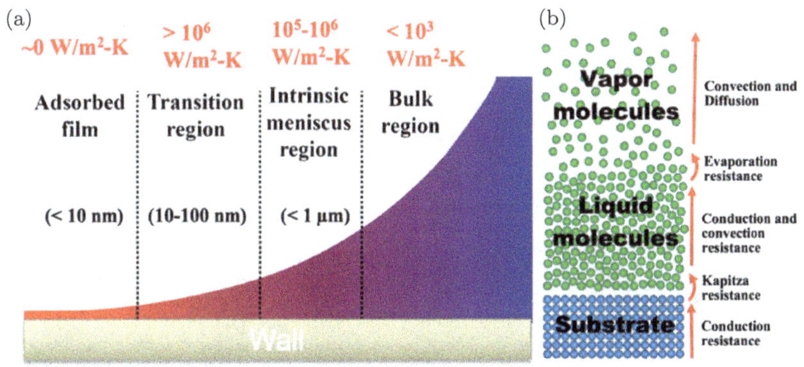

Figure 2. (a) The variation of local film thickness and heat transfer coefficient in different regions of an evaporating thin film (e.g., water). The heat transfer coefficient at the transition region is over 1,000 times greater than that at the bulk region. (b) Schematic illustration of different thermal resistances involved in the thin-film evaporation process.

between the solid substrate and the liquid effectively prevent evaporation and create a flat liquid–vapor interface. Next is the transition region, where the liquid–vapor interface and the heated solid substrate are increasingly separated, and the attractive forces between the liquid molecules at the interface and those at the heated surface are significantly reduced. In this region, a sharp change in the droplet curvature of the liquid–vapor interface creates large capillary forces at the liquid–vapor interface that drive liquid flow toward the contact line region during evaporation. The total thermal resistance in each region can be composed of several components: (1) solid–liquid interfacial thermal resistance (Kapitza resistance), (2) conduction and convection thermal resistances across the liquid body, which are functions of the film thickness, (3) evaporation resistance across the liquid–vapor interface, and (4) vapor diffusion mass transport and convection resistance in the gas/vapor domain. The magnitude of these resistances varies locally along an evaporating meniscus, which gives rise to spatial variation in the evaporation rate. Figure 3 shows an example of the total thermal resistance of the film thickness, and the contributions of each thermal resistance to the total resistance, with water as the evaporating liquid. Here, the convection thermal resistances inside the liquid film and in the gas domain have been neglected due to the small Peclet number, Marangoni number, and Rayleigh number at the micron and sub-micron length scales [184].

The evaporation rate from a meniscus is governed by the competition between two factors: a high superheat at the liquid–vapor interface and a suppressing liquid pressure caused by the capillary pressure p_c and/or the disjoining pressure Π. p_c depends on the meniscus curvature and decreases as the meniscus becomes flatter. Π is dominant in the transition region, where the liquid film thickness, $h_l < 100$ nm. For a film thickness larger than 150 nm, the total disjoining pressure balances the capillary pressure and becomes increasingly smaller for larger film thicknesses [185, 186].

Π arises due to long-range intermolecular forces between a liquid film and its solid substrate. This pressure is a manifestation of surface forces acting in thin liquid layers and decreases rapidly with increasing film thickness. The variations of Π and p_c in the meniscus lead to a gradient in the liquid pressure, which drives the liquid flow needed to maintain the evaporation [187]. The concept of disjoining pressure was first introduced by Derjaguin [188]. This concept has been widely applied to interactions in colloid systems, such as suspensions, emulsions, foams, and liquid films on solid surfaces. Disjoining pressure is also dominant in the vicinity of a three-phase contact line in the case of wetting/spreading. According to Gibbs, the intensive values of pressure at each point inside the volume of each phase of a heterogenous system, which, at equilibrium, are equal, except at the transition layer closer to the interfaces. In the transition zones, pressure becomes anisotropic, exhibiting a tensor characteristic which varies along the surface normal. For a flat interface, only the normal component of this tensor, $p_N = p_{zz}$, remains constant. In the equilibrium state, the condition $p_T = p_N = p$ is fulfilled only outside the transition zone, where p is the isotropic pressure in the bulk of adjacent phases. In the equilibrium state, the normal component is constant throughout the cross-section of a plane-parallel interlayer. The tangential component p_T changes in a complex way depending on the nature of the forces acting both in the interlayer and its neighborhood in adjacent phases.

When liquid gets in contact with a solid substrate, the liquid molecules will inevitably experience an attractive force from the solid atoms. For a sufficiently thin liquid layer, such an attractive force can become dominant over the intermolecular forces between the liquid molecules and therefore introduce an overwhelming pressure at the liquid surface. This pressure, known as the disjoining pressure, was first introduced by Derjaguin as the difference between the pressure of the interlayer on the surfaces confining it and the pressure in the bulk phase, the interlayer being part of this phase and/or in equilibrium with it.

According to the classical Derjaguin–Landau–Verwey–Overbeek (DLVO) theory, the disjoining pressure consists of two important components, including the electrostatic double-layer force and the van der Waals force. The electrostatic double-layer force originates from the charged surface and the neutralizing diffuse layer of counterions formed due to electrostatic repulsion. Based on the

Poisson–Boltzmann equation and Gouy–Chapman theory [189], the electrostatic double-layer interaction between charged surfaces in an electrolyte can be expressed by

$$\Pi = k_b T(\rho_m - \rho_\infty) \tag{7}$$

where k_b is Boltzmann's constant, T is temperature, ρ_m and ρ_∞ are the counterion concentrations at mid-plane of the charged surface and the far end in the bulk, respectively. Assuming a 1:1 electrolyte, one can have

$$\Pi = 2RTc_i \left[\cosh\left(\frac{e\Psi_{\text{mid}}}{k_b T}\right) - 1\right] \tag{8}$$

where R is the gas constant, c_i is the electrolyte concentration, e is the electric charge, and Ψ_{mid} is the potential at the mid-plane of the liquid film. Since the midplane potential Ψ_{mid} is very small, such a value can be considered the sum of potentials of each surface at $x = \delta/2$. Therefore, the electrostatic force between two planar surfaces can be estimated using weak overlap approximation, which yields

$$\Pi = 64c_i RT\tanh^2\left(\frac{e\Psi_{\text{mid}}}{4k_b T}\right)\exp\left(-\kappa\delta\right) \tag{9}$$

where κ is the Debye length that characterizes the distance of the diffuse electric double layer. The van der Waals interaction contribution of a non-polar liquid can be estimated using the Derjaguin approximation as

$$p_{vdw} = \frac{A}{6\pi\delta^3} \tag{10}$$

where A is the Hamaker constant (typically $\sim 10^{-20}$ J) and δ is the film thickness. The Hamaker constant is defined as

$$A = \pi^2 C\rho_1\rho_2 \tag{11}$$

where ρ_1 and ρ_2 are the number of atoms per unit volume in the two materials and C is the coefficient in the atom–atom pair potential. Equation (1) provides a simple yet effective method to estimate the disjoining pressure in a thin-film liquid. For example, Wayner *et al.* [190] applied Eq. (10) to calculate the evaporative heat transfer coefficient of a thin water film at the interline region. Wang *et al.* [191] used the same equation to calculate the distribution of evaporative heat flux along different regions of a thin film. However, it should be noted that the use of the non-polar disjoining pressure equation (Eq. (10)) to describe a polar liquid (e.g., water) has resulted in different thicknesses of the non-evaporating region as well as thin-film profiles in past literature studies [192–194]. Several attempts have been made to model the disjoining pressure more accurately for polar liquids. For example, a logarithmic model was presented by Holm and Goplen [194] based on Deryagin and Zorin's [195] experimental work as

$$p_{vdw} = \rho_l R_g T_v \ln(a\delta^b) \tag{12}$$

where ρ_l is the liquid density, R_g is the gas constant, T_v is the temperature, and a and b are dimensionless parameters obtained empirically from experimental measurements. Churaev and Derjaguin [196] proposed an exponential model of disjoining pressure by considering the effect of the structural force

$$p_{vdw} = \frac{A}{\delta^3} + Kexp\left(-\frac{\delta}{l}\right) \tag{13}$$

where K and l are experimental parameters defined for specific conditions. Basu and Sharma [197] proposed a revised equation to calculate the disjoining pressure as

$$p_{vdw} = \frac{A(15.96\delta/\lambda + 2)}{12\pi\delta^3(1 + 5.32\delta/\lambda)^2} \tag{14}$$

Combining Eq. (9) with Eqs. (10)–(14) provides a complete expression of disjoining pressure based on the DLVO theory. It should be noted that while the effect of electrostatic force has been typically neglected in past studies when analyzing the evaporation transport from a thin liquid film, some studies [198] have demonstrated a significant change in the shape of the interface due to electrostatic forces that extended the contact line, especially within the nanoscale-sized pores. As a result, the net rate of evaporation was promoted due to an increase in the total free surface area, as well as due to an enlarged thin-film region in the meniscus. Furthermore, a number of studies [199, 200] have shown significant discrepancies between the DLVO theory and the experimental measurement results from thin soap film, which suggests that the DLVO forces alone may not provide a full picture of the disjoining pressure. Therefore, it may be necessary to include non-DLVO forces, such as the steric, hydration, and hydrophobic forces, when analyzing the topology and evaporative transport behavior of complex fluids. The Kapitza resistance is the thermal resistance across the solid–liquid interface caused by the difference in the electronic and vibrational properties between the liquid and the solid substrate. Such a resistance was first discovered by Kapitza [201] in 1941, who observed a temperature discontinuity at the solid–liquid interface when investigating the superfluidity of helium at a temperature below 2 K. Later, this temperature jump was confirmed to be a universal phenomenon for all types of interfaces [202]. Traditionally, the Kapitza resistance has been considered to be significant mostly in cryogenic systems due to the following two reasons: (1) liquids generally possess very high thermal conductivities at ultralow temperature; (2) the Kapitza resistance scales approximately inversely with the cube of the absolute temperature [203]. With advances in micro and nanoengineering in recent years, however, the impact of the Kapitza resistance has become quite significant for certain nanostructured systems, along with a weak intermolecular interaction between the solid and the liquid. As a typical example, several studies have demonstrated that the heat flow in carbon nanotube composite systems is strongly impeded by the Kapitza resistance [204–206]. The Kapitza resistance across the solid–liquid interface is highly dependent on the material properties since such a resistance arises from a balance of the intermolecular forces between the two sides; when an energy carrier (phonon or electron, depending on the material) attempts to traverse the interface, it will scatter at the interface [207]. In general, however, the Kapitza resistance is smaller than 10^{-7} m^2K/W [208–212]. Therefore, it is often reasonable to neglect such resistance in an evaporative system at larger length scales ($\sim$1 μm), where conduction, evaporation, and diffusion resistances have values equal to or greater than $\sim 10^{-6}$ m^2K/W. However, as the film thickness becomes smaller, the magnitude of the

Kapitza resistance starts to become comparable with that of other resistances [213].

Conduction resistance also varies with film thickness. For example, the total thermal resistance reaches a minimum in the transition region. In this region, the comparatively thin liquid film thickness (<1 μm) [134, 183] imposes a moderate conduction resistance (<1.67×10^{-6} m^{2}K/W), yet the film thickness is large enough that the interfacial resistance (Kapitza resistance) decreases rapidly as the film thickens. Near the contact line, where the liquid thickness decreases to ~1 μm, the conduction resistance of water at ambient conditions reaches about 1.67×10^{-6} m^{2}K/W [214, 215].

When the film thickness increases from 0.1 to 10 μm, the percentage of conduction resistance in the total thermal resistance increases from 43% to 98% (Figure 3). The convection resistance inside the

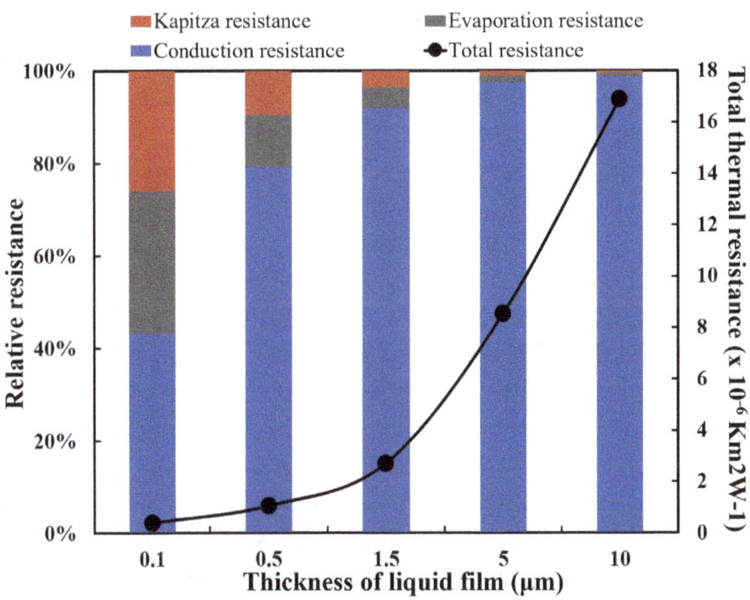

Figure 3. The variation of total thermal resistance and the percentage of each individual resistance with different evaporating film thicknesses. The percentages of each individual resistance are shown on the left vertical axis, while the total thermal resistance is shown on the right vertical axis. The contribution from the Kapitza resistance exceeds 25% of the total resistance when the film thickness becomes 100 nm.

liquid film also becomes significant at larger scales (>10 μm). The magnitude of the convection resistance is dependent on the strength of the convection current inside the liquid film, which can be represented by the Péclet number (Pe). Using the Péclet number, the relative strength of the convective heat transfer with respect to diffusive heat transfer can be assessed. In general, convection is only significant if the Péclet number is greater than 1. Inside the liquid body, different convection processes can happen, such as buoyancy-induced convection (Rayleigh convection) and thermo-capillary convection (Marangoni convection). Buoyancy-driven convection can occur due to the density difference created by temperature gradients inside the liquid body. Marangoni or thermo-capillary convection can be induced by surface tension gradient along the liquid–vapor interface due to temperature or solutal concentration gradient. Specifically, a temperature gradient can arise due to two factors: (1) self-cooling effect [142] and (2) conduction resistance across the liquid body on a heated substrate. The magnitudes of the buoyancy-driven and Marangoni convections can be characterized, respectively, by the Rayleigh number (Ra) and the Marangoni number (Ma). A critical Rayleigh number between 1,000 and 2,000 has been observed to be associated with the onset of buoyancy-driven convection [90]. Many studies on droplet evaporation have neglected the effect of natural convection due to the low Rayleigh number reported in their research [134, 216, 217]. It has also been shown that the strength of the natural convection scales with the cubic relationship of the characteristic length of the liquid film [218]. The strength of the convective flow induced by self-cooling is further analyzed in detail by Xu and Ma [219]. They proposed a dimensionless evaporative cooling number, E_c, to characterize the importance of the self-cooling effect on an evaporating microdroplet. The critical value for E_c is 1 [219–222], which means that when $E_c < 1$, the self-cooling effect is not important for evaporative heat transfer behavior. At room temperature (300 K), E_c for a water droplet is only 0.11, which is significantly smaller than the critical value but can be very large for droplets on heated substrates. For the onset of Marangoni convection, a minimum Marangoni number of 8,000 is required. On an easily contaminated surface, the Marangoni stress of a water body will be significantly reduced by even a small concentration, $\leq$300 molecules/μm^2, of an insoluble surfactant on the free surface. Moreover, other studies [142, 223, 224] have shown that the effect of Marangoni flow on evaporation is negligible for liquid films with a characteristic length of less than 60 μm. Although, typically, the micro-convection effects can be negligible at small scales, changing the droplet from a capped spherical to a non-axisymmetric shape and increasing the temperature can yield stronger micro-convection, even for droplets with smaller sizes. In particular, the convection current will flow in the circumferential direction in a non-axisymmetric droplet due to the non-uniform temperature and evaporation rate at different azimuthal angles [137], where the azimuthal angles are the local angles between the radial direction and the location of interest.

For evaporation of droplets, the evaporation resistance increases as the thickness of the droplet radius decreases and dominates the heat and mass transfer for droplets with radii below 1 μm [174]. The contribution of the evaporation resistance to the total thermal resistance also decreases with increasing liquid thickness. For example, for a film thickness of 0.1 μm, the evaporation resistance constitutes about 15% of the total resistance, while it becomes less than 1% of the total resistance when the film thickness reaches 10 μm (Figure 3). Lowering the system pressure can also lower the evaporation resistance. For example, the evaporation resistance reported in Figure 3 can be decreased by 80% when the pressure is lowered from 1 atm to 0.2 atm. The evaporation resistance also becomes very dominant in a pure vapor medium. In a pure vapor medium, where diffusion does not play any role in the vapor, the heat and mass transport can be completely governed by the evaporation resistance across the Knudsen layer, which is a few-molecules thick layer adjacent to the interface [161, 166].

In the case of high atmospheric pressure conditions or where non-condensable gases are present, diffusion and convection resistances dominate the mass and heat transport in the gas medium. Here, mass diffusion is governed by the concentration gradient of the vapor molecules and mainly depends on environmental conditions, such as temperature and pressure, rather than the dimension of the liquid film. Convection in the gas domain originates from two sources: (1) natural convection, and (2) the Stefan flow driven by the vapor evaporating from the gas–liquid interface [225–227]. Natural convection is composed of two components of buoyancy force: (1) an upward, vapor concentration-dependent, buoyancy force arising from the density difference between water vapor and air, and (2) a downward thermal buoyancy force due to the self-cooling near the liquid–gas interface. The overall natural convection pattern is then governed by the balance of these two factors. The magnitudes of both natural convection and Stefan flow are highly dependent on the length scale of the liquid film. Recently, Zhang *et al.* [228] characterized the effects of Stefan flow on the heat transfer performance of evaporating droplets. They showed that Stefan flow suppressed natural convection and became dominant with a Stefan flow Reynolds number (Re_{V0}) of >150. Such conditions are often satisfied when the characteristic length of the liquid film is greater than 1 mm, while the effect of Stefan flow is completely negligible [224].

One important characteristic of microscale evaporation is that the thermal resistance across the bulk region of the droplet is at least 10 times larger than that across the thin-film region of the droplet, which limits the evaporation rate in the bulk region. Thus, the overall heat transfer rate is restricted by the limited fraction of the total evaporating area that is occupied by the thin-film region [229, 230]. There, the rate of heat transfer depends on the conduction/convection resistance across the liquid film body and the mass transfer resistance to vapor transport from the evaporation interface to the ambient environment. If the liquid film thickness is larger than 100 μm, it suffers from a high conduction resistance, which is inversely proportional to the local film thickness. This high conduction resistance creates a lower liquid–vapor interfacial temperature, which in turn creates a higher diffusion resistance [139, 231].

2.3. Microscale evaporation in microchannels and microgap devices

As we discussed in the introduction section of this chapter, among different flow boiling regimes in microchannels, the annular flow regime is associated with two distinct two-phase heat transfer modes: boiling and thin-film evaporation [68–71]. In this regime, a thin liquid layer flows along the channel periphery and a thick vapor core flows in the center. The thin liquid layer has a very small conduction resistance and can transport heat very effectively. Harirchian and Garimella [232] studied the flow patterns using flow visualization in microchannels with a height of 400 μm and widths ranging from 100 to 5,850 μm. They found the annular flow regime to dominate over other regimes in the smallest channels. In a study of the two-phase flow of FC-72 in a microchannel with a height of 276 μm and a width of 225 μm, Hassan found annular flow to dominate the flow behaviors of channels operating at vapor qualities of 0.25 and above. Lee *et al.* [67] also reported annular flow to dominate the hydrodynamic behavior of R113 in a rectangular channel with a hydraulic diameter ranging from 400 μm to 3.6 mm. In the channel with the hydraulic diameter of 400 μm, the heat transfer coefficient reached a maximum value of 5 kW/m^{2}K in the annular flow regime. In a similar experiment, Yang and Fujita [71] observed that for the channels with hydraulic diameters ranging from 396 μm to 3.636 mm, flow was in the annular flow regime for vapor qualities of 0.2 and larger. In a channel with a hydraulic diameter of 2.55 mm, Kuznetsov and Shamirzaev [70] studied the hydraulic and thermal behavior of R21 fluid. They found an annular flow regime to begin at a vapor quality of 0.1 when heat transfer coefficient started to increase continuously with superficial gas velocity. The heat transfer coefficient displayed a very steep rise after a vapor quality of 0.5, and at a vapor quality of 0.98, it reached a maximum value of 18 kW/m^{2}K. They credited this high heat transfer coefficient to thin-film evaporation as the governing heat transfer mode. In an experimental study conducted by Rahim and Bar-Cohen [233] with R113 as the coolant fluid, it was concluded that the variation of heat transfer coefficient with vapor quality is very sensitive to channel diameter, increasing far more steeply and reaching a significantly higher value for a small-diameter (100 μm) channel compared to a large-diameter (1 cm) channel. Particularly, the maximum heat transfer coefficient increased from 8 to 20 kW/m^{2}K when the diameter decreased from 100 μm to 1 cm. The increase in the heat transfer coefficient in the 100 μm diameter channel is attributed to the thinner liquid film and stronger shear forces in the flowing gas. Additionally, they found a strong relation between heat transfer coefficient and coolant mass flux, irrespective of the vapor quality. For the highest mass flux of 1,000 kg/m^{2}s, the heat transfer coefficient reached a peak value of 25 kW/m^{2}K, at a vapor quality of 0.85.

The principle of heat transfer from a thin liquid film in the annular flow regime in microchannels is mimicked in microgap or microgroove devices [132–134, 234]. There are two compartments in the device: The bottom compartment holds the liquid film, and the top compartment is occupied by the flowing gas. The partition between the two compartments has an opening created at the middle section, which is exposed to the flowing gas medium. After being pumped into the reservoir channel, the liquid also flows along the channel as a film due to shear forces at the liquid–gas interface exerted by the flowing gas. The thickness of the film can be controlled by placing the partition at a certain height. A similar device concept was used by Orlova *et al.* [234]. In their device, the bottom wall of the liquid compartment was heated, and the applied heat was dissipated by the evaporating thin liquid film. The resulting vapor was carried away by the gas passing over the liquid film. Stepanov *et al.* [133] could remove a heat flux of 1 kW/cm^2 from a heated surface of 1×1 cm^2 area using thin-film evaporation assisted by gas flow in a microchannel.

The high heat removal capability of devices utilizing shear-driven thin-film evaporation is due to two factors: less conduction resistance of the thin liquid film and reduced diffusion resistance due to vapor removal from the evaporating meniscus by the flowing gas. While the conduction resistance is primarily controlled by the film thickness, the diffusion resistance is influenced by three factors: diffusion coefficient, which is influenced by the velocity and flow mode of the gas/vapor medium; the relative humidity or dryness of the gas/vapor medium; and the saturation concentration of the liquid that is being evaporated. Since the conduction and diffusion resistances act in series, in order to maximize the evaporative heat transfer, it is critical to minimize both resistances. This is achieved by sweeping the gas over the evaporating thin liquid film in the microgap devices so that evaporated molecules are taken away to maintain a large concentration gradient around the evaporating meniscus. One issue with these devices is the difficulty of maintaining a stable liquid film. Oftentimes, at some locations, liquid is fully evaporated and not replenished, thus creating dry spots. These dry spots cause a decrease in the heat transfer coefficient locally, resulting in an increase in the surface temperature and, eventually, failure of the cooling system. This challenge has been addressed by Narayanan *et al.* [128, 144, 198], who used a nanoporous membrane to utilize capillary wicking in order to maintain a stable film. More details of this method are discussed in the following section.

2.4. *Implementation of micro- and nanostructures to facilitate microscale evaporative cooling*

Micro- and nanoscale roughness features can be introduced to modify the surface energy of the substrate, altering the wettability of the surface and macroscopic properties, such as the contact angle. These structures can also facilitate liquid feeding to the evaporating liquid body and can change the shape of an evaporating liquid meniscus. In particular, changing the surface wettability can significantly affect phase-change heat transfer, especially for evaporative processes.

Microscale surface roughness alters the interfacial forces at the solid and liquid–vapor interface meniscus and lowers the contact angle, thus making the surface more hydrophilic. Roughness at the nanoscale was found to exhibit distinctive features which affect interfacial thermal transport in several different ways. For example, a study by Mittal and Hummer demonstrated that introducing nanoscale roughness (amplitude = ∼3 Å and wavelength = ∼11 Å) can make a hydrophilic surface hydrophobic with a higher interfacial energy [235]. This increase is attributed to the distortion of the hydrogen bond network created as the interfacial water follows the bumpy surface. As the free energy penalty of the distorted hydrogen bond network overwhelms the energetic gain from the larger accessible area, the surface becomes more hydrophobic, which impedes interfacial thermal transport. This behavior was further validated by Acharya *et al.*, who observed an increase in the Kapitza resistance per solvent-accessible area for a nanogrooved surface [236]. However, for a sinusoidal surface with a larger wavelength, hydrogen bonds were able to follow the surface topology, so the interfacial energy decreased, which enhanced the interfacial thermal transport.

Creating a nanoscale surface topology has also been reported to augment interfacial thermal transport by modifying the degree of vibrational coupling between the solid surface and the compressed liquid layer adsorbed at the interface. For example, using a nanopillar array, Goicochea *et al.* investigated thermal transport between water and a quartz surface [237]. They found that the difference in the vibrational density of states (VDOSs) between oxygen atoms in water near the quartz surface and the oxygen atoms of the quartz substrate is reduced significantly with increasing pillar length and density, which reduces the Kaptiza resistance by a factor of three. Similar results have been reported by Issa and Mohamad, who found that nanopatterning creates a noticeable shift in the VDOS of atoms at both the solid surface and the adsorbed liquid layer [238]. The alteration of VDOS in the solid and liquid domains provides better vibrational coupling between the two phases. In summary, the design of a surface nanostructure can potentially improve thermal transport across the solid–liquid interface by tailoring both the surface wetting characteristics and the vibrational coupling.

A liquid body with a lower apparent contact angle has a larger transition region where maximum evaporation occurs [114]. Because it increases the surface area, nanoscale surface roughness can alter the wettability of a surface and can be realized macroscopically as a change in the apparent contact angle [114, 120, 127, 239]. Roughness or microscale structures on a surface affect the solid–liquid interfacial forces and alter the interfacial shape. Particularly, increased roughness on hydrophilic surfaces leads to improved wetting, forming Wenzel droplets. With an increase in the surface roughness, these Wenzel droplets eventually transition to thin liquid films (via imbibition), and the droplets propagate onto the surface due to the enhanced capillarity generated by the surface roughness. Modulating the surface wettability improves evaporative heat transfer performance by decreasing drag and increasing wetting [114, 127] as well as extending the contact line region [240]. However, a microstructured surface heated to a critical temperature beyond the saturation point has been shown to behave as a non-wetting surface, with an apparent contact angle as high as 160° for water droplets [241].

Capillary evaporation, a very efficient heat transfer mode, has been widely implemented in heat exchangers and heat pipes [242–246], most extensively by using wicking structures [114, 121, 122, 145, 247]. In these devices, the movement of fluid is driven by capillary and viscous forces. Other than in wicking structures [145, 247, 248], capillary evaporation has also been implemented to enhance heat transfer in spray cooling methods that use micro–nano porous structures, such as porous coatings and micro- and nanoporous membranes [128, 129, 249]. One disadvantage of using wicking structures is that, as the pore size is reduced in the wicking material to achieve a higher evaporation rate, the viscous resistance increases and, eventually, fluid flow stops. For example, the flow resistances in microscale mono-porous structures can be very high, resulting in a low CHF ($\sim$83 W/cm^2) due to the limited liquid supply. To overcome this problem, new studies are pursuing the development of biporous or even hierarchically structured wicking materials that allow for partially raising the limit imposed by momentum transport. In this approach, the viscous pressure of fluid flow is divided into two parts: (1) a long length of the fluid flow in large-size channels with small hydraulic resistances, and (2) a short length of the fluid flow in small-size channels with high hydraulic resistances. In this way, the overall small hydraulic resistance allows for raising the momentum transport limit. The different types of hierarchical surfaces include bi-porous coatings, biphilic surfaces, and bi-textured titania surfaces (BTSs), which can provide better capillary feeding capabilities to the device.

To precisely control the liquid film thickness and reduce conduction resistance, Narayan *et al.* proposed a novel cooling design, known as the "perspiration nanopatch," in which a liquid-filled channel is placed beneath a suspended porous membrane, with pore sizes as small as 50–100 nm [128]. A coolant fluid is pumped into the channel of such dimensions that the film thickness is imposed by the precise structure of the cavity, rather than by flow conditions that are not easily controllable. This design allows for making the liquid film as thin as possible ($\sim$1 μm) to minimize thermal resistance, yet it reduces the possibility of local dry-out because a sufficient pressure head keeps the cavity filled with the coolant fluid driven by negative capillary pressure created by the small pores. Experimental heat transfer coefficients approached 10^5 W/m^{2}K, while the surface temperature was maintained well below the saturation temperature of the working fluid. Using similar principles, capillary evaporation has been utilized in micro- and nanomembrane devices [241, 250]. Lu *et al.* [129] reported a maximum heat flux of 500 W/cm^2, with a heat transfer coefficient of 8×10^4 W/m^{2}K, using a membrane device with a pore diameter of 130 nm. A very high heat flux of 1,400 W/cm^2, with a heat transfer coefficient of 6.66×10^5 W/m^{2}K, was reported by Hanks *et al.* [250] for their nanoporous membrane device, with pentane as the coolant.

The perspiration nanopatch and other designs proposed in the past two decades primarily rely on wicking structures to form concave evaporating menisci with ultrasmall thicknesses [94, 109, 125, 247, 251–259]. This approach suffers from a negative Laplace pressure at the liquid–vapor interface, which reduces the liquid pressure. Compared to a flat droplet, the equilibrium vapor pressure for a droplet with a concave meniscus, will decrease exponentially [260]. This reduced pressure can be estimated using Kelvin's equation (Eq. (14)). This low pressure gives rise to a reduced saturation temperature, which limits the maximum allowable superheat temperature because of the risk associated with early boiling [261] (Figure 4). The Laplace pressure ($\Delta p = 2\sigma/R$) for water reaches 14 kPa when the radius of curvature, R, is smaller than 10 μm. The liquid pressure for a 200 nm droplet can be reduced by 95% compared to a 10 μm

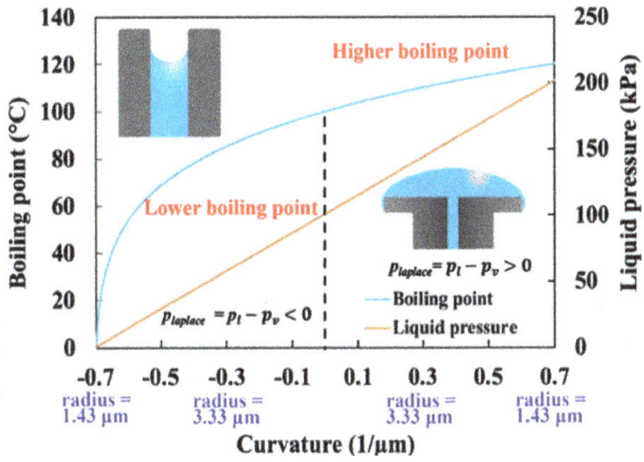

Figure 4. Effect of curvature on the boiling point of water: Due to the interface curvature, the liquid pressure is reduced for a concave meniscus and enhanced for a convex meniscus. Thus, a concave meniscus has a lower boiling point.

droplet. The decrease in equilibrium pressure significantly suppresses evaporation [260] and thus reduces the evaporative heat flux, which can be predicted using Eqs. (16) and (17). In these equations, p_l is the equilibrium liquid pressure, p_v is the vapor pressure near the interface, corresponding to liquid–vapor interfacial temperature, T_l, and vapor temperature, T_v, respectively. $\gamma_{lv}, M, h_{fg}, R, r$, and ρ are the surface tension coefficient, molar mass, the latent heat of vaporization, the universal gas constant, the droplet radius, and the density of the liquid, respectively. For a droplet with 200 nm diameter, the evaporative heat flux decreases by 98% to that of 10 μm diameter. Mitigating this issue requires a convex evaporating meniscus, i.e., a microdroplet, whose positive Laplace pressure will suppress the early onset of boiling.

$$p_{l,\text{corrected}} = p_l \exp\left(\frac{2\gamma_{lv}M}{rR\rho}\right) \tag{15}$$

$$q'' = \frac{2\sigma}{2-\sigma} h_{fg} \sqrt{\frac{M}{2\pi R}} \left(\frac{p_l}{\sqrt{T_l}} - \frac{p_v}{\sqrt{T_v}}\right) \tag{16}$$

$$q''_{\text{corrected}} = \frac{2\sigma}{2-\sigma} h_{fg} \sqrt{\frac{M}{2\pi R}} \left(\frac{p_{l,\text{corrected}}}{\sqrt{T_l}} - \frac{p_v}{\sqrt{T_v}}\right) \tag{17}$$

The decreased evaporative heat flux in nanoporous membrane devices as a result of the Kelvin effect was experimentally demonstrated by Lu *et al.* [166]. They showed that depending on the pore wall temperature or the liquid supply pressure, the liquid–vapor interface self-regulates its curvature. As the radius of curvature reduces with increasing wall temperature or heat flux, the evaporation resistance increases. As a result, the evaporation rate decreases. In extreme operating conditions, this can lead to a complete dry-out since the liquid–vapor meniscus can recede to the bottom of the pore.

Besides modulating the Kapitza resistance, changing surface wettability, and facilitating capillary feeding, another role that micro- and nanostructures can play is the manipulation of droplet shapes. Microscale evaporation can be enhanced by changing the droplet shape to increase the perimeter-to-wetted area ratio. As described earlier, the ultrasmall conduction thermal resistance at the contact line region results in an evaporative heat transfer coefficient more than hundred times greater than that of the bulk fluid. Consequently, the evaporative transport near the contact line region often dominates the total phase-change heat transfer from a thin meniscus [94, 125, 134, 138, 139, 142, 180, 231]. Thus, it is desirable to design surfaces that maximize the ratio of the perimeter to the fluid volume or the perimeter to the interfacial area. Non-axisymmetric droplets also have larger average curvatures, k, which induce a higher vapor concentration gradient in the vicinity of the liquid–vapor interface that can lead to a higher evaporation rate. Saenz *et al.* [141] performed evaporation experiments on non-heated substrates and concluded that droplets with a higher average curvature, k_{avg}, evaporate faster. Beyond this finding, the heat transfer mechanism for asymmetric droplets evaporating on a heated substrate has not been explored in detail.

Though there are various techniques to create non-axisymmetric droplets, which include the use of external forces [262, 263] such as wind, gravity [264, 265], mechanical oscillation of substrates [266], and an AC electric field [267], they are not practically very convenient. For example, while these studies were able to form asymmetric droplets by deforming a capped spherical geometry, these deformed shapes represent a higher energy state and are inherently unstable. Once the external force is removed, these droplets resume a capped spherical shape, neglecting gravitational effects (i.e., a low Bond number). Thus, it is difficult to incorporate these asymmetric droplets in practical applications. An alternative is to pin the droplet meniscus by using microstructures (i.e., microposts or micropillars), whose sharp edges impose energy barriers [239]. One such technique is the use of hollow micropillars to form stable microdroplets pinned atop micropillar surfaces. This technique has been developed by Agonafer *et al.* [136, 137, 139, 142, 184, 231] at the Nanoscale Energy and Interfacial Transport Laboratory (NEIT Lab) at Washington University in St. Louis. The thermal characteristics of the NEIT technology are discussed in Section 3.

The improved thermal performance achieved using the NEIT technology can be clearly seen in Figure 5, which plots heat flux versus heat transfer coefficient for various cooling technologies. All relevant data for this plot are shown in Table 1. The single-phase

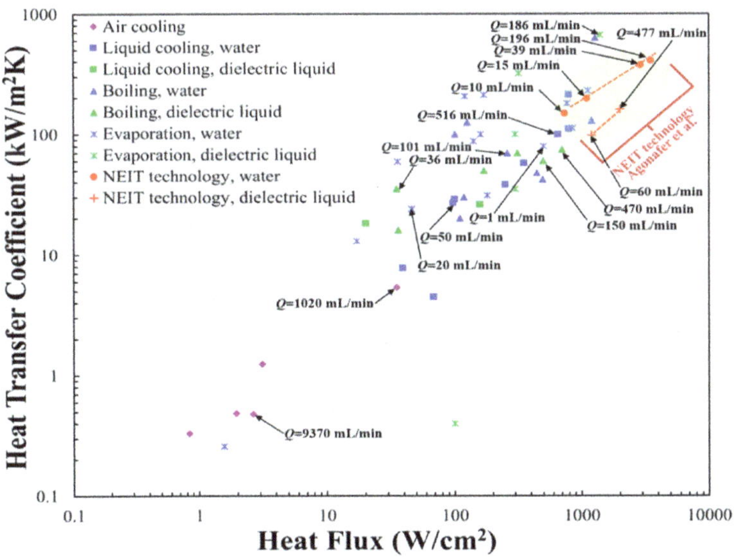

Figure 5. Heat flux vs. heat transfer coefficient for different cooling technologies summarized in Table 1.

Table 1. Thermal performances of different cooling technologies.

Heat transfer mode	Method	Year	Source	Heat flux (W/cm^2)	h (10^5 W/m^{2}K)	Δp (kPa)	Flow rate (mL/min)	Pumping power, P (W)	Exit quality χ_e
Air cooling	Rectangular plate fin heat sink	2001	[270]	3.1	0.01	0.08	1.2×10^6	1.630	—
	Microjet cooling	2008	[269]	35	0.05	207	—	—	—
	Heat spreader	2003	[272]	2.6	0.005	—	—	—	—
	Heat sink with fins	2007	[268]	1.9	0.005	0.22	9,370	0.040	—
Single phase, water	Microchannel heat sink	1981	[41]	790	1.11	213	516	1.870	—
	Pin-fin-enhanced microgap cooler	2014	[273]	100	0.29	—	—	—	—
	Microchannel	2016	[274]	350	0.58	0.25	400	0.002	—
	Jet impingement	2009	[275]	250	1.15	100	4,000	6.800	—
	Pin fin arrays in microchannel	2005	[276]	790	2.57	203	—	—	—
	Microchannel flow	2008	[39]	39	0.08	448	800	5.970	—
	Microchannel flow	2015	[277]	97	0.27	90	50	0.075	—
	Microchannel flow	2016	[278]	6.7	0.66	7	1,200	0.140	—
	Microchannel flow	2009	[279]	650	1.00	50	350	0.298	—
	Microchannel flow	2005	[280]	68	0.05	26.30	7.5	0.003	—
	Microchannel with pin-fin arrays	2019	[281]	250	0.38	2.36	100	0.004	—
Single phase flow, R245fa fluid	Staggered fins in microchannels	2018	[66]	158	0.26	500	150	1.275	—
	Micropin-fin array in microchannel	2019	[282]	20	0.18	47	7.4	0.350	—

Flow boiling, water	Diamond microchannel with microporous copper	2017	[283]	1,280	6.30	42	186	0.133	—
	Microgap heat sink	2012	[72]	110	0.20	2	281	0.010	—
	Staggered, micropin-fin arrays	2014	[284]	118	0.40	45	—	—	0.04
	Parallel microchannels	2008	[67]	260	0.70	6.50	101	0.011	—
	Bi-porous wicking structures	2008	[285]	494	0.42	4.32	—	—	—
	Micropin-fin array in microgap	2016	[274]	444	0.48	448	800	6.093	0.07
	Superhydrophilic nanowires on microchannel walls	2013	[286]	125	1.26	—	—	—	0.52
Pool boiling, water	Super-biphilic surface	2013	[287]	100	1.0	—	—	—	—
Capillary-fed boiling, water	Channel coated with copper inverse opal wicks	2018	[288]	1,200	1.30	—	—	—	—
Flow boiling of HFE7000	Microchannel with "Piranha Pin Fins"	2015	[289]	700	0.75	120	470	0.940	0.20
Flow boiling, R245fa	Pin fin in microgap channel	2012	[290]	35	0.35	67	40	0.045	0.28
	Microgap with staggered fins	2018	[66]	498	0.60	500	150	1.250	0.18
Flow boiling, R-123	Hydrofoil-based micropin-fin heat sink	2007	[73]	312	0.70	0.25	400	0.002	0.40

(Continued)

Table 1. (*Continued*)

Heat transfer mode	Method	Year	Source	Heat flux (W/cm^2)	h (10^5 W/m^2K)	Δp (kPa)	Flow rate (mL/min)	Pumping power, P (W)	Exit quality χ_e
Film boiling of R134a	Spray cooling, with microstructured surface	2013	[126]	170	0.50	480	492	4.015	—
Capillary evaporation, water	Micromembrane-enhanced evaporating surface	2013	[145]	140	0.88	—	—	—	1
	Biphilic nanoporous surface	2014	[127]	170	2.13	—	—	—	1
	Ultrathin nanoporous membrane Membrane pore diameter: 130 nm Porosity: 0.23	2017	[129]	500	0.80	—	1	—	1
	Nano Cu wick	2013	[147]	180	0.31	—	—	—	1
Capillary evaporation of pentane	Nanoporous membrane device	2018	[250]	1,400	6.66	460	—	—	1

Thin-film evaporation, water	Carbon nano tube (CNT) wicking structures	2012	[291]	770	1.80	—	—	—	1
	Micropillar wick	2016	[241]	46	0.24	17	20	4.08	1
	Capillary wicking using micropillars Micropillar diameter: 10 μm, No. of pillars in 1 cm^2 area: 10×10^5	2017	[143]	35.7	0.56	17	36	7.34	1
	Shear-driven film flow in channel	2018	[292]	850	1.13	—	350	—	1
	Shear-driven flow in a minichannel	2011	[132]	160	1.0	—	—	—	1
Evaporation dominated but nucleate boiling present, water	Bi-porous media consisting of microscale pin fins separated by microchannels	2012	[251]	119.6	2.07	—	—	—	1

(Continued)

Table 1. (*Continued*)

Heat transfer mode	Method	Year	Source	Heat flux (W/cm^2)	h (10^5 W/m^2K)	Δp (kPa)	Flow rate (mL/min)	Pumping power, P (W)	Exit quality χ_e
Evaporation of sessile droplets, water	Copper substrate with microposts	2017	[293]	1.56	0.003	—	3.25	—	1
Nucleate boiling and evaporation	Mono-porous silicon wick structures	2014	[271]	1,127	2.30	—	—	—	1
Thin-film evaporation of FC72	Gas-assisted thin-film evaporation	2009	[128]	300	0.35	20	36	0.00027	1
Thin-film evaporation, octane	Microgrooved heatsink	2016	[134]	320	3.20	—	—	—	1
Evaporation of HFE-8401 HT	Embedded jet impingement	2001	[294]	100	0.004	—	600	—	1
Evaporation of HFE-7100	Forced fed using enhanced surfaces	2008	[295]	300	1.0	—	—	—	1
NEIT technology, evaporation, water	0.5–10 μm hollow micropillar array No. of pillars in 1 cm^2 area: $2.44 \times 10^3 - 9.77 \times 10^5$	2020	[315]	735–14,700	2.20–83.50	29–621	10–196	1.3×10^{-4} -1.1×10^{-1}	1

cooling technologies reported in the table can operate at a device temperature of 85°C [268–270]. Although a maximum heat flux of 35 W/cm^2 can be dissipated with microjet cooling [269], the required flow rate can be as high as 1.2×10^6 mL/min. Single-phase cooling with water as a coolant can remove a maximum heat flux of 790 W/cm^2 using microchannels [41]. The heat transfer coefficient can reach up to 111 kW/m^{2}K, for a flow rate of 516 mL/min and a pressure drop of 213 kPa. Palko *et al.* reported a maximum heat flux of 1,280 W/cm^2 for flow boiling of water in microchannels coated with copper microporous coating. The heat transfer coefficient (630 kW/m^{2}K) is also significantly higher than that in single-phase cooling. One disadvantage of boiling is that the required mass flow rate and exit vapor quality maintain an inverse relationship. Therefore, at a higher flow rate, the vapor quality is lower, and vice versa. Since the heat transfer coefficient also diminishes with decreasing vapor exit quality, increasing flow rate means decreased heat transfer coefficients. Maximum vapor quality is also associated with flow instabilities [67, 72, 73]. Therefore, as shown in Table 1, most of the experiments performed on boiling heat transfer have very low exit qualities. These limitations are overcome by evaporative cooling. In general, the thermal performances of evaporative heat transfer techniques are comparable to that of boiling heat transfer, but with the added advantage of requiring less coolant consumption as well as lower pumping power. Cai *et al.* [271] reported a maximum heat flux of 1,127 W/cm^2, with a heat transfer coefficient of 230 kW/m^{2}K. They used mono-porous silicon wicking structures to facilitate fluid capillarity, which enhanced thin-film evaporation. The porous structures also assisted in nucleate boiling. Lu *et al.* [129] obtained a maximum heat flux of 500 W/cm^2 using a ultrathin nanoporous membrane device with a very small flow rate (1 mL/min). Particularly attractive thermal performance can be achieved by using evaporative cooling technology developed at the NEIT Lab. This technology utilizes a constant pressure microfluidic system to form stable evaporating microdroplets by feeding the working fluid to the top of a micropillar through a central hole. As can be clearly seen from Figure 5, the maximum allowable heat flux using this technology can exceed the values reported in the current literature. For example, an array of micropillars with a diameter of 0.5 μm can accommodate 9.78×10^7 droplets, covering an area of 1 cm^2, and can remove a maximum heat flux of 14.7 kW/cm^2 while maintaining the operating temperature below 100°C on the substrate surface. The maximum heat transfer coefficient can reach up to 830 kW/m^2K, with a flow rate of only 196 mL/min. Unlike other evaporative cooling technologies, since NEIT technology uses a constant pressure microfluidic system, the liquid flow rate is completely governed by the evaporation rate: The amount of liquid fed into the system is equal to the liquid being evaporated. This low flow rate results in extremely low pumping power requirements (i.e., 0.11 W). A comparison can also be made between the NEIT technology and flow boiling in microchannels. By using the NEIT technology with an array of micropillars of 0.5 um diameter, a maximum heat flux of 73.52 W/mm^2 can be dissipated, with a required flow rate of 97 mL/min and a pressure drop of 620 kPa for a device with 20×20 mm^2 area. For microchannel flow boiling, by providing the same flow rate through a channel of 200 μm diameter and 20 mm length while maintaining an annular flow regime with an exit vapor quality of 0.8, a critical heat flux of 22.10 W/mm^2 can be obtained with a significantly larger pressure drop of 3.71×10^7 kPa. Therefore, for the same flow rate, the NEIT technology can remove 3.3 times more heat flux with a 59,000 times smaller pressure drop, compared to flow boiling.

3. Evaporation of Asymmetric Droplets Formed Using Hollow Micropillars

3.1. *Implementation of micro- and nanostructures to facilitate microscale evaporative cooling*

As described earlier, the ultrasmall conduction thermal resistance at the contact line region results in an evaporative heat transfer coefficient more than hundred times greater than that of the bulk fluid. Consequently, evaporative transport near the contact line region often dominates the total phase-change heat transfer from a thin meniscus [94, 134, 137, 142, 191, 231, 296]. Thus, it is desirable to design surfaces that maximize the ratio of the perimeter to the fluid volume or the perimeter to the interfacial area. Such a structured surface causes a large fraction of the heat transfer to occur along the triple-phase contact line, enhancing the phase-change heat transfer performance of the entire system.

Designing hollow micropillar structures with sharp edge features is a promising approach to tuning the geometry of an evaporating meniscus, where the advance of the contact line is obstructed by the topographic discontinuity [137, 142, 231]. In particular, the droplet shape can be tuned into non-axisymmetric geometries, which, compared to capped spherical droplets, exhibit a larger ratio between the perimeter and interfacial solid–liquid area. As an example, given the same liquid–vapor interfacial area and solid–liquid contact line length, the perimeter-to-area ratio of a droplet resting on a triangular micropillar is 29% greater than that on a circular micropillar, as shown in Figure 6. As a result, a greater fraction of heat is transported along the contact line. In addition, the non-axisymmetric droplet also exhibits a larger average curvature than a capped spherical droplet. The larger curvature causes the iso-concentration line to experience a sharp change in direction, which yields a higher vapor concentration gradient and therefore faster vapor diffusion near the droplet surface [118, 141]. Another distinctive feature of droplet evaporation atop a hollow micropillar structure is that vapor molecules can diffuse partially downward. For a pinned droplet evaporating on a non-planar substrate, such as that supported on a micropillar structure at a finite height above the bottom substrate, there is open space for vapor molecules to diffuse below the droplet itself [184]. Therefore, a greater number of vapor molecules can be added to an already partially occupied vapor space. Finally, a stable evaporating droplet with a constant size and shape can be maintained on a hollow micropillar structure by replenishing the evaporated liquid through the central micropillar hole using a constant-pressure microfluidic system [137]. All these features together demonstrate enhanced evaporation from a non-axisymmetric droplet atop a hollow micropillar structure.

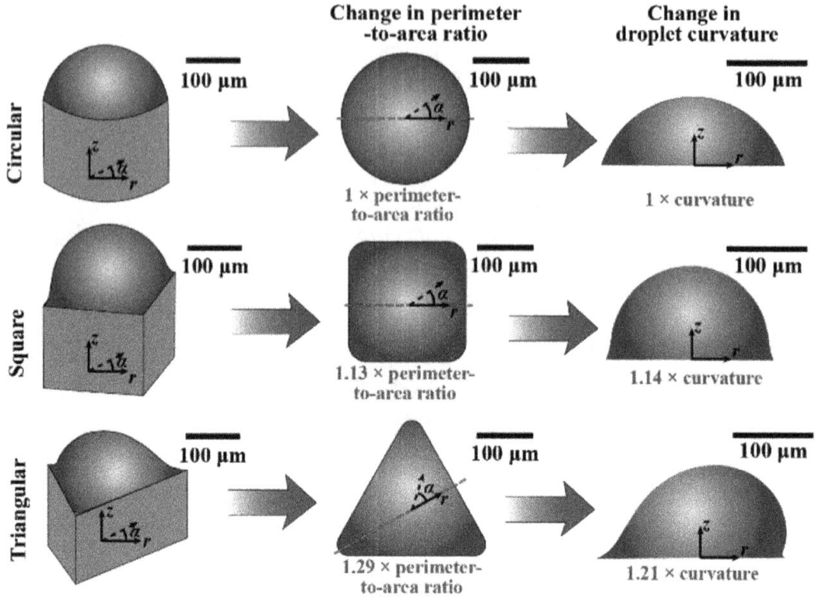

Figure 6. Illustration of the increased perimeter-to-area ratio and higher surface curvature as the solid–liquid contact area of a droplet changes from a circle to a square and a triangle. All three droplets have the same solid–liquid contact perimeter and liquid–vapor interfacial area. Changing the solid–liquid contact shape from a circle to a square and a triangle, respectively, results in 13% and 29% increases in the perimeter-to-area ratio, as well as 14% and 21% enhancements in the surface curvature k.
Source: Reprinted from Ref. [137], Copyright (2019), with permission from Elsevier.

3.2. *Major characteristics of droplets atop a hollow micropillar structure*

3.2.1. *Asymmetry in droplet morphology and diffusion transport*

Li *et al.* investigated the morphological features of different non-axisymmetric droplets with the help of Surface Evolver [231]. The algorithm embedded in Surface Evolver allows for the surface to evolve toward a minimum energy state by the gradient descent method, during which each vertex is shifted in the desired direction within preset constraints. In this way, the final equilibrium

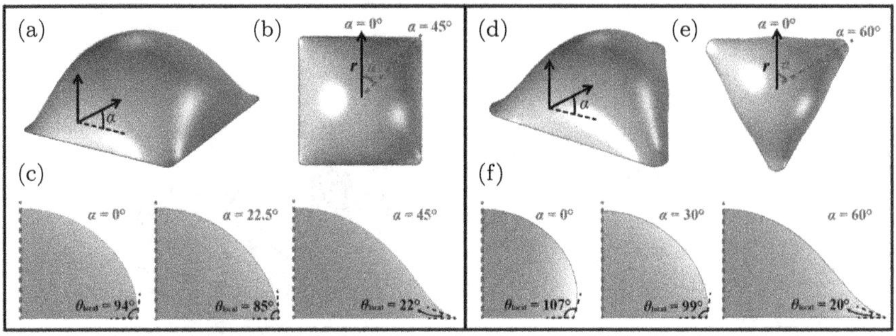

Figure 7. (a)–(c) Isometric view, top view, and cross-section of the equilibrium droplet profile on a square micropillar structure. (d)–(f) Isometric view, top view, and cross-section of the equilibrium droplet profile on a triangular micropillar structure. The cross-sections shown in (c) and (f) are taken at three different azimuthal angles, α. The definition of the azimuthal angle is given in (b) and (e). The maximum and minimal local apparent contact angles, $\theta_{a,\text{local}}$, are found at the locations with the smallest and largest azimuthal angles, respectively.
Source: Reprinted from Ref. [137], Copyright (2019), with permission from Elsevier.

shape can be obtained by iterating the successive energy minimization processes for many steps. Figure 7 shows the cross-sectional shapes of droplets atop triangular and square micropillars obtained by increasing the volume of the microdroplets in Surface Evolver until the liquid–vapor interfacial area, A_{lv}, of the equilibrated microdroplets reached 1,200 μm^2. The corresponding liquid volumes for microdroplets on circular, square, and triangular microdroplets were $1.68 \times 10^{-4}, 1.69 \times 10^{-4}$, and $1.62 \times 10^{-4} \mu$m^3, respectively. As shown, the equilibrium droplet profiles on square and triangular micropillars are thicker near the center of the side edge (i.e., locations with a small azimuthal angle, α) and much thinner near the corner regions (i.e., locations with a large azimuthal angle, α). In other words, the local contact angle, θ_{local}, decreases monotonically with increasing α. This unique non-axisymmetric characteristic is attributed to a uniform Laplace pressure over the entire droplet under equilibrium conditions, which is determined by the two principal radii of curvature at the droplet surface as

$$\Delta p = \gamma_{\text{H}_2\text{O}} \left(\frac{1}{R_1} + \frac{1}{R_2} \right) \tag{18}$$

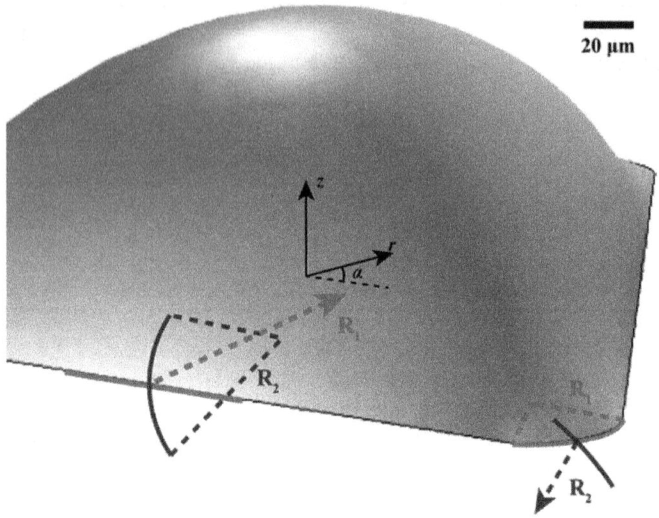

Figure 8. Schematic representation of the two principal radii of curvature at the center of the side edge and at the corner of a droplet on a triangular micropillar. The principal radius of curvature (R_1) at the center of the side edge is much larger than that at the corner region of the droplet. To ensure that the mean curvature is constant over the droplet interface, the secondary radius of curvature (R_2) at the center of the side edge must be much smaller than that at the corner region of the droplet.

Source: Reprinted from Ref. [137], Copyright (2019), with permission from Elsevier.

where R_1 and R_2 are the two principal radii of curvature and γ_{H_2O} is the surface tension of water. If R_1 and R_2 are defined on the planes parallel and perpendicular to the top surface of the micropillar, respectively, as shown in Figure 8, then R_1 equals the local radius of the curved contact line at the corner region. At the center of the side edge, however, the magnitude of R_1 goes to infinity since the edge is a straight line. Therefore, to ensure that $1/R_1 + 1/R_2$ are the same at these two different locations, the magnitude of R_2 at the corner region must be much larger than that at the center of the side edge. In other words, the meniscus is much thinner at the corner region, measured in the direction normal to the top surface of the micropillar. Thus, the local apparent contact angle is the smallest in the corner region.

While the average curvature is uniform along the non-axisymmetric droplet, the two principal curvatures, k_1 and k_2, have different degrees of impact on the interfacial vapor diffusion transport. We can decompose the diffusion transport into two components: the planes parallel to and normal to the substrate. In this case, the first component is governed by the primary curvature, k_1, while the second component is governed by the secondary curvature, k_2. However, the vapor diffusion transport in the plane normal to the substrate is, to a certain extent, confined by the impenetrable solid substrate and is therefore more dependent on the local contact angle than the secondary curvature. Contrariwise, the diffusion transport in the plane parallel to the substrate is dominated by the primary curvature k_1 only. Therefore, the distribution of the local vapor diffusion transport is generally more affected by the change in the primary curvature k_1 along the droplet surface. As a consequence, the higher primary curvature will induce a stronger evaporation rate at locations with higher azimuthal angles. In general, the smaller contact angle and large primary curvature at locations with high azimuthal angles will both contribute to a significantly higher local evaporation rate. Such a characteristic will result in highly non-uniform evaporation over the whole droplet, where the local evaporation is enhanced near the edge corners and suppressed near the edge center locations.

3.2.2. *Enhanced vapor diffusion from a suspended droplet*

When droplets evaporate on a planar substrate, the downward diffusion of vapor molecules is blocked by the impenetrable substrate. Hence, a sessile droplet has a smaller evaporation rate than a free droplet. However, for a pinned droplet evaporating on a non-planar substrate, such as that supported on a micropillar, vapor diffusion is substantially less confined because the vapor molecules escaping from the contact line can now diffuse partially downward. Using the simulation results shown in Figure 9, this behavior can also be identified from the iso-concentration contours and the local diffusion flux vectors for droplets evaporating on micropillars with different

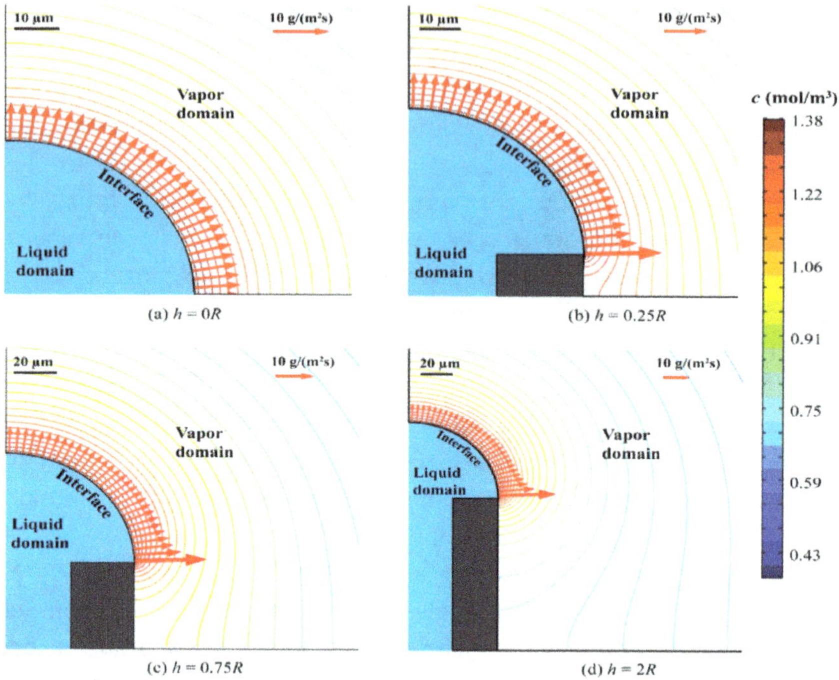

Figure 9. The iso-concentration contour with the vector plot of the local evaporative flux, J, along the liquid–air interface at four different pillar heights: (a) $0R$, (b) $0.25R$, (c) $0.75R$ and (d) $2R$. The concentration of the vapor domain is represented by the different colored contour lines. The length and direction of the red arrows represent the magnitude and direction vector of the evaporative flux. *Source*: Reprinted from Ref. [184], Copyright (2019), with permission from Elsevier.

heights. In this figure, the red arrow shows the magnitude and direction of the local diffusion flux along the liquid–vapor interface. With a zero micropillar height, the iso-concentration lines form concentric circular rings around the center of the droplet, and the magnitude of the diffusion flux is constant along the liquid–vapor interface. However, with a finite micropillar height, the iso-concentration line close to the liquid–vapor interface must end on the side wall of the micropillar in the horizontal direction in order to satisfy the impenetrable boundary condition. This geometric constraint results in a sharp bending of the iso-concentration line near the contact line region, creating a high local concentration gradient and therefore a strong local diffusion flux near the contact line region. In addition, for iso-concentration lines ending on the x-axis, the location of the ending points shifts slightly toward the left-hand side, which represents a closer packing of the iso-concentration lines. In other words, the diffusion flux in the ambient air is intensified by the stronger diffusion at the liquid–vapor interface. With increasing micropillar height, more iso-concentration lines land on the side wall of the micropillar, and there is a more compact distribution of iso-concentration lines near the contact line region. Therefore, the local diffusion flux near the contact line becomes higher with the increasing pillar height, resulting in a more non-uniform distribution of the local diffusion flux along the liquid–vapor interface.

3.3. *Evaporative heat transfer performance from non-axisymmetric droplet atop hollow micropillar structure*

3.3.1. *Conduction resistance for a non-axisymmetric droplet*

Given that the conduction resistance inside a droplet is inversely proportional to the liquid film thickness, the variation of the liquid–vapor interfacial area for different film thicknesses (shown in Figure 10) results in a substantial change in the total thermal resistance of the droplet. Changing the base shape of the droplet from a circle to a square or a triangle yields a larger contact line length and therefore a greater portion of the thin-film region, owing to the larger perimeter-to-solid liquid interfacial area ratio of triangles and squares compared to circles. To quantify the distribution of film thicknesses along different regions of the droplet, the meniscus profiles of square, triangular, and circular droplets can be sectioned into 20 segments along the z-axis direction with a uniform spacing. Subsequently, the total surface area within each segment can be calculated, and their fractions in the total meniscus area can be obtained accordingly. These fractions reveal the percentage of the total evaporating surface

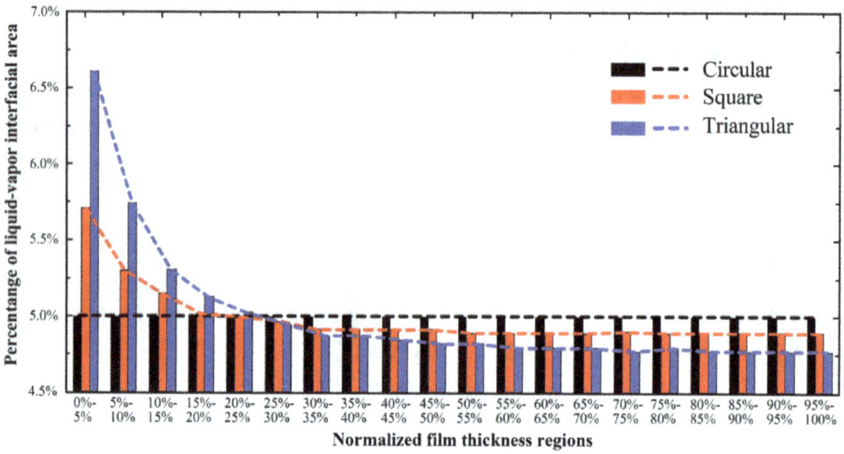

Figure 10. Film thickness distributions of triangular, square, and circular micro-droplets: The fraction of the total liquid–vapor interfacial area evaporating with a large liquid thickness (e.g., greater than 95% of the maximum droplet height) is reduced as the shape of the droplet base changes from a circle to a square and a triangle.

Source: Reprinted from Ref. [231], Copyright (2019), with permission from Elsevier.

(liquid–vapor interface) that falls in the thin-film region. Figure 10 shows the percentage distributions of meniscus profiles falling in different film thickness ranges.

For droplets evaporating on a circular micropillar, the total evaporating area is distributed uniformly along the liquid–vapor interface with different film thicknesses. However, the fraction of the total evaporating area possessing a smaller film thickness is significantly higher for non-axisymmetric droplets. For droplets evaporating on a square micropillar, around 5.7% of the total evaporating area exhibits a film thickness of less than 5% of the maximum droplet height. This fraction increases to 6.6% for droplets evaporating on a triangular micropillar. Correspondingly, the fraction of the total liquid–vapor interfacial area evaporating with a large liquid thickness (e.g., greater than 95% of the maximum droplet height) is also reduced as the shape of the droplet base changes from a circle to a square and a triangle. In a nutshell, the larger perimeter-to-area ratio associated with the non-axisymmetric droplet causes a greater fraction of the phase-change heat transfer to occur in the thin-film region. In addition, if we approximate the heat transport in the droplet as 1D conduction behavior, then the conduction thermal resistances associated with each of the small liquid–vapor interfacial areas are integrated in a parallel network configuration since the total evaporation heat is distributed over the entire droplet meniscus. Therefore, to quantify the change in the total thermal resistance caused by the varied liquid–vapor areas for different film thicknesses, we can estimate the effective thickness of the entire meniscus by taking the weighted harmonic mean of the different film thicknesses:

$$\frac{1}{t^*_{\text{eff}}} = \sum_{i=0}^{19} \frac{\emptyset_{A,i}}{t^*_i}, \tag{19}$$

Here, t^*_i is the normalized film thickness of each meniscus segment and i is the index for each segment, ranging from 0 to 19. In this study, the meniscus is sectioned into 20 segments with a uniform spacing; therefore, t^*_i is approximately $2.5\% + i \times 5\%$. Further, $\emptyset_{A,i}$ is the fraction of the liquid–vapor interfacial area associated with each thickness range, and t^*_{eff} is the effective thickness of the entire meniscus. Based on the data shown in Figure 10, t^*_{eff} is calculated to be 0.202, 0.190, and 0.176 for droplets evaporating on circular, square, and triangular micropillars, respectively. In other words, the effective film thickness and conduction resistance of the evaporating droplet are reduced by 5.9% and 12.8%, respectively, when the base shape of the droplet is changed from a circle to a square and from a circle to a triangle.

3.3.2. *Heat transfer coefficient for non-axisymmetric droplets*

Experimental and numerical studies have been performed to investigate the evaporation heat transfer performance of non-axisymmetric droplets. A thermal test platform was designed and developed to form microdroplets with non-axisymmetric geometries. A liquid feed system was incorporated into the test platform to maintain a constant droplet volume during the evaporation. Two resistance temperature detectors (RTDs) were fabricated on the backside of the samples to maintain a constant substrate temperature during the experiments. A pressure-driven liquid feed system was installed to keep replenishing the evaporated liquid during the experiment and ensure that the shapes of the droplets were maintained constantly, regardless of the evaporation rate. The evaporation heat transfer coefficient was calculated using the following equation:

$$h = \frac{m_v h_{fg}}{A_{sl}(T_l - T_\infty)}, \tag{20}$$

where m_v is the total evaporation rate, A_{sl} is the solid–liquid interfacial area, T_l is the liquid–vapor interfacial temperature, and T_∞ is the ambient temperature.

The evaporative heat transfer performance of microdroplets evaporating on the circular, square, and triangular micropillars is summarized in Figure 11. The heat transfer coefficients of microdroplets atop the triangular micropillars are 8% and 45% larger than those of droplets on the square micropillars at 60°C and 98°C, respectively, and 46% and 71% larger than those of droplets pinned on the circular micropillars. This enhancement is attributed to the smaller solid liquid area compared to the contact line perimeter ratio of triangular microdroplets. Since all three microdroplets evaporate from micropillars with the same contact line perimeter, the solid–liquid area for the microdroplet on the triangular micropillar is the smallest (4,683.5 μm^2), followed by the square (6,162.3 μm^2) and circular (7,854.0 μm^2) micropillars. In other words, a microdroplet evaporating on a triangular micropillar exhibits the longest contact line for a given solid–liquid area. Given that the evaporation heat transfer is most efficient at the contact line region, the overall heat transfer coefficient is highest for microdroplets evaporating on triangular micropillar structures. All the experimental measurements were validated by numerical simulations in COMSOL Multiphysics, which demonstrated a close match in the evaporation rate, with a maximum difference of 24% at a substrate temperature of 98°C.

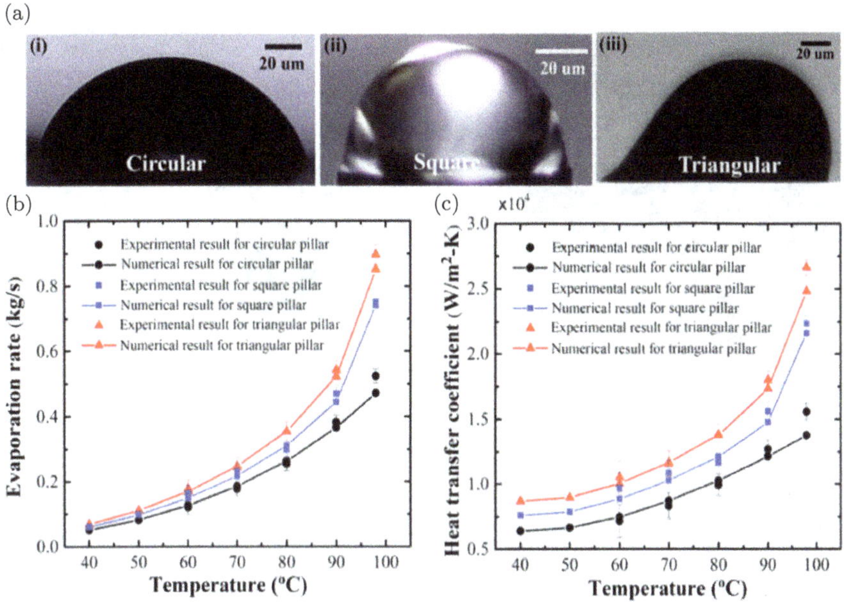

Figure 11. (a) Microdroplet profiles, (b) a comparison of the evaporation rates, and (c) a comparison of the heat transfer coefficients. For all three micropillar shapes, the evaporative flux and heat transfer coefficient obtained from simulation results agree well with the experimental measurement, with the relative error ranging from 4% to 11% as temperature decreases from 98°C to 60°C.

Source: Reprinted from Ref. [231], Copyright (2019), with permission from Elsevier.

3.3.3. *Vapor diffusion from suspended droplets*

Experimental and numerical studies have shown that increasing the micropillar height can reduce the vapor diffusion confinement of the evaporating droplet. This trend is reflected in both higher local and higher total evaporation rates. Specifically, we have shown that raising the micropillar height from 0 to $2R$, $5R$, and $20R$ led respectively to 26%, 35%, and 42% enhancement in the total evaporation rate of a hemispherical droplet under adiabatic conditions. In addition, we found that the confinement of vapor diffusion at the contact line for droplets with a contact angle greater than 90° was alleviated with increasing pillar height due to the greater vapor diffusion space available beneath the droplet. When the pillar height was sufficiently

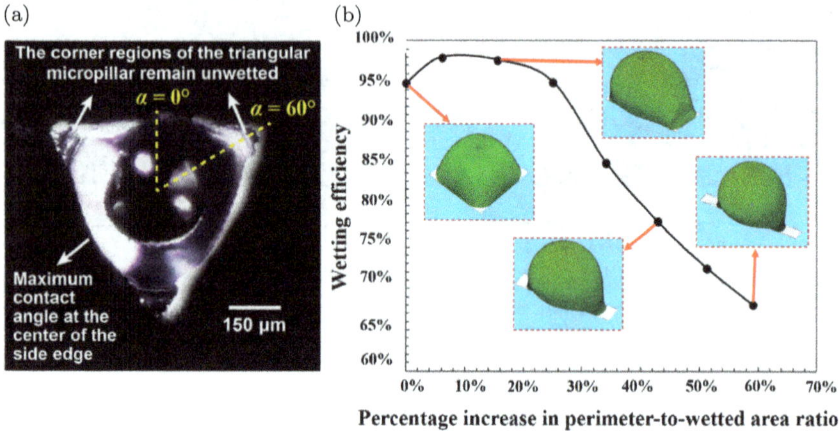

Figure 12. (a) The corner region of the triangular micropillar structure remains unwetted. (b) The variation in the wetting efficiency with increasing perimeter-to-area ratio for a droplet on a rectangular micropillar.

large (e.g., $h > 20R$), the local evaporation rate increased from the top center to the contact line region of these large droplets, contrary to the trend observed for the evaporation of large droplets on a flat substrate. For droplets evaporated on heated micropillars, however, we found that evaporation transport was affected by competition between the effects of increasing conduction resistance and reduced vapor diffusion confinement as the pillar height increased, as shown in Figure 12. Thus, with increasing pillar height, the total evaporation rate first increases and later decreases (except for the substrate temperature of 98°C). Specifically, for substrate temperatures of 60°C, 80°C, and 98°C, the pillar heights providing the largest evaporation rates were $\overline{h} = 5.5$, 3.5, and 1.5, respectively (where, $\overline{h} = h/R$ represents the non-dimensional pillar height).

3.3.4. *Wetting efficiency*

To achieve better heat transfer performance, the goal is not only to maximize the ratio of the droplet's contact line perimeter to the solid–liquid area. It is also important to consider the wetting efficiency of a droplet atop the hollow micropillar structures. Wetting efficiency is defined as the percentage of the micropillar top surface wetted by the droplet. The wetting efficiency for a droplet on a triangular micropillar is less than 100% since the liquid meniscus might not wet the entire top surface of a non-axisymmetric micropillar. For example, a droplet atop a triangular micropillar will not fully wet the three corner regions (Figure 13(a)). The wetting efficiency and pinning/depinning behavior of high and low surface tension liquids expanding on these micropillars would be different. Additionally, such a wetting efficiency will decrease with an increasing perimeter-to-area ratio of the micropillar, creating a challenge in analyzing the equilibrium profile of a non-axisymmetric droplet. Figure 13(b) shows the preliminary results obtained from Surface Evolver, in which the wetting efficiency decreases with increasing perimeter-to-area ratio for droplets on rectangular micropillars. As the perimeter-to-area ratio increases, the overall non-wetted area increases, and the wetting

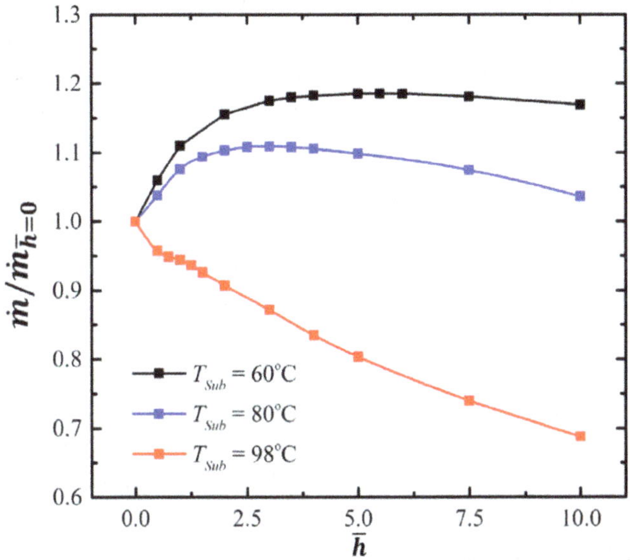

Figure 13. Simulation results of evaporation rates for various pillar heights at three different substrate temperatures of 60° C, 80° C, and 98° C. The evaporation rates for different pillar heights are normalized with respect to the evaporation rate for a pillar height of zero.

Source: Reprinted from Ref. [184], Copyright (2019), with permission from Elsevier.

efficiency decreases. Compared to a square micropillar, a pillar with a 5% increased perimeter-to-wetted area ratio, provides the maximum wetting efficiency. However, a further increase in the perimeter-to-wetted area ratio decreases the wetting efficiency to as low as 68% with a 60% increase in the perimeter-to-wetted ratio. Based on preliminary results, micropillars should be designed to yield a minimum of 95% wetting efficiency.

4. Fabrication Techniques for Evaporative Cooling Devices

4.1. *Fabrication of wicking structures and surface coatings*

To increase evaporative surface areas and capillary forces, researchers have developed various mono-porous structures, such as sintered particles/powders [297] and woven copper mesh laminates [242–246]. Under surface tension, a working fluid forms an ultrathin evaporating meniscus along these wicking structures. Traditionally, wicking structures have been fabricated by sintering Cu microparticles [121, 147]. Due to its high thermal conductivity, Cu is a highly effective material for phase-change processes, but the poor water wettability of the Cu surface is a significant drawback. However, a Cu wicking structure has better wettability due to enhanced capillary wetting [258]. Su *et al.* deposited layers of 50 μm Cu particles and sintered them to form a 150 μm thick layer [121]. Compared to pool boiling, they achieved 4.6 times faster cooling using their wicking structure for thin-film evaporation. Taking another approach, Nam *et al.* used electrochemical deposition (ECD) and chemical oxidation to fabricate dense arrays of superhydrophilic Cu microposts with diameters ranging from 30 to 100 μm, solid fractions as high as 58%, and aspect ratios as high as four [147]. To enhance the wettability of the Cu microposts, they deposited CuO layers on them using chemical and thermal oxidation processes. In yet another approach exploiting capillary evaporation, Amon *et al.* [48] proposed a jet-impingement cooling technique using droplet evaporation. To enhance the spreading of coolant droplets, they used DRIE to fabricate microstructured silicon surfaces covered with islands and studs of various dimensions. Plasma etching was used to create surface roughness on fused silica microscope slides, and then PECVD was used to grow a silicon-oxide film that increased the wetting properties of the substrate. Along similar lines, Ojha *et al.* [114] investigated the effects of surface roughness on phase change by using silica surfaces with various roughness features [94].

Different approaches have been adopted by researchers to fabricate hierarchical surfaces. For example, in 2014, Dai *et al.* [127] fabricated a hybrid surface with both hydrophilic and hydrophobic properties. The hydrophilic properties improved liquid spreading and augmentation in the thin-film evaporation area, while the hydrophobic properties reduced the pinning force of the contact line. To reduce flow friction and enhance capillary evaporation, Dai *et al.* also synthesized biphilic nanoporous surfaces by creating hydrophilic functional groups in the defect areas of hydrophobic pristine carbon nanotubes (CNTs). They used commercially available CNTs as the hydrophobic microstructures and oxidized them in a concentrated nitric acid solution to create hydrophilic carboxyl and hydroxyl groups on the walls of the CNTs. Finally, they bonded the functionalized CNTs to a substrate. In 2009, Ding *et al.* fabricated a thin layer of titania nanostructures on Ti microstructures, forming a BTS [248]. The BTS helps increase the capillary pressure and avoid dry-out in the evaporator region. Using titania micromachining and surface modification techniques, they fabricated BTS with two different topologies: 5/5 μm (diameter/gap) and 100/20 μm (diameter/gap), with an etch depth of 50 μm. The fabrication process for making BTS begins with the deep etching of a 300 μm thick Ti substrate through inductively coupled plasma (ICP) etching. After the deep etching process, the sample is oxidized at 83°C in a 30% solution of hydrogen peroxide for 15 min to create the BTS and the bottom floor surface. In general, fabricating bi-porous or hierarchically structured wicking materials is challenging because it requires precise control of the pore size distribution. A number of different techniques, such as colloidal assembly, are available to build highly ordered porous wicking materials [298].

4.2. *Fabrication of micro- and nanoporous membrane devices*

Membrane-based devices have been fabricated using various techniques to enhance capillary forces during evaporation [128, 129, 144, 198, 250, 299]. Narayanan *et al.* created a "MEMS-NEMS perspiration nanopatch device" using bulk and surface micromachining and low-temperature processes compatible with CMOS. They co-fabricated a nanoporous membrane (porous anodic alumina, PAA) with a silicon substrate using an established PAA etching process. The process started with electron beam deposition of a 10 μm uniform layer of pure aluminum (99.99% or higher) onto the silicon substrate, followed by a two-step electrochemical process that transformed pure aluminum into self-organized nanoporous alumina. The nominal pore diameter was found to be 60 nm. Hanks *et al.* [250] fabricated an ultrathin nanoporous membrane device with a membrane thickness of 200 nm and a pore radius of 65 nm using MEMS fabrication processes. Both sides of a double-side-polished Si wafer were coated with Si_3N_4 ($\sim$300 nm thick) using low-pressure CVD. Through interference lithography and RIE, they created non-through nanoporous patterns on the Si_3N_4 layer on its frontside. During this step, the pores were not yet etched into the Si substrate. Later, they performed KOH etching on the back, while protecting the frontside Si, with the remaining Si_3N_4 layer. Finally, they performed additional RIE to open up the nanopores.

4.3. *Fabrication of hollow micropillar structures*

Hollow micropillar structures are microfabricated on a 4″ diameter silicon wafer ($\langle 100 \rangle$, 300 μm thick, University Wafer) through SiO_2 thin-film deposition, photolithography, and RIE. First, a thin oxide layer ($\sim$450 nm) is grown on both sides of the silicon wafer by heating the wafer in an O_2 atmosphere in a tube furnace (Lindberg/Blue M) at 1,100°C for 14 h. This oxide layer prevents potential current leakage from the heating and temperature sensor units that will be fabricated in the next step (Figure 14). Next, two

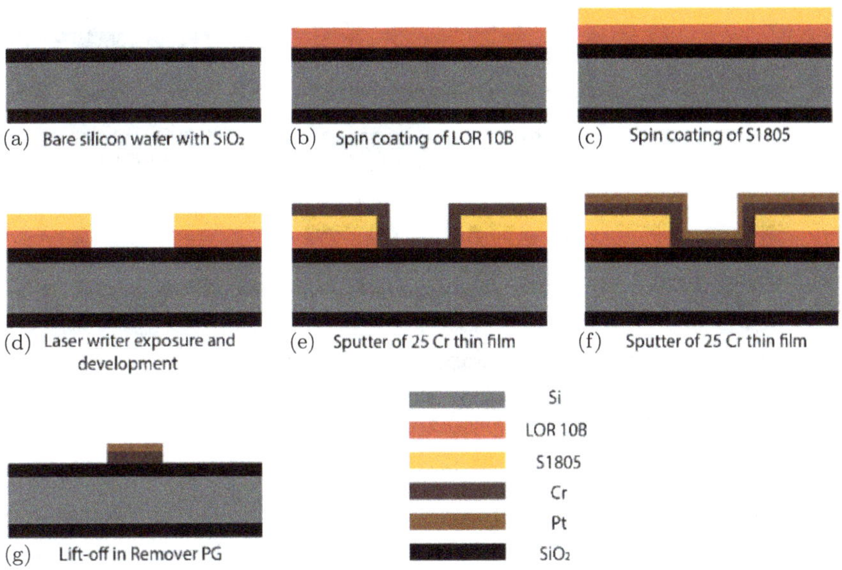

Figure 14. Schematic of fabrication process of Pt thin film heater and RTD elements.

125 nm thick platinum RTDs are deposited on the backside of the wafer by sputtering (Kurt J. Lesker, PVD 75) and lift-off. During this process, the redundant metal layers are removed by submerging the wafers in Remover PG at 60°C for 12 h. A 25 nm thick chromium (Cr) layer is used as an adhesion layer to ensure strong bonding between the platinum and silicon surfaces. After the fabrication of the RTD elements, a deep hole is etched at the center of the silicon wafer by first removing the oxide layer with RIE and then removing the silicon with DRIE. Both processes take place in the same RIE chamber (Oxford Instruments, Plasma Lab System 100). Finally, the micropillar structures are fabricated using SiO_2 etching and DRIE as well (Figure 15).

The fabrication of porous micropillar arrays and micromembrane devices faces at least four challenges. First, as the micropillar size approaches 10 μm, additional efforts or techniques may be required to obtain accurate mask alignment. Second, it may also be challenging to fabricate a through hole with a large aspect ratio (e.g., a diameter of 10 μm and a depth of 300 μm). Alternatively,

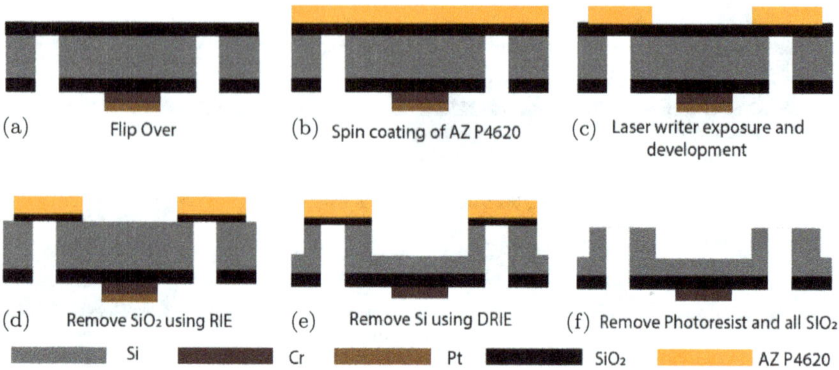

Figure 15. Schematic of fabrication process of the frontside micropillar.

advanced microfabrication techniques are available for realizing a deep and narrow hollow channel, such as parameter ramping [300, 301], the multiplexed Bosch process, and the oxygen plasma irradiation inserted DRIE process [302]. Third, while the diameter of the proposed micropillar structure ranges between 10 and 500 μm, to realize an ultrahigh heat transfer coefficient (exceeding 10^6 W/m^{2}K) may require further scaling down of the micropillar size to a few microns or even the sub-micron scale. However, it is challenging to create these features through conventional photolithography and DRIE. An alternative solution is to use electron beam lithography in conjunction with ion etching to create the porous nanopillar structures.

Last and most important, fabrication defects can ruin the phase separation performance of the entire hollow micropillar array or porous membrane. A single point defect can be either a pore absence in a membrane or a defect that alters the hollow channel size in a micropillar. Missing pores can potentially increase flow resistance but still allow the porous membrane to function properly. On the other hand, a variation in pore size, merging of pores, or destruction of pore edge features can directly impact the phase separation selectivity of the hollow micropillar array because it reduces the maximum pinning pressure of liquid attempting to penetrate the membrane. Therefore, it is critical to ensure a high success rate and yield when fabricating the porous micropillar structures. Postprocessing approaches, such as Piranha etching, can help fix the limited level of clogging or contamination, which may often occur in the fabrication process.

4.4. *Challenges to integrating micro- and nanoengineered surfaces in an embedded cooling system for 3D stacked chips*

In recent years, 3D chips have been gaining attention as a way to improve processor performance. By stacking the chips vertically, the computational and memory resources can be brought closer to each other, which can reduce latency and I/O power by up to 50%. However, the vertical stacking compounds the dissipated heat over each stacked layer, making thermal management extremely challenging. As one of the best performing solutions to date, embedded cooling systems using micro-/nanoengineered surfaces can effectively remove heat directly from each layer of the stacked chips. Section 4 briefly reviewed several types of micro-/nanoporous coatings and wicking structures, nanoporous membrane devices, microposts, and micropillars that have been utilized to enhance evaporative cooling. These cooling technologies can all be integrated into embedded-cooling platforms to achieve better cooling of 3D chip stacks.

This integration presents three aspects for consideration. First, electrical and fluidic delivery channels must be fabricated in the layers of the 3D stacked chips. TSVs, conductive interconnections fabricated through the silicon wafer, are widely used in 3D packaging. TSVs create low-latency, high-bandwidth paths for transporting information and power between layers in 3D stacked chips. In an embedded-cooling system for 3D stacked chips, structures similar to TSVs can be used to deliver the coolant to different layers. TSVs that transport electrical signals are called through-silicon electrical vias (TSEVs), and similar structures for fluidic delivery are called through-silicon fluidic vias (TSFVs). A second consideration is the bonding and connection of the electrical and fluidic channels in the assembly of the embedded cooling system. In the commonly used flip-chip technique, also known as controlled collapse chip connection, solder bumps are precisely positioned on the upper surfaces of 3D stacked chip layers. Mating layers are then flipped over and aligned, and the solder bumps are remelted. A flip-chip assembly is both compact and short, so it provides better electrical performance and causes less stress on an active area than traditional wire bonding methods [303]. As with TSEVs, in TSFVs, the flip-chip method can be used to interconnect aligned fluidic channels between layers. By using a similar method, the bonding process for the embedded cooling systems can be simplified. The third consideration is that the micro- and nanostructures and their interconnection method must be optimized for thermal and electrical performance. Co-design optimization methods are being used to design and simulate embedded-cooling systems. Properly implementing these three challenging aspects is essential to realizing embedded-cooling systems for 3D stacked chips.

The NEIT lab has developed a concept for embedded two-phase evaporative cooling, as shown in Figure 16. Here, liquid is delivered by a constant-pressure microfluidic system. It initially flows horizontally through an array of micropin fins and then vertically up (or down) through an array of hollow micropillar structures. The liquid is pinned on the top surface of each hollow micropillar structure, where evaporative heat transfer occurs. In this embedded-cooling system, there are three functional layers: the evaporation layers, the liquid delivery layers, and the chip layers. The evaporation layers are

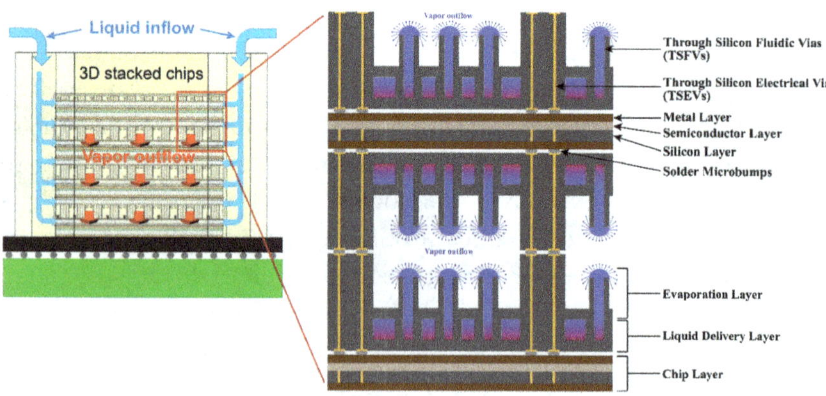

Figure 16. Schematic diagram of an embedded two-phase cooling package developed at the NEIT Lab.

composed of silicon micropillar arrays, silicon layers, and TSEVs. The liquid delivery layers are composed of silicon microposts, silicon layers, and microbumps. The chip layers are composed of metal, silicon, and semiconductor layers. The evaporation layers and the liquid delivery layers are coupled through direct silicon bonding techniques to connect the TSFVs and the liquid delivery layers. Opposing pairs of evaporation layers are bonded facing each other through the solder microbumps so that electrical signals can transfer through different chip layers. Solder microbumps are also used to connect the liquid delivery layers and the chip layers.

4.5. *Challenges in fabricating TSVs*

Although different fabrication methods are employed in producing TSEVs with high electrical performance, the process begins with using photolithography and DRIE to etch holes through the silicon wafer. Next, to create electrical interconnections, metallic materials with high electrical conductivity are either coated on the interior walls of the holes or completely filled in [148, 304–307]. TSFVs that carry coolant have structures that vary with the specific cooling technique, but most are fabricated by etching the silicon substrate. In the NEIT's design, the TSFVs are the through holes inside the micropillar arrays, which deliver dielectric coolant from the liquid layer to the vapor domain. In order for the coolant to emerge from all the holes uniformly, the pressure restriction of each TSFV must be the same. The exit pressure of the TSFVs is dictated by their size. Therefore, fabrication accuracy becomes challenging while scaling down the size. Yu *et al.* [304] used microfins to create liquid coolant channels in their embedded cooling system. They sputtered thick copper layers on both faces of the wafer, which also coated the interior walls of the TSEVs with a thin metal film. However, during the coating processes, it is hard to ensure uniform conductivity due to the roughness of the holes' sidewalls. In addition, the thick copper layers sputtered on the faces of the silicon wafers make the wet etching process more difficult. An alternative solution is to fill the through holes with a conductive material. Civale *et al.* [305] deposited a Ti/Cu seed layer, applied a 10 μm thick negative resist mask layer, and then used ECD to fill the

TSEVs with Cu. Compared to coating the holes, filling them makes it easier to ensure uniform results. However, as the dimensions of engineered surfaces become smaller than 10 μm, the fabrication processes become more challenging. The Bosch DRIE process is widely used in fabricating micro-/nanoengineered structures and TSVs. It can be used to make TSEVs with diameters less than 5 μm; however, the sidewalls of the through holes are typically tapered and scalloped [308]. Going forward, the Bosch etching process needs to be improved, and alternative methods, such as laser drilling, need to be studied.

4.5.1. *Wafer bonding methods and challenges*

After the TSVs are created, microstructures for interconnections are fabricated on the side of the wafer opposite the TSFVs. One of the most widely used interconnection techniques in 3D stacked chips is called flip chip. This technique uses projecting solder microbumps deposited onto the chip pads as bonds and interconnections between the different layers of 3D chips.

Microbumps bond different layers of 3D stacked chips together and transport electrical signals [148, 306, 309, 310]. Zheng *et al.* [310] fabricated microbumps that were 25 μm in diameter and 50 μm in pitch. First, a Ti/Cu seed layer was deposited on the backside of the silicon wafer, and copper pads for the electrical microbumps were electroplated on the seed layer. Next, Ni and solder were electroplated on the copper pads, in that order. In a different approach, Schultz *et al.* [307] used silicon and sealing structures to interconnect the different layers. They used a handler attach and release process [311] to fabricate silicon column interconnects with TSEVs inside the silicon. Similarly, microbump structures have also been used to interconnect TSFVs, where microbumps at the ends of the TSFVs contain channels for a liquid coolant to flow through different layers [310]. Micropipes are also used as interconnections of TSFVs in embedded-cooling systems [309, 312]. Dang *et al.* [152] spin-coated a thick layer of photoimageable polymer and then used wet etching to open the passivation layer inside the micropipes so that fluid could circulate. After fabricating the microbumps, Zheng *et al.* [310] used the flip-chip method to bond the different layers. A force of approximately 7 N was applied to the two layers, and the temperature was increased to 230°C, with a ramp of 2°C/s. Madhour *et al.* [313] used AgSn as the solder and Ni and Au materials in the process of fabricating the microbumps.

One of the challenges in bonding the TSVs is alignment accuracy, which directly affects the electrical and fluidic performance of the cooling devices and the 3D chips. High alignment accuracy requires both accurate fabrication of the microstructures and accurate physical assembly of the different stacks. The fabrication accuracy challenges were discussed in Section 4.5. The assembly challenges increase with the size and decreasing pitch of the microbumps. Thermo-mechanical concerns also arise when bonding the 3D chips and the embedded-cooling device. The peak temperature of the flip-chip method is normally higher than the operating temperature of the chips, so some thermo-mechanical issues can occur during the bonding. The small size of the microbumps can also cause stress concentration problems during the assembly process.

4.5.2. *Optimized designs for improved thermal and electrical performance of embedded-cooling systems*

Since the material used in TSEVs is mainly copper, the TSEVs conduct heat better than the silicon wafer. Although TSEVs enable high electrical performance in 3D stacked chips, they inevitably decrease the thermal performance of the embedded cooling system. Therefore, the number and positions of the TSEVs must be optimized to balance their electrical and thermal performance. In addition, the thermal performance is also influenced by the pressure drop and flow rate of the TSFVs. Therefore, optimization methods for both TSEVs and TSFVs are needed to achieve both the desired thermal and electrical performance. Several optimization methods are used to arrange the position and number of the TSEVs and TSFVs. Cong *et al.* [314] developed an algorithm to refine the 3D placement of the TSVs, which successfully reduced the maximum on-chip temperature by 37%, with only 6% more TSVs and an 8% greater wirelength. Shi *et al.* [37] proposed a co-optimization method for the allocation of microchannels and TSVs to achieve maximum cooling effectiveness at minimum cooling power.

The challenging part of the co-design method is balancing thermal performance with electrical performance. Due to the limited area of 3D stacked chips, the number of TSEVs and TSFVs is limited. Therefore, an efficient and effective optimization algorithm is very important for balancing the number and position of different kinds of TSVs. However, it is difficult to find a universal algorithm for different embedded-cooling systems since many aspects influence their performance.

5. Conclusion

This chapter briefly discussed cooling technologies focused on removing high heat fluxes from microelectronic devices by implementing micro- and nanostructures to enhance evaporative cooling. These technologies aim to achieve mainly two purposes: maximizing the region with the least resistance to evaporation and supplying the liquid to that region in the required quantity. The three-phase contact line has been identified as the region with the least resistance and maximum evaporation rate. Therefore, most of the recent evaporative cooling technologies have implemented methods to increase the contact line region. Many of them have used wicking structures, surface coatings, and other micro–nano porous structures, such as membranes, microposts, and hollow micropillars, to maximize the contact line region and facilitate liquid delivery. Micro- and nanoscale surface roughness modifies the intermolecular forces between the solid and the liquid–vapor interface and reduces the apparent contact angle, thus making the surface more wetting. Increased surface wettability helps the liquid film thin out more at the contact line. Among different surface structures, wicking materials have been extensively used in numerous passive cooling devices, such as heat pipes and jet-impingement cooling devices. In these devices, under surface tension, the working fluid forms an ultrathin evaporating meniscus along the wicking structure. The fluid propagates along the wicking surface under the action of capillarity. Most of these wicking structures have been fabricated from sintered Cu microparticles. One drawback of mono-porous wicking structures is their increased viscous flow resistance at smaller pore sizes, which causes dry-out. To avoid this, researchers have developed hierarchical surfaces, such as bi-porous, bi-textured, and biphilic surfaces, which can simultaneously reduce the viscous flow resistance and increase fluid supply by facilitating fluid imbibition. Another category of devices using surface structures to maximize evaporative cooling performance are the micro–nano porous membrane devices, which mimic biological perspiration as a strategy to cool electronic devices. In these devices, a very thin liquid film ($\sim$1 μm) can be maintained in a porous membrane stretched atop a constantly replenished liquid reservoir. This technique achieved a very high heat flux removal (heat flux $\sim$1,400 W/cm^2), with a heat transfer coefficient of 6.66×10^5 W/m^{2}K [250]. A novel technique for maximizing evaporative cooling performance is the use of micropillars or microposts to change the shape of the evaporating interface. Sharp features of these structures obstruct an advancing evaporating meniscus by providing an energy barrier which forces the liquid body to conform to a particular shape. A comprehensive investigation of the evaporation dynamics and heat transfer characteristics of asymmetric droplets has been carried out by Agonafer *et al.* [136, 139, 184]. In their technology (the NEIT technology), a hollow micropillar with a triangular top is used to form droplets with different shapes. A 71% increase (at a substrate temperature of 98°C) in heat transfer coefficient has been reported when the droplet footprint was changed from circular to triangular. There are several aspects of this cooling scheme which increases the heat transfer rate:

- Compared to a capped spherical droplet, a triangular droplet has a larger contact line region. The liquid layer thins out more, particularly at the corner regions of the micropillar, which results in a lower average effective liquid layer thickness. As a result, the conduction resistance across the liquid layer is reduced, which increases heat transfer.
- A triangular droplet also possesses a larger average meniscus curvature due to its non-uniform and lower contact angle. The larger

liquid–vapor meniscus curvature creates steeper density gradients in the gas/vapor domain, which reduces diffusion resistance, and the evaporated vapor molecules are taken away from the vicinity of the interface to be replenished by newly evaporated molecules.

- Another factor contributing to the higher evaporation rate by confining evaporating droplets atop micropillars is the reduced confinement effect. For sessile droplets evaporating on a solid substrate, the diffusion of vapor molecules is restricted in the direction toward the substrate due to the physical barrier posed by the substrate. By placing the droplets on top of the micropillar surfaces, vapor molecules are allowed to diffuse in the downward direction, which increases heat transport.

Although microscale evaporative cooling technologies using micro–nano surface structures are very attractive for high heat flux applications using 3D embedded-cooling systems, the current fabrication and integration of different components of these devices, such as micro-/nanostructures, porous membranes structures, microposts/micropillars, and heating and temperature-sensing elements, and TSVs pose many challenges which need to be overcome.

Another bottleneck in reaching the fundamental limit of evaporative heat transfer performance is removing vapor molecules from the evaporating interface since evaporation may finally be limited due to the accumulated vapor molecules in the vicinity of the liquid meniscus. Some innovative strategies have been adopted to address this issue, for example, by using vapor-venting [64, 65, 154], flowing gas [128, 299], and passive methods, such as increasing vapor diffusion by increased meniscus curvature, and by reducing substrate confinement effect (e.g., in the NEIT technology). Further research is needed to gain more in-depth understanding of the gas-phase fluid physics and mass transfer to minimize the diffusion resistance to realize the full potential of micro–nano-engineered surfaces to provide enhanced evaporative cooling.

The discussion in this chapter recognizes the strategies using micro- and nanostructures to assist heat dissipation in the next generation of ultrahigh heat flux microelectronic applications. It also identifies ways to increase the region of an evaporating meniscus with the least thermal resistance, namely the contact line region. By comparing the performances of different cooling technologies and reviewing their capabilities and limitations, this chapter provides a guideline for choosing a suitable cooling solution for high heat flux microelectronic applications.

References

[1] A.W. Topol, *et al.*, Three-dimensional integrated circuits. *IBM J. Res. Dev.*, 50 (4.5), 2006, 491–506.

[2] T. Simonite, Moore's law is dead. Now what? Microchips 2016, May 13. Available at: https://www.technologyreview.com/2016/05/13/245938/moores-law-is-dead-now-what/ (accessed August 1, 2020).

[3] Tech Design Forum. 2.5D-IC, 3D-IC, and 5.5D-IC — Stacked-die integration. Available at: https://www.techdesignforums.com/practice/guides/2-5dic-3dic-5-5dic-integration/ (accessed July 30, 2020).

[4] S. Anthony, 7nm, 5nm, 3nm: The new materials and transistors that will take us to the limits of Moore's law, July 26, 2013. Available at: https://www.extremetech.com/computing/162376-7nm-5nm-3nm-the-new-materials-and-transistors-that-will-take-us-to-the-limits-of-moores-law (accessed July 29, 2020).

[5] A. Shilov, TSMC: 5nm on track for Q2 2020 HVM, will ramp faster than 7nm, October 23, 2019 Available at: https://www.anandtech.com/show/15016/tsmc-5nm-on-track-for-q2-2020-hvm-will-ramp-faster-than-7nm (accessed August 2, 2020).

[6] D. Liu, S.J.J.O.E.P. Park, Three-dimensional and 2.5 dimensional interconnection technology: State of the art, 136 (1), 2014.

[7] J. Knickerbocker, *et al.* 2.5 D and 3D technology challenges and test vehicle demonstrations, in: *2012 IEEE 62nd Electronic Components and Technology Conference*, 2012, IEEE.

[8] A.L. Moore, L.J.M.T. Shi, Emerging challenges and materials for thermal management of electronics, *Materials today*, 17 (4), 2014, 163–174.

[9] S.S. Iyer, The evolution of dense embedded memory in high performance logic technologies, in: *2012 International Electron Devices Meeting*, 2012, IEEE.

[10] H.-S.P. Wong, Beyond the conventional transistor, *IBM J. Res. Dev.*, 46 (2.3), 2002, 133–168.

[11] S.S. Wong, A. El Gamal, The prospect of 3D-IC, in: *2009 IEEE Custom Integrated Circuits Conference*, 2009, IEEE.

[12] V. Solberg, 2.5 D and 3D semiconductor package technology: Evolution and innovation. *IPC APEX EXPO Proceedings, 2016.*

[13] W.-W. Shen, K.-N.J.N.R.L. Chen, Three-dimensional integrated circuit (3D IC) key technology: Through-silicon via (TSV), *Nanoscale research letters*, 12 (1), 2017, 1–9.

[14] Semiconductor Engineering. 2.5D. Available at: https://semiengineering.com/knowledge_centers/packaging/advanced-packaging/2-5d-ic/ (accessed August 1, 2020).

[15] Y. Zhang, *et al.* 3D stacked microfluidic cooling for high-performance 3D ICs, in: *2012 IEEE 62nd Electronic Components and Technology Conference*, 2012, IEEE.

[16] OpenSilicon, 2.5D technology. Available at: https://www.open-silicon.com/2-5d-technology/ (accessed July 3, 2020).

[17] A. Desai, 2.5D and 3D ICs: New paradigms in ASIC, October 18, 2017. Available at: https://www.einfochips.com/blog/2-5d-3d-ics-new-paradigms-in-asic/ (accessed July 21, 2020).

[18] M. Maxfield, 2D vs. 2.5D vs. 3D ICs 101, April 8, 2012. Available at: https://www.eetimes.com/2d-vs-2-5d-vs-3d-ics-101/# (accessed July 26, 2020).

[19] R. Grigalunas, Design engineers! 8 benefits of die stacking, October 3, 2017. Available at: https://www.escomponents.com/blog/2017/10/3/design-engineers-8-benefits-of-die-stacking (accessed August 1, 2020).

[20] MonolithIC3D, The monolithic 3D Advantage. Available at: http://www.monolithic3d.com/3d-ic-edge1.html (accessed August 1, 2020).

[21] GSA, 3D IC architecture: A natural evolution, 2012, 1–27.

[22] C.-C. Yang, *et al.* Enabling monolithic 3D image sensor using large-area monolayer transition metal dichalcogenide and logic/memory hybrid 3D+ IC, in: *2016 IEEE Symposium on VLSI Technology*, 2016, IEEE.

[23] Z. Or-Bach, The monolithic 3D advantage: Monolithic 3D is far more than just an alternative to 0.7× scaling, in: *2013 IEEE International 3D Systems Integration Conference (3DIC)*, 2013, IEEE.

[24] L. Zhang, L. Liu, Polymeric self-assembled monolayers anomalously improve thermal transport across graphene/polymer interfaces, *ACS Applied Materials & Interfaces*, 9 (34), 2017, 28949–28958.

[25] P. Jain, *et al.*, Thermal and power delivery challenges in 3D ICs, in: *Three Dimensional Integrated Circuit Design*, 2010, Springer, pp. 33–61.

[26] P. Shukla, *et al.* An overview of thermal challenges and opportunities for monolithic 3D ICs, in: *Proceedings of the 2019 on Great Lakes Symposium on VLSI*, 2019.

[27] C.S. Sharma, *et al.*, A novel method of energy efficient hotspot-targeted embedded liquid cooling for electronics: An experimental study, *Int. J. Heat Mass Transfer*, 88, 2015, 684–694.

[28] M. Li, *et al.*, Joule heating induced thermomigration failure in un-powered microbumps due to thermal crosstalk in 2.5 D IC technology, *Journal of Applied Physics*, 120 (7), 2016, 075105.

[29] K.J. Geisler, A. Bar-Cohen. Passive immersion cooling of 3-D stacked dies, in: *International Electronic Packaging Technical Conference and Exhibition*, 2007.

[30] F. Buja, *et al.*, High heat flux dissipation via interposer active micro-cooling, *Japanese Journal of Applied Physics*, 58 (SB), 2019, SBBB11.

[31] V. Venkatadri, *et al.*, A review of recent advances in thermal management in three dimensional chip stacks in electronic systems, *J. Electron. Packag.*, 133 (4), 2011.

[32] M. Ebadian, C.-X. Lin, A review of high-heat-flux heat removal technologies, *J. Electron. Packag.*, 133 (11), 2011.

[33] S. Islam, I.A. Motaleb, Investigation of the dynamics of liquid cooling of 3D ICs, in: *2019 8th International Symposium on Next Generation Electronics (ISNE)*, 2019, IEEE.

[34] H. Coles, M. Herrlin, *Immersion Cooling of Electronics in DoD Installations*, California University Berkeley, Berkeley, USA, 2016.

[35] A. Sinha, *et al.*, Novel immersion cooling technique for a 3D chip stack, in: *International Electronic Packaging Technical Conference and Exhibition*, 2013, American Society of Mechanical Engineers.

[36] D. Kearney, T. Hilt, P. Pham, A liquid cooling solution for temperature redistribution in 3D IC architectures, *Microelectron. J.*, 43 (9), 2012, 602–610.

[37] B. Shi, A. Srivastava, A. Bar-Cohen. Hybrid 3D-IC cooling system using micro-fluidic cooling and thermal TSVs, in: *2012 IEEE Computer Society Annual Symposium on VLSI*, 2012, IEEE.

[38] M.M. Sabry, *et al.*, Greencool: An energy-efficient liquid cooling design technique for 3-d mpsocs via channel width modulation, *IEEE Transactions on Computer-Aided Design of Integrated Circuits and Systems*, 32 (4), 2013, 524–537.

[39] F. Mei, *et al.*, Fabrication, assembly, and testing of Cu-and Al-based microchannel heat exchangers, *Journal of Microelectromechanical systems*, 17 (4), 2008, 869–881.

[40] 3DInCites. What is 3D integration? Available at: https://www.3dincites.com/3d-incites-knowledge-portal/what-is-3d-integration/ (accessed July 27, 2020).

[41] D.B. Tuckerman, R.F.W. Pease, High-performance heat sinking for VLSI, *IEEE Electron device letters*, 2 (5), 1981, 126–129.

[42] P.-S. Lee *et al.*, Investigation of heat transfer in rectangular microchannels, *Int. J. Heat Mass Transfer*, 48 (9), 2005, 1688–1704.

[43] D. Copeland, M. Behnia, W. Nakayama, Manifold microchannel heat sinks: Isothermal analysis, in: *InterSociety Conference on Thermal Phenomena in Electronic Systems, I-THERM V*, 1996, IEEE.

[44] J. Ryu, *et al.*, Three-dimensional numerical optimization of a manifold microchannel heat sink, *Int. J. Heat Mass Transfer*, 46 (9), 2003, 1553–1562.

[45] A. Bar-Cohen, F.L. Robinson, D.C. Deisenroth. Challenges and opportunities in Gen3 embedded cooling with high-quality microgap flow, in: *2018 International Conference on Electronics Packaging and iMAPS All Asia Conference (ICEP-IAAC)*, 2018, IEEE.

[46] E. Kermani, Manifold microchannel cooling of photovoltaic cells for high efficiency solar energy conversion systems, University of Maryland, College Park, 2008.

[47] A. Bar-Cohen, M. Arik, M. Ohadi, Direct liquid cooling of high flux micro and nano electronic components, *Proceedings of the IEEE*, 94 (8), 2006, 1549–1570.

[48] C.H. Amon, *et al.*, MEMS-enabled thermal management of high-heat-flux devices EDIFICE: Embedded droplet impingement for integrated cooling

of electronics, *Thermal Sciences 2000. Proceedings of the International Thermal Science Seminar. Volume 1.*, 25 (5), 2001, 231–242.

[49] S.G. Kandlikar, Review and projections of integrated cooling systems for three-dimensional integrated circuits, *Journal of Electronic Packaging*, 136 (2), 2014.

[50] A. Robinson *et al.*, A single phase hybrid micro heat sink using impinging micro-jet arrays and microchannels, *Appl. Thermal Eng.*, 136, 2018, 408–418.

[51] C.R. King Jr., *Thermal Management of Three-Dimensional Integrated Circuits Using Inter-Layer Liquid Cooling*, Georgia Institute of Technology, 2012.

[52] S. Wang, *et al.*, 3D integrated circuit cooling with microfluidics, *Micromachines*, 9 (6), 2018, 287.

[53] A. Bar-Cohen, Gen 3 "embedded" cooling: key enabler for energy efficient data centers, *IEEE Transactions on Components, Packaging and Manufacturing Technology*, 7 (8), 2017, 1206–1211.

[54] M. Schultz, *et al.*, Embedded two-phase cooling of large three-dimensional compatible chips with radial channels, *Journal of Electronic Packaging*, 138 (2), 2016.

[55] D.G. Bae, *et al.* Embedded two-phase cooling of high heat flux electronics on silicon carbide (SiC) using thin-film evaporation and an enhanced delivery system (FEEDS) manifold-microchannel cooler, in: *2017 16th IEEE Intersociety Conference on Thermal and Thermomechanical Phenomena in Electronic Systems (ITherm)*, 2017, IEEE.

[56] L.S. Wu, *et al.* Experimental research on phase-change cooling for high-power chip. *Appl. Mech. Mater.*, 628, 2014, 306–310.

[57] S. Sengupta, *et al.*, Rotational instabilities in microchannel flows, *Physics of Fluids*, 31 (5), 2019, 054101.

[58] C.-J. Kuo, Y. Peles, Flow boiling instabilities in microchannels and means for mitigation by reentrant cavities, *J. Heat Transfer*, 130 (7), 2008.

[59] Y.K. Prajapati, P.J.E.T. Bhandari, F. Science, Flow boiling instabilities in microchannels and their promising solutions: A review, *Experimental Thermal and Fluid Science*, 88, 2017, 576–593.

[60] M. Bruder, G. Bloch, and T. Sattelmayer, Critical heat flux in flow boiling: Review of the current understanding and experimental approaches, *Heat Transfer Engineering*, 38 (3), 2017, 347–360.

[61] W. Qu, I. Mudawar, Measurement and correlation of critical heat flux in two-phase micro-channel heat sinks, *Int. J. Heat Mass Transfer*, 47 (10–11), 2004, 2045–2059.

[62] S.-M., Kim, I. Mudawar, Universal approach to predicting saturated flow boiling heat transfer in mini/micro-channels–Part I. *Dryout Incipience Qual.*, 64, 2013, 1226–1238.

[63] M.P. David, *et al.*, Adiabatic and diabatic two-phase venting flow in a microchannel, *International Journal of Multiphase Flow*, 37 (9), 2011, 1135–1146.

[64] M.P. David, *et al.*, Hydraulic and thermal characteristics of a vapor venting two-phase microchannel heat exchanger, *Int. J. Heat Mass Transfer*, 54 (25–26), 2011, 5504–5516.

[65] M.P. David, *et al.* Vapor-venting, micromachined heat exchanger for electronics cooling, in: *ASME International Mechanical Engineering Congress and Exposition*, 2007.

[66] P. Asrar, *et al.*, Flow boiling of R245fa in a microgap with staggered circular cylindrical pin fins, *Int. J. Heat Mass Transfer*, 121, 2018, 329–342.

[67] P.-S. Lee, S.V. Garimella, Saturated flow boiling heat transfer and pressure drop in silicon microchannel arrays, *Int. J. Heat Mass Transfer*, 51 (3–4), 2008, 789–806.

[68] A. Bar-Cohen, E.J.H.T.E. Rahim, Modeling and prediction of two-phase microgap channel heat transfer characteristics, *Heat Transfer Engineering*, 30 (8), 2009, 601–625.

[69] E. Rahim, A. Bar-Cohen. Parametric dependence of annular flow heat transfer in microgaps, in: *International Heat Transfer Conference*, 2010.

[70] V. Kuznetsov, A. Shamirzaev. Flow boiling heat transfer mechanism in minichannels, in: *International Conference on Nanochannels, Microchannels, and Minichannels*, 2007.

[71] Y. Yang, Y. Fujita. Flow boiling heat transfer and flow pattern in rectangular channel of mini-gap, in: *International Conference on Nanochannels, Microchannels, and Minichannels*, 2004.

[72] T. Alam, *et al.*, Experimental investigation of local flow boiling heat transfer and pressure drop characteristics in microgap channel, *Int. J. Multiphase Flow*, 42, 2012, 164–174.

[73] A. Koşar, Y. Peles, Boiling heat transfer in a hydrofoil-based micro pin fin heat sink, *Int. J. Heat Mass Transfer*, 50 (5–6), 2007, 1018–1034.

[74] C. Green, *et al.*, A review of two-phase forced cooling in three-dimensional stacked electronics: Technology integration, *Journal of Electronic Packaging*, 137 (4), 2015.

[75] C. Huh, M.H. Kim, Pressure drop, boiling heat transfer and flow patterns during flow boiling in a single microchannel, *Heat Transfer Engineering*, 28 (8–9), 2007, 730–737.

[76] P. Zhou, K. Goodson, J. Suntiago, Vapor escape microchannel heat exchanger, US Patent 6,994,151 B2, Feb. 7, 2006.

[77] W. Li, *et al.*, Enhanced flow boiling in microchannels through integrating multiple micro-nozzles and reentry microcavities, *Appl. Phys. Lett.*, 110 (1), 2017, 014104.

[78] L. Zhang, K.E. Goodson, T.W. Kenny, Enhanced nucleate boiling in microchannels, in: *Silicon Microchannel Heat Sinks*, 2004, Springer, pp. 95–118.

[79] A. Mukherjee, S.G. Kandlikar, The effect of inlet constriction on bubble growth during flow boiling in microchannels, *Int. J. Heat Mass Transfer*, 52 (21–22), 2009, 5204–5212.

[80] H.Y. Erbil, Evaporation of pure liquid sessile and spherical suspended drops: A review, *Advances in colloid and interface science*, 170 (1–2), 2012, 67–86.

[81] W.-L. Cheng, *et al.*, Spray cooling and flash evaporation cooling: The current development and application, *Renewable and Sustainable Energy Reviews*, 55, 2016, 614–628.

[82] A.-M. Cazabat, G. Guena, Evaporation of macroscopic sessile droplets, *Soft Matter*, 6 (12), 2010, 2591–2612.
[83] E.Y. Gatapova, *et al.*, Evaporation dynamics of a sessile droplet on glass surfaces with fluoropolymer coatings: Focusing on the final stage of thin droplet evaporation, *Soft Matter*, 14 (10), 2018, 1811–1821.
[84] H. Hu, R.G. Larson, Evaporation of a sessile droplet on a substrate, *The Journal of Physical Chemistry B*, 106 (6), 2002, 1334–1344.
[85] J. Oh, *et al.*, Jumping-droplet electronics hot-spot cooling, *Appl. Phys. Lett.*, 110 (12), 2017, 123107.
[86] N. Kumari, S.V. Garimella, Characterization of the heat transfer accompanying electrowetting or gravity-induced droplet motion, *Int. J. Heat Mass Transfer*, 54 (17–18), 2011, 4037–4050.
[87] V.K. Pamula, K. Chakrabarty. Cooling of integrated circuits using droplet-based microfluidics, in: *Proceedings of the 13th ACM Great Lakes Symposium on VLSI*, 2003.
[88] J. Prakash, B.S. Sikarwar, Modeling of sessile droplet evaporation on engineered surfaces, *Journal of Thermal Science and Engineering Applications*, 11 (6), 2019.
[89] S.K. Singh, *et al.*, Combined effect of substrate wettability and thermal properties on evaporation dynamics of a sessile droplet, *Interf. Phenom. Heat Transfer*, 5.4 (2017), 2017.
[90] S.Y. Misyura, Evaporation of a sessile water drop and a drop of aqueous salt solution, *Scientific reports*, 7 (1), 2017, 1–11.
[91] Y.C. Kim, Evaporation of nanofluid droplet on heated surface, *Advances in Mechanical Engineering*, 7 (4), 2015, 1687814015578358.
[92] Z. Yan, *et al.*, Droplet-based microfluidic thermal management methods for high performance electronic devices, *Micromachines*, 10 (2), 2019, 89.
[93] B. Deryagin, S. Nerpin, N.J.K.Z. Churaev, To the theory of liquid evaporation from capillaries, *Kolloidn. Zh*, 26, 1964, 301–307.
[94] J. Plawsky, *et al.*, Nano-and microstructures for thin-film evaporation: A review, *Nanoscale Microscale Thermophys. Eng.*, 18 (3), 2014, 251–269.
[95] M. Mochizuki, *et al.* Advanced micro-channel vapor chamber for cooling high power processors, in: *International Electronic Packaging Technical Conference and Exhibition*, 2007.
[96] B. Agostini, *et al.*, State of the art of high heat flux cooling technologies, *Heat transfer engineering*, 28 (4), 2007, 258–281.
[97] A. Bar-Cohen, J.J. Maurer, J.G. Felbinger. DARPA's intra/interchip enhanced cooling (ICECool) program, in: *CS MANTECH Conference*, May 13–16, 2013.
[98] S.G. Kandlikar, *et al.*, Heat transfer in microchannels: 2012 status and research needs, *Journal of Heat Transfer*, 135 (9), 2013.
[99] L. Cao, H.-H. Hu, D.J.L. Gao, Design and fabrication of micro-textures for inducing a superhydrophobic behavior on hydrophilic materials, *Langmuir*, 23 (8), 2007, 4310–4314.
[100] L. Cao, *et al.*, Super water-and oil-repellent surfaces on intrinsically hydrophilic and oleophilic porous silicon films, *Langmuir*, 24 (5), 2008, 1640–1643.

[101] X.-M. Li, D. Reinhoudt, M. Crego-Calama, What do we need for a superhydrophobic surface? A review on the recent progress in the preparation of superhydrophobic surfaces, *Chemical Society Reviews*, 36 (8), 2007, 1350–1368.
[102] Y. Nam, *et al.*, Fabrication and characterization of the capillary performance of superhydrophilic Cu micropost arrays, *Journal of Microelectromechanical Systems*, 19 (3), 2010, 581–588.
[103] J.L. Plawsky, J.K. Kim, E.F. Schubert, Engineered nanoporous and nanostructured films, *Materials Today*, 12 (6), 2009, 36–45.
[104] X. Deng, *et al.*, Candle soot as a template for a transparent robust superamphiphobic coating, *Science*, 335 (6064), 2012, 67–70.
[105] J.L. Plawsky, *et al.*, Review of the effects of surface topography, surface chemistry, and fluid physics on evaporation at the contact line, *Chemical Engineering Communications*, 196 (5), 2008, 658–696.
[106] R. Xiao, E.N. Wang, Microscale liquid dynamics and the effect on macroscale propagation in pillar arrays, *Langmuir*, 27 (17), 2011, 10360–10364.
[107] W. Liu, *et al.*, Capillary rise of liquids over a microstructured solid surface, *Langmuir*, 27 (23), 2011, 14260–14266.
[108] L. Courbin, *et al.*, Imbibition by polygonal spreading on microdecorated surfaces, *Nature materials*, 6 (9), 2007, 661–664.
[109] R. Ranjan, J.Y. Murthy, S.V. Garimella, Analysis of the wicking and thin-film evaporation characteristics of microstructures, *J. Heat Transfer*, 131 (10), 2009, 101001.
[110] R.M. Manglik, On the advancements in boiling, two-phase flow heat transfer, and interfacial phenomena, 2006.
[111] S.C. Maroo, J. Chung, A possible role of nanostructured ridges on boiling heat transfer enhancement, *Journal of Heat Transfer*, 135 (4), 2013.
[112] D. Quéré, Wetting and roughness, *Annu. Rev. Mater. Res.*, 38, 2008, 71–99.
[113] S.J. Gokhale, J.L. Plawsky, P.C. Wayner, Spreading, evaporation, and contact line dynamics of surfactant-laden microdrops, *Langmuir*, 21 (18), 2005, 8188–8197.
[114] M. Ojha, *et al.*, Role of solid surface structure on evaporative phase change from a completely wetting corner meniscus, *Physics of Fluids*, 22 (5), 2010, 052101.
[115] F. Barberis, M. Capurro, Wetting in the nanoscale: A continuum mechanics approach, *Journal of colloid and interface science*, 326 (1), 2008, 201–210.
[116] K. Kamrin, M.Z. Bazant, H.A. Stone, Effective slip boundary conditions for arbitrary periodic surfaces: The surface mobility tensor, *Journal of Fluid Mechanics*, 658, 2010, 409–437.
[117] J.P. Rothstein, Slip on superhydrophobic surfaces, *Annual review of fluid mechanics*, 42, 2010, 89–109.
[118] P. Sáenz, *et al.*, Evaporation of sessile drops: A three-dimensional approach, *Journal of Fluid Mechanics*, 772, 2015, 705–739.

[119] A. Faghri, Y. Zhang, 5-solid-liquid-vapor phenomena and interfacial heat and mass transfer, *Transport Phenomena in Multiphase Systems*, 2006, 331–420.

[120] N. Miljkovic, *et al.*, Jumping-droplet-enhanced condensation on scalable superhydrophobic nanostructured surfaces, *Nano letters*, 13 (1), 2013, 179–187.

[121] F. Su, *et al.*, Ultra-high cooling rate utilizing thin film evaporation, *Appl. Phys. Lett.*, 101 (11), 2012, 113702.

[122] Y. Nam, Y.S. Ju, A comparative study of the morphology and wetting characteristics of micro/nanostructured Cu surfaces for phase change heat transfer applications, *Journal of Adhesion Science and Technology*, 27 (20), 2013, 2163–2176.

[123] F.D. Espinosa, *et al.*, Effect of fabrication parameters on the thermophysical properties of sintered wicks for heat pipe applications, *Int. J. Heat Mass Transfer*, 55 (25–26), 2012, 7471–7486.

[124] F.L. Chang, Y.M. Hung, Dielectric liquid pumping flow in optimally operated micro heat pipes, *Int. J. Heat Mass Transfer*, 2017, 108, 257–270.

[125] R. Ranjan, J.Y. Murthy, S.V. Garimella, A microscale model for thin-film evaporation in capillary wick structures, *Int. J. Heat Mass Transfer*, 54 (1), 2011, 169–179.

[126] J. Xie, *et al.*, Study of heat transfer enhancement for structured surfaces in spray cooling, *Appl. Thermal Eng.*, 59 (1–2), 2013, 464–472.

[127] X. Dai, *et al.*, Biphilic nanoporous surfaces enabled exceptional drag reduction and capillary evaporation enhancement, *Appl. Phys. Lett.*, 105 (19), 2014, 191611.

[128] S. Narayanan, *et al.*, Gas-assisted thin-film evaporation from confined spaces for dissipation of high heat fluxes, 13 (1), 2009, 30–53.

[129] Z. Lu, *et al.*, An ultrathin nanoporous membrane evaporator, 17 (10), 2017, 6217–6220.

[130] A. Bar-Cohen, *et al.*, Two-phase thermal transport in microgap channels: Theory, experimental results, and predictive relations, 24 (1), 2012, 1–15.

[131] S. Bigham, A. Fazeli, S. Moghaddam, Physics of microstructures enhancement of thin film evaporation heat transfer in microchannels flow boiling, 7, 2017, 44745.

[132] O.A. Kabov, *et al.*, Evaporation and flow dynamics of thin, shear-driven liquid films in microgap channels, *Exp. Thermal Fluid Sci.*, 35 (5), 2011, 825–831.

[133] K. Stepanov, *et al.* An experimental model of evaporative cooling system with forced circulation of coolant for high-performance electronic components, in: *EPJ Web of Conferences*, 2017, EDP Sciences.

[134] R. Mandel, A. Shooshtari, M. Ohadi, Thin-film evaporation on microgrooved heatsinks, *Numer. Heat Transfer A Appl.*, 71 (2), 2017, 111–127.

[135] J. Kim, *et al.*, Evaporative spray cooling of plain and microporous coated surfaces, 47 (14–16), 2004, 3307–3315.

[136] D.D. Agonafer, *et al.*, Porous micropillar structures for retaining low surface tension liquids, 514, 2018, 316–327.

[137] L. Shan, *et al.*, Experimental investigation of evaporation from asymmetric microdroplets confined on heated micropillar structures, 109, 2019, 109889.

[138] H. Wang, S.V. Garimella, J.Y. Murthy, Characteristics of an evaporating thin film in a microchannel, *Int. J. Heat Mass Transfer*, 50 (19), 2007, 3933–3942.

[139] B. Ma, *et al.*, Evolution of microdroplet morphology confined on asymmetric micropillar structures, 35 (37), 2019, 12264–12275.

[140] N. Farokhnia, *et al.*, Rational micro/nanostructuring for thin-film evaporation, 120 (16), 2016, 8742–8750.

[141] P. Sáenz, *et al.*, Dynamics and universal scaling law in geometrically-controlled sessile drop evaporation, 8 (1), 2017, 1–9.

[142] L. Shan *et al.*, Numerical investigation of shape effect on microdroplet evaporation, 141 (4), 2019.

[143] S. Adera, *et al.*, Hotspot thermal management via thin-film evaporation—Part I: Experimental characterization. *IEEE Trans. Compon. Packag. Manuf. Technol.*, 8 (1), 2017, 88–98.

[144] S. Narayanan, A.G. Fedorov, Y.K. Joshi. Experimental characterization of a micro-scale thin film evaporative cooling device, in: *2010 12th IEEE Intersociety Conference on Thermal and Thermomechanical Phenomena in Electronic Systems*, 2010, IEEE.

[145] X. Dai, *et al.*, Micromembrane-enhanced capillary evaporation, *Int. J. Heat Mass Transfer*, 64, 2013, 1101–1108.

[146] J.-N. Chen, *et al.*, Dropwise evaporative cooling of heated surfaces with various wettability characteristics obtained by nanostructure modifications, 11 (1), 2016, 158.

[147] Y. Nam, Y.S. Ju, A comparative study of the morphology and wetting characteristics of micro/nanostructured Cu surfaces for phase change heat transfer applications, *J. Adhesion Sci. Technol.*, 27 (20), 2013, 2163–2176.

[148] T. Brunschwiler, *et al.* Dual-side heat removal by micro-channel cold plate and silicon-interposer with embedded fluid channels, in: *2016 6th Electronic System-Integration Technology Conference (ESTC)*, 2016, IEEE.

[149] G. Van der Plas, *et al.*, Design issues and considerations for low-cost 3-D TSV IC technology, 46 (1), 2010, 293–307.

[150] H.M. Wyss, *et al.*, Mechanism for clogging of microchannels, 74 (6), 2006, 061402.

[151] R. van Zwieten, *et al.*, From cooperative to uncorrelated clogging in cross-flow microfluidic membranes, 8 (1), 2018, 1–10.

[152] B. Dang, *et al.* Integration and packaging of embedded radial micro-channels for 3D chip cooling, in: *2016 IEEE 66th Electronic Components and Technology Conference (ECTC)*, 2016, IEEE.

[153] A. Bar-Cohen, *et al.*, Thermal management of high heat flux nanoelectronic chips, 19 (3), 2007, 48–52.
[154] A. Bar-Cohen, *et al.*, Gas-assisted evaporative cooling of high density electronic modules, 18 (3), 1995, 502–509.
[155] W. Deng, A. Gomez, M. Transfer, Electrospray cooling for microelectronics, 54 (11–12), 2011, 2270–2275.
[156] Y.O. Popov, Evaporative deposition patterns: Spatial dimensions of the deposit. *Phys. Rev. E*, 71, 2005, 036313-1–036313-17.
[157] J.M. Stauber, *et al.*, On the lifetimes of evaporating droplets with related initial and receding contact angles, 27 (12), 2015, 122101.
[158] G. Dunn, *et al.*, The strong influence of substrate conductivity on droplet evaporation, 623, 2009, 329–351.
[159] M.A. Saada, *et al.*, Evaporation of a sessile drop with pinned or receding contact line on a substrate with different thermophysical properties, 58 (1–2), 2013, 197–208.
[160] M. Ait Saada, S. Chikh, L. Tadrist, Numerical investigation of heat and mass transfer of an evaporating sessile drop on a horizontal surface, 22 (11), 2010, 112115.
[161] N.H. Aly, A.K.J.D. El-Figi, Mechanical vapor compression desalination systems: A case study, 158 (1–3), 2003, 143–150.
[162] S.-S. Hsieh, *et al.*, Thermal performance of flat vapor chamber heat spreader, 49 (6), 2008, 1774–1784.
[163] D.F. Hanks, *et al.*, Nanoporous membrane device for ultra high heat flux thermal management, 4 (1), 2018, 1–10.
[164] A.K. El-Feky, Mechanical vapor compression (MVC) desalination system optimal design, 94, 2016, 1–13.
[165] Y. Sone, Kinetic theoretical studies of the half-space problem of evaporation and condensation, 29 (3–5), 2000, 227–260.
[166] Z. Lu, *et al.*, A unified relationship for evaporation kinetics at low Mach numbers, 10 (1), 2019, 1–8.
[167] J. Barrett, Kinetic evaporation and condensation rates and their coefficients, 150 (2), 1992, 352–364.
[168] R.W. Schrage, *A Theoretical Study of Interphase Mass Transfer*, Columbia University Press, 1953.
[169] R. Hołyst, M. Litniewski, D.J.S.M. Jakubczyk, A molecular dynamics test of the Hertz–Knudsen equation for evaporating liquids, 11 (36), 2015, 7201–7206.
[170] A.H. Persad, C.A. Ward, Expressions for the evaporation and condensation coefficients in the Hertz-Knudsen relation, 116 (14), 2016, 7727–7767.
[171] P. Jafari, *et al.*, Evaporation mass flux: A predictive model and experiments, 34 (39), 2018, 11676–11684.
[172] J. Young, The condensation and evaporation of liquid droplets in a pure vapour at arbitrary Knudsen number, 34 (7), 1991, 1649–1661.
[173] J. Julin, *et al.*, Mass accommodation of water: Bridging the gap between molecular dynamics simulations and kinetic condensation models, 117 (2), 2013, 410–420.

[174] P.L. Barclay, J.R. Lukes, Curvature dependence of the mass accommodation coefficient, 35 (18), 2019, 6196–6202.

[175] F. Duan, B. He, Evaporation of liquid droplet in nano and micro scales from statistical rate theory, 15 (4), 2015, 3011–3016.

[176] G. Fang, C.A. Ward, Examination of the statistical rate theory expression for liquid evaporation rates, 59 (1), 1999, 441.

[177] V. Ryazanov, Statistical theory of evaporation and condensation processes in liquid droplets, 68 (2), 2006, 217–227.

[178] A. Kapoor, J.A. Elliott, Statistical rate theory insight into evaporation and condensation in multicomponent systems, 114 (46), 2010, 15052–15056.

[179] L. Zhang, *et al.* Comparison study on the calculation formula of evaporation mass flux through the plane vapour-liquid interface, in: *Journal of Physics: Conference Series*, IOP Publishing, 2017.

[180] P.A. Raghupathi, S.G. Kandlikar, Contact line region heat transfer mechanisms for an evaporating interface, *Int. J. Heat Mass Transfer*, 95, 2016, 296–306.

[181] H. Wang, S.V. Garimella, J.Y. Murthy, An analytical solution for the total heat transfer in the thin-film region of an evaporating meniscus, *Int. J. Heat Mass Transfer*, 51 (25), 2008, 6317–6322.

[182] M. Hanchak, *et al.*, Profile measurements of thin liquid films using reflectometry, *Appl. Phys. Lett.*, 103 (21), 2013, 211607.

[183] V.S., Ajaev, O.A. Kabov, Heat and mass transfer near contact lines on heated surfaces, 108, 2017, 918–932.

[184] J. Li, *et al.*, Investigation of the confinement effect on the evaporation behavior of a droplet pinned on a micropillar structure, *J. Colloid Interface Sci.*, 555, 2019, 583–594.

[185] S.I. Karakashev, A.V. Nguyen, R.J.C.J. Tsekov, Effect of the adsorption component of the disjoining pressure on foam film drainage, 75 (2), 2013, 176–180.

[186] J. Zhang, *Theory of Curvature-Dependent Disjoining Pressure and Its Application to Liquid Films.* Louisiana State University and Agricultural & Mechanical College, 2006.

[187] X. Wang, *et al.*, Assessing the impact of disjoining pressure on thin-film evaporation with atomistic simulation and kinetic theory, 116 (21), 2020, 213701.

[188] B. Derjaguin, N. Churaev, On the question of determining the concept of disjoining pressure and its role in the equilibrium and flow of thin films, 66 (3), 1978, 389–398.

[189] J.N. Israelachvili, *Intermolecular and Surface Forces*, Academic Press, 2011.

[190] P. Wayner Jr., *et al.*, The interline heat-transfer coefficient of an evaporating wetting film, 19 (5), 1976, 487–492.

[191] H. Wang, *et al.*, Characteristics of an evaporating thin film in a microchannel, 50 (19–20), 2007, 3933–3942.

[192] K.H. Do, *et al.*, A mathematical model for analyzing the thermal characteristics of a flat micro heat pipe with a grooved wick, 51 (19–20), 2008, 4637–4650.

[193] K. Park, *et al.*, Transport phenomena in the thin-film region of a micro-channel, 46 (13), 2003, 2381–2388.
[194] F. Holm, S. Goplen, Heat transfer in the meniscus thin-film transition region. *Journal Heat Transfer*, 101 (3), 1979, 543–547.
[195] B. Derjaguin, Z. Zorin, Optical study of the adsorption and surface condensation of vapours in the vicinity of saturation on a smooth surface, 40 (1–4), 1992, 83–117.
[196] N. Churaev, B.J. Derjaguin, Inclusion of structural forces in the theory of stability of colloids and films, 103 (2), 1985, 542–553.
[197] S. Basu, M.M. Sharma, Measurement of critical disjoining pressure for dewetting of solid surfaces, 181 (2), 1996, 443–455.
[198] S. Narayanan, A.G. Fedorov, Y.K.J.L. Joshi, Interfacial transport of evaporating water confined in nanopores, 27 (17), 2011, 10666–10676.
[199] D. Exerowa, *et al.*, Direct measurement of disjoining pressure in black foam films. I, *Films Ionic Surfactant*, 22 (2), 1987, 161–169.
[200] S. Tchaliovska, *et al.*, Interactions in equilibrium free films of aqueous dodecylammonium chloride solutions, 168 (1), 1994, 190–197.
[201] P. Kapitza, The study of heat transfer in helium II, *J. Phys. (Moscow)*, 4, 1941, 181.
[202] I.M. Khalatnikov, *An Introduction to the Theory of Superfluidity*, CRC Press, 2018.
[203] N. Snyder, The Kapitza conductance of the (100) surface of copper, *J. Low Temp. Phys.*, 22 (3–4), 1976, 257–284.
[204] S. Shenogin, *et al.*, Role of thermal boundary resistance on the heat flow in carbon-nanotube composites, *J. Appl. Phys.*, 95 (12), 2004, 8136–8144.
[205] S.T. Huxtable, *et al.*, Interfacial heat flow in carbon nanotube suspensions, *Nat. Mater.*, 2 (11), 2003, 731–734.
[206] N. Singh, D. Shin, D. Banerjee, Nanoscale effects in multiphase flows and heat transfer, *Microelectron. Nanoelectron. Mater. Dev. Manuf.*, 2012, 309.
[207] G.D. Mahan, Kapitza resistance at a solid–fluid interface, 12 (4), 2008, 294–310.
[208] G. Nagayama, *et al.*, On the evaporation rate of ultra-thin liquid film at the nanostructured surface: A molecular dynamics study, *Int. J. Thermal Sci.*, 49 (1), 2010, 59–66.
[209] A.T. Pham, M. Barisik, B. Kim, Molecular dynamics simulations of Kapitza length for argon-silicon and water-silicon interfaces, *Int. J. Precis. Eng. Manuf.*, 15 (2), 2014, 323–329.
[210] B.H. Kim, A. Beskok, T. Cagin, Molecular dynamics simulations of thermal resistance at the liquid–solid interface, *J. Chem. Phys.*, 129 (17), 2008, 174701.
[211] H. Han, *et al.*, Solid–liquid interface thermal resistance affects the evaporation rate of droplets from a surface: A study of perfluorohexane on chromium using molecular dynamics and continuum theory, *Langmuir*, 33 (21), 2017, 5336–5343.

[212] B. Ramos-Alvarado, S. Kumar, G. Peterson, solid–liquid thermal transport and its relationship with wettability and the interfacial liquid structure, *J. Phys. Chem. Lett.*, 7 (17), 2016, 3497–3501.
[213] B. Ma, *et al.* Molecular dynamic simulation of the effect of nanocoating on two-phase evaporative heat and mass transfer, in: *2019 18th IEEE Intersociety Conference on Thermal and Thermomechanical Phenomena in Electronic Systems (ITherm)*, 2019, IEEE.
[214] A. Faghri, Y. Zhang, *Transport Phenomena in Multiphase Systems*, Academic Press, 2006.
[215] M. Zientara, *et al.*, Temperature dependence of the evaporation coefficient of water in air and nitrogen under atmospheric pressure: Study in water droplets, *J. Phys. Chem. A*, 112 (23), 2008, 5152–5158.
[216] Y. Wang, *et al.*, Combined effects of underlying substrate and evaporative cooling on the evaporation of sessile liquid droplets, 11 (28), 2015, 5632–5640.
[217] Z. Zheng, *et al.*, Numerical investigation on conjugate heat transfer of evaporating thin film in a sessile droplet, 101, 2016, 10–19.
[218] Z. Pan, J.A. Weibel, S.V. Garimella, Influence of surface wettability on transport mechanisms governing water droplet evaporation, 30 (32), 2014, 9726–9730.
[219] X. Xu, L. Ma, Analysis of the effects of evaporative cooling on the evaporation of liquid droplets using a combined field approach, *Sci. Rep.*, 5, 2015, 8614.
[220] A.A. Guünay, *et al.*, Steady method for the analysis of evaporation dynamics, *Langmuir*, 33 (43), 2017, 12007–12015.
[221] T.A. Nguyen, S.R. Biggs, A.V.J.L. Nguyen, Analytical model for diffusive evaporation of sessile droplets coupled with interfacial cooling effect, 34 (23), 2018, 6955–6962.
[222] Y. Zhang, *et al.*, Three-dimensional numerical simulation on Marangoni convection in a sessile water droplet evaporating in its vapor at low pressure, 31 (2), 2019, 231–240.
[223] F. Girard, M. Antoni, K. Sefiane, On the effect of Marangoni flow on evaporation rates of heated water drops, *Langmuir*, 24 (17), 2008, 9207–9210.
[224] S. Semenov, *et al.*, Computer simulations of evaporation of pinned sessile droplets: Influence of kinetic effects, *Langmuir*, 28 (43), 2012, 15203–15211.
[225] J. Mitrovic, Josef Stefan and his evaporation–diffusion tube—The Stefan diffusion problem, 75, 2012, 279–281.
[226] S. Tonini, G. Cossali, M. Transfer, A multi-component drop evaporation model based on analytical solution of Stefan–Maxwell equations, 92, 2016, 184–189.
[227] F. Carle, *et al.*, Contribution of convective transport to evaporation of sessile droplets: Empirical model, 101, 2016, 35–47.
[228] L. Zhang, *et al.*, Stefan flow induced natural convection suppression on high-flux evaporators, 110, 2020, 104255.

[229] J.C. Johnson, Measurement of the surface temperature of evaporating water drops, 21 (1), 1950, 22–23.
[230] G. Langstroth, C. Diehl, E.J.C. Winhold, The evaporation of droplets in still air, 28 (6), 1950, 580–595.
[231] L. Shan, B. Ma, J. Li, B. Dogruoz, D. Agonafer, Investigation of the evaporation heat transfer mechanism of a non-axisymmetric droplet confined on a heated micropillar structure, *Int. J. Heat Transfer*, 141, 2019, 191–203.
[232] T. Harirchian, S.V. Garimella. An investigation of flow boiling regimes in microchannels of different sizes by means of high-speed visualization, in: *2008 11th Intersociety Conference on Thermal and Thermomechanical Phenomena in Electronic Systems*, 2008, IEEE.
[233] E. Rahim, *et al.*, Characterization and prediction of two-phase flow regimes in miniature tubes, 2011. 37 (1), 12–23.
[234] E.G. Orlova, A.G. Islamova, O.A. Kabov. Evaporation rate of a liquid layer streamlined by gas flow in minichannel, in: *MATEC Web of Conferences*, 2017, EDP Sciences.
[235] J. Mittal, G. Hummer, Interfacial thermodynamics of confined water near molecularly rough surfaces, *Faraday Discus.*, 146, 2010, 341–352.
[236] H. Acharya, *et al.*, How chemistry, nanoscale roughness, and the direction of heat flow affect thermal conductance of solid–water interfaces, *Ind. Eng. Chem. Res.*, 51 (4), 2012, 1767–1773.
[237] J.V. Goicochea, *et al.*, Surface functionalization mechanisms of enhancing heat transfer at solid-liquid interfaces, *J. Heat Transfer*, 133 (8), 2011.
[238] K. Issa, A. Mohamad, Lowering liquid-solid interfacial thermal resistance with nanopatterned surfaces, *Phys. Rev. E*, 85 (3), 2012, 031602.
[239] R. Raj, *et al.*, High-resolution liquid patterns via three-dimensional droplet shape control, 5 (1), 2014, 1–8.
[240] A.-L. Ljung, T.S. Lundström, Heat and mass transfer boundary conditions at the surface of a heated sessile droplet, 53 (12), 2017, 3581–3591.
[241] S. Adera, *et al.*, Design of micropillar wicks for thin-film evaporation, *Int. J. Heat Mass Transfer*, 101, 2016, 280–294.
[242] Y.F. Maydanik, Loop heat pipes, 25 (5–6), 2005, 635–657.
[243] G. Hwang, *et al.*, Multi-artery heat pipe spreader: Experiment, 53 (13–14), 2010, 2662–2669.
[244] G. Hwang, *et al.*, Multi-artery heat-pipe spreader: Lateral liquid supply, 54 (11–12), 2011, 2334–2340.
[245] G. Hwang, *et al.*, Modulated wick heat pipe, 50 (7–8), 2007, 1420–1434.
[246] T. Zhao, Q. Liao, On capillary-driven flow and phase-change heat transfer in a porous structure heated by a finned surface: Measurements and modeling, 43 (7), 2000, 1141–1155.
[247] K.K. Bodla, J.Y. Murthy, S.V. Garimella, Evaporation analysis in sintered wick microstructures, *Int. J. Heat Mass Transfer*, 61, 2013, 729–741.
[248] C. Ding, *et al.* Super wetting of micro &nano structured titania surfaces, in: *TRANSDUCERS 2009–2009 International Solid-State Sensors, Actuators and Microsystems Conference*, 2009, IEEE.

[249] Z. Lu, *et al.*, Design and modeling of membrane-based evaporative cooling devices for thermal management of high heat fluxes, 6 (7), 2016, 1056–1065.

[250] D.F. Hanks, *et al.*, Nanoporous membrane device for ultra high heat flux thermal management, *Microsyst. Nanoeng.*, 4 (1), 2018, 1–10.

[251] D. Ćoso, *et al.*, Enhanced heat transfer in biporous wicks in the thin liquid film evaporation and boiling regimes, *J. Heat Transfer*, 134 (10), 2012, 101501.

[252] B.D. Iverson, *et al.*, Heat and mass transport in heat pipe wick structures, *J. Thermophys. Heat Transfer*, 21 (2), 2007, 392–404.

[253] M. Kole, T. Dey, Thermal performance of screen mesh wick heat pipes using water-based copper nanofluids, *Appl. Thermal Eng.*, 50 (1), 2013, 763–770.

[254] C. Li, G. Peterson, Y. Wang, Evaporation/boiling in thin capillary wicks (l): Wick thickness effects, *J. Heat Transfer*, 128 (12), 2006, 1312–1319.

[255] Q. Liang, *et al.* Experiment and modeling of microstructured capillary wicks for thermal management of electronics, in: *Electronics Packaging Technology Conference (EPTC 2013)*, 2013 IEEE 15, 2013, IEEE.

[256] T. Semenic, I. Catton, Experimental study of biporous wicks for high heat flux applications, *Int. J. Heat Mass Transfer*, 52 (21), 2009, 5113–5121.

[257] J.A. Weibel, *et al.*, Design of integrated nanostructured wicks for high-performance vapor chambers, *IEEE Trans. Compone. Packag. Technol.*, 1 (6), 2011, 859–867.

[258] J.A. Weibel, S.V. Garimella, M.T. North, Characterization of evaporation and boiling from sintered powder wicks fed by capillary action, *Int. J. Heat Mass Transfer*, 53 (19), 2010, 4204–4215.

[259] J.A. Weibel, *et al.* Characterization and nanostructured enhancement of boiling incipience in capillary-fed, ultra-thin sintered powder wicks, in: *Thermal and Thermomechanical Phenomena in Electronic Systems (ITherm), 2012 13th IEEE Intersociety Conference on*, 2012, IEEE.

[260] R. Xiao, S.C. Maroo, E.N. Wang, Negative pressures in nanoporous membranes for thin film evaporation, 102 (12), 2013, 123103.

[261] M. Nosonovsky, B. Bhushan, Phase behavior of capillary bridges: Towards nanoscale water phase diagram, *Phys. Chem. Chem. Phys.*, 10 (16), 2008, 2137–2144.

[262] M.J. Hancock, M.C. Demirel, Anisotropic wetting on structured surfaces, 38 (5), 2013, 391.

[263] E. Portuguez, *et al.*, Study of the contact and the evaporation kinetics of a thin water liquid bridge between two hydrophobic plates, 7 (4), 2017, 99.

[264] A. Hooshanginejad, S. Lee, Droplet depinning in a wake, 2 (3), 2017, 031601.

[265] J.Y. Kim, I.G. Hwang, B.M. Weon, Evaporation of inclined water droplets, 7, 2017, 42848.

[266] H. Lu, *et al.*, Controllable transport of water through nanochannel by rachet-like mechanism, 136 (17), 2012, 174511.
[267] H.-D. Xi, *et al.*, AC electric field induced droplet deformation in a microfluidic T-junction, 16 (16), 2016, 2982–2986.
[268] T. Açıkalın, *et al.*, Characterization and optimization of the thermal performance of miniature piezoelectric fans, 28 (4), 2007, 806–820.
[269] Y. Joo, *et al.*, Air cooling of a microelectronic chip with diverging metal microchannels monolithically processed using a single mask, *J. Micromech. Microeng.*, 18 (11), 2008, 115022.
[270] M. Iyengar, A. Bar-Cohen, Design for manufacturability of SISE parallel plate forced convection heat sinks, *IEEE Trans. Compon. Packag. Technol.*, 24 (2), 2001, 150–158.
[271] S.Q. Cai, A. Bhunia, Characterization of phase change heat and mass transfers in monoporous silicon wick structures, *J. Heat Transfer*, 136 (7), 2014.
[272] S. Tonapi, *et al.* An overview of thermal management for next generation microelectronic devices, in: *Advanced Semiconductor Manufacturing Conference and Workshop, 2003 IEEEI/SEMI*, 2003, IEEE.
[273] Z. Wan, Y. Joshi. Pressure drop and heat transfer characteristics of square pin fin enhanced microgaps in single phase microfluidic cooling, in: *Fourteenth Intersociety Conference on Thermal and Thermomechanical Phenomena in Electronic Systems (ITherm)*, 2014, IEEE.
[274] G. Tang, *et al.*, Development of a compact and efficient liquid cooling system with silicon microcooler for high-power microelectronic devices, *IEEE Trans. Compone. Packag. Technol.*, 6 (5), 2016, 729–739.
[275] A.J. Robinson, A thermal–hydraulic comparison of liquid microchannel and impinging liquid jet array heat sinks for high-power electronics cooling, *IEEE Trans. Compon. Packag. Technol.*, 32 (2), 2009, 347–357.
[276] Y. Peles, *et al.*, Forced convective heat transfer across a pin fin micro heat sink, *Int. J. Heat Mass Transfer*, 48 (17), 2005, 3615–3627.
[277] L. Zheng, Y. Zhang, M.S. Bakir, A silicon interposer platform utilizing microfluidic cooling for high-performance computing systems, *IEEE Trans. Compon. Packag. Manuf. Technol.*, 5 (10), 2015, 1379–1386.
[278] R.S. Andhare, *et al.*, Heat transfer and pressure drop characteristics of a flat plate manifold microchannel heat exchanger in counter flow configuration, *Appl. Thermal Eng.*, 96, 2016, 178–189.
[279] T. Brunschwiler, *et al.*, Interlayer cooling potential in vertically integrated packages, *Microsyst. Technol.*, 15 (1), 2009, 57–74.
[280] J.-M. Koo, *et al.*, Integrated microchannel cooling for three-dimensional electronic circuit architectures, *J. Heat Transfer*, 127 (1), 2005, 49–58.
[281] K.W. Jung, *et al.*, Embedded cooling with 3D manifold for vehicle power electronics application: Single-phase thermal-fluid performance, *Int. J. Heat Mass Transfer*, 130, 2019, 1108–1119.
[282] D. Kong, *et al.*, Single-phase thermal and hydraulic performance of embedded silicon micro-pin fin heat sinks using R245fa, *Int. J. Heat Mass Transfer*, 141, 2019, 145–155.

[283] J.W. Palko, *et al.*, Extreme two—Phase cooling from laser—Etched diamond and conformal, template—Fabricated microporous copper, *Adv. Funct. Mater.*, 27 (45), 2017, 1703265.

[284] A. Reeser, A. Bar-Cohen, G. Hetsroni, High quality flow boiling heat transfer and pressure drop in microgap pin fin arrays, *Int. J. Heat Mass Transfer*, 78, 2014, 974–985.

[285] T. Semenic, *et al.*, Use of biporous wicks to remove high heat fluxes, *Appl. Thermal Eng.*, 28 (4), 2008, 278–283.

[286] F. Yang, *et al.*, Flow boiling phenomena in a single annular flow regime in microchannels (I): Characterization of flow boiling heat transfer, *Int. J. Heat Mass Transfer*, 68, 2014, 703–715.

[287] A.R. Betz, J. Jenkins, D. Attinger, Boiling heat transfer on superhydrophilic, superhydrophobic, and superbiphilic surfaces, *Int. J. Heat Mass Transfer*, 57 (2), 2013, 733–741.

[288] C. Zhang, *et al.*, Enhanced capillary: Fed boiling in copper inverse opals via template sintering, *Adv. Funct. Mater.*, 28 (41), 2018, 1803689.

[289] C. Woodcock, *et al.*, Piranha pin fin (PPF): Advanced flow boiling microstructures with low surface tension dielectric fluids, *Int. J. Heat Mass Transfer*, 90, 2015, 591–604.

[290] S.A. Isaacs, *et al.* Two-phase flow and heat transfer in pin-fin enhanced micro-gaps, in: *13th InterSociety Conference on Thermal and Thermomechanical Phenomena in Electronic Systems*, 2012, IEEE.

[291] J.A. Weibel, S.V. Garimella, Visualization of vapor formation regimes during capillary-fed boiling in sintered-powder heat pipe wicks, *Int. J. Heat Mass Transfer*, 55 (13–14), 2012, 3498–3510.

[292] K.S. Eloyan, D.V. Zaitsev. Thin film evaporative cooling system for high heat flux applications, in: *Journal of Physics: Conference Series*, IOP Publishing Ltd., 2018.

[293] S. Chakraborty, M.A. Rosen, B.D. MacDonald, Analysis and feasibility of an evaporative cooling system with diffusion-based sessile droplet evaporation for cooling microprocessors, *Appl. Thermal Eng.*, 125, 2017, 104–110.

[294] C.H. Amon, *et al.*, MEMS-enabled thermal management of high-heat-flux devices EDIFICE: Embedded droplet impingement for integrated cooling of electronics, *Exp. Thermal Fluid Sci.*, 25 (5), 2001, 231–242.

[295] T. Baummer, *et al.*, Force-fed evaporation and condensation utilizing advanced micro-structured surfaces and micro-channels, *Microelectron. J.*, 39 (7), 2008, 975–980.

[296] H.J.L. Wang, From contact line structures to wetting dynamics, 35 (32), 2019, 10233–10245.

[297] S.-C. Wong, *et al.*, Evaporation resistance measurement with visualization for sintered copper-powder evaporator in operating flat-plate heat pipes, 53 (19–20), 2010, 3792–3798.

[298] Y. Li, Z.Y. Fu, B.L. Su, Hierarchically structured porous materials for energy conversion and storage, 22 (22), 2012, 4634–4667.

[299] X. Dai, *et al.*, Capillary evaporation on micromembrane-enhanced microchannel wicks with atomic layer deposited silica, 103 (15), 2013, 151602.

[300] M.K. Hooda, *et al.*, A systematic study of DRIE process for high aspect ratio microstructuring, 84 (9), 2010, 1142–1148.
[301] Y. Tang, A. Sandoughsaz, K. Najafi. Ultra high aspect-ratio and thick deep silicon etching (UDRIE), in: *2017 IEEE 30th International Conference on Micro Electro Mechanical Systems (MEMS)*, 2017, IEEE.
[302] J. Ohara, *et al.*, Improvement of high aspect ratio Si etching by optimized oxygen plasma irradiation inserted DRIE, 19 (9), 2009, 095022.
[303] P. Elenius, L. Levine, Comparing flip-chip and wire-bond interconnection technologies, 4 (6), 2000.
[304] A. Yu, *et al.*, Fabrication of silicon carriers with TSV electrical interconnections and embedded thermal solutions for high power 3-D packages, 32 (3), 2009, 566–571.
[305] Y. Civale, *et al.* Die stacking using 3D-wafer level packaging copper/polymer through-Si via technology and Cu/Sn interconnect bumping, in: *2009 IEEE International Conference on 3D System Integration*, 2009, IEEE.
[306] R. Agarwal, *et al.* Cu/Sn microbumps interconnect for 3D TSV chip stacking, in: *2010 Proceedings 60th Electronic Components and Technology Conference (ECTC)*, 2010, IEEE.
[307] M. Schultz, *et al.* Column interconnects: A path forward for embedded cooling of high power 3D chip stacks, in: *2018 IEEE 68th Electronic Components and Technology Conference (ECTC)*, 2018, IEEE.
[308] T. Lu, *et al.*, TSV-based 3-D ICs: Design methods and tools, 36 (10), 2017, 1593–1619.
[309] C.R. King, *et al.*, 3D stacking of chips with electrical and microfluidic I/O interconnects, in: *2008 58th Electronic Components and Technology Conference*, 2008, IEEE.
[310] L. Zheng, Y. Zhang, M.S. Bakir. Design, fabrication and assembly of a novel electrical and microfluidic I/Os for 3-D chip stack and silicon interposer, in: *2013 IEEE 63rd Electronic Components and Technology Conference*, 2013, IEEE.
[311] P. Andry, *et al.* Advanced wafer bonding and laser debonding, in: *2014 IEEE 64th Electronic Components and Technology Conference (ECTC)*, 2014, IEEE.
[312] B. Dang, M.S. Bakir, J.D. Meindl, Integrated thermal-fluidic I/O interconnects for an on-chip microchannel heat sink, 27 (2), 2006, 117–119.
[313] Y. Madhour, *et al.* Patterned die-to-die thin film bonding for 3D chip stacks with integrated microfluidic cooling, in: *2012 13th International Conference on Electronic Packaging Technology & High Density Packaging*, 2012, IEEE.
[314] J. Cong, *et al.* Thermal-aware 3D IC placement via transformation, in: *2007 Asia and South Pacific Design Automation Conference*, 2007, IEEE.
[315] M. M. Nahar, *et al.* Microscale evaporative cooling technologies for high heat flux microelectronics devices: Background and recent advances. *Appl. Thermal Eng.*, 194, 2021, 117109.

https://doi.org/10.1142/9789811279379_0003

Chapter 3

Numerical Modeling of Embedded Two-Phase Cooling in Silicon Microelectronics

Daniel Lorenzini and Yogendra Joshi*

G.W. Woodruff School of Mechanical Engineering
Georgia Institute of Technology, Atlanta, GA, USA
**yogendra.joshi@me.gatech.edu*

Abstract

The two-phase cooling of microelectronics is studied by adapting a mechanistic phase change model that can be used with commercial computational fluid dynamics and heat transfer (CFD–HT) codes. Silicon microdevices with variable density of pin fins and hotspots are simulated, and results are validated with in-house experimental data generated for the purpose of studying the two-phase flow regimes and their thermal/hydraulic implications. The CFD–HT modeling approach is found to constitute a valuable tool in the design and analysis of heterogeneous microfluidic cooling devices under two-phase operation.

Keywords: Microfluidic cooling, silicon, microelectronics, hotspots, numerical modeling

1. Introduction

The heat transfer phenomena associated with flow boiling in microsystems are not as well understood as those associated with single-phase cooling. There are still a number of challenges related to the control of flow instabilities, critical heat flux (CHF), flow regime transitions, and two-phase pressure drops. The literature reviews in Refs. [1–5] contain useful guidelines and discussion regarding the design, analysis, and areas of opportunity for two-phase cooling in microsystems.

Krishnamurthy and Peles [6] investigated the features of flow boiling of water in a silicon device with a flow channel 250 μm high and 1.8 mm wide, populated with cylindrical pin fins of 100 μm diameter. The device was powered at the bottom with heat fluxes ranging from 20 to 350 W/cm^2 and mass fluxes in the range of 346–794 kg/m^{2}s. The reported wall temperatures ranged between 100°C and 140°C due to the saturation properties of water above atmospheric pressures. Koo *et al.* [7] investigated a 1D thermal model to predict the cooling performance of a 3D IC with an integrated microchannel network using flow boiling for the removal of heat fluxes up to 135 W/cm^2, with a maximum temperature of 85°C. In the investigation by Kim *et al.* [8], the single- and two-phase cooling of a 3D-stacked IC was analyzed with a reduced-order model for a non-uniform power map, corresponding to a dual-core microprocessor, with localized heat fluxes ranging from 11.3 to 305 W/cm^2. The implications for heat transfer, microchannel layout, and flow directions were reported, and suggestions for the optimization of such devices were provided.

Han *et al.* [9] investigated the flow boiling of water at sub-atmospheric pressures within a silicon microgap of 200 μm height and 10 mm width over a 1 cm^2 heated area, populated with cylindrical pin fins of 90 μm diameter in a staggered array. The device was tested for heat fluxes ranging from 198 to 444 W/cm^2 and mass fluxes from 1,351 to 1,784 kg/m^{2}s. In an effort to reduce the operating temperatures, subcooled water was supplied to the device, and results indicated a maximum wall superheat of 60°C at a heat flux of 444 W/cm^2. Although water has desirable thermophysical properties, reliability concerns, such as the direct contact of the coolant with the device in interlayer configurations, have prompted the use of dielectric refrigerants instead. In the numerical investigation by

Asrar *et al.* [10], the flow boiling of the dielectric refrigerant R245fa was studied in a silicon microgap 200 μm high and 10 mm wide, with pin fins of circular cross-section shape covering a 1 cm^2 heated area. The cooling performance was assessed for heat fluxes up to 498 W/cm^2, mass flux values up to 7,896 kg/m^{2}s, and inlet temperatures of 13°C and 18°C, while also discussing the observed two-phase flow regimes.

Modeling is a key aspect that allows for a better understanding of the underlying details of boiling phenomena and is therefore a powerful tool for the development of more efficient cooling technologies. Although there are a number of investigations with experimental evidence that suggest two-phase cooling as a potential candidate for the 3D stacking of microelectronics, the detailed thermal modeling of this mechanism has not advanced as much due to its intrinsic complexity. Although this type of modeling (which involves a microscale, three-dimensional, transient, and phase-change problem) is computationally intensive, the development and refinement of new and current models can eventually lead to more cost-effective tools for optimizing and analyzing these systems, which would be advantageous considering that the microfabrication and testing of experimental prototypes are very expensive, time-consuming, and complex. Several challenges remain to date in order to improve the capabilities of the known multiphase models, in particular for incorporating the effect of phase change. The CFD analysis of boiling requires the coupling of a multiphase flow model with a phase-change model to estimate the heat and mass transport across the interface. Despite inherent modeling limitations and assumptions, a few modeling efforts regarding this topic have been made. Wu *et al.* [11] and Yang *et al.* [12] investigated the induced two-phase flow regimes due to boiling of the refrigerant R141B in a coiled tube evaporator using the mechanistic phase-change model proposed by Lee [13], which was coupled to the volume of fluid (VOF) method for interface tracking. The presented results indicated a remarkable capability to predict the flow regime evolution when compared to experimental results. De Schepper *et al.* [14] presented a similar phase-change model, comprehensively describing its features and limitations in order to simulate the flow boiling of a hydrocarbon feedstock; the authors indicated that the predicted flow regimes were in accordance with experimental results available in the literature. Fang *et al.* [15] studied the flow boiling inside a microchannel with a hydrophobic vapor-venting membrane in an effort to reduce flow instabilities and vapor phase accumulation, where the phase-change model was adapted from the work of Yang *et al.* [12]. The bubble dynamics and condensation process due to the subcooled flow of the refrigerant HFE-7100 were numerically analyzed by Zhuan and Wang [16] using the model proposed by Plesset and Zwick [17] to predict bubble growth. The authors indicated that their results were in good agreement with a range of experimental data. More recently, Magnini and Thome [18] presented the numerical simulation of saturated flow boiling in a circular microchannel using the VOF method with additional functions to improve the surface tension calculation and include phase-change effects based on the kinetic theory of gases.

Based on the aforementioned challenges, there are a number of opportunities to expand research in this emerging area. In this work, the physics of two-phase flows, specifically flow boiling in microgap cooling layers, is studied by means of proposing and adapting a mechanistic phase-change model that is compatible with general-purpose CFD–HT codes.

In addition, the modeling results are validated with experimental measurements of flow boiling in silicon microgaps with a variable density of pin fins and with spatially variable heat fluxes specifically designed for the vertical integration of microelectronics. The results of this comparison provide evidence that such a modeling technique is capable of predicting the flow regimes, temperature distribution, and two-phase pressure drop in complex cooling layers, with the ultimate goal of constituting a valuable tool in the design of advanced, two-phase cooling microsystems.

2. Numerical Model

2.1. *Coupled level-set volume of fluid model*

The liquid–vapor interface can be efficiently tracked with the coupled level-set volume of fluid (CLSVOF) method [19], a hybrid technique that combines the volume-conservative features of the VOF method [20] and the accurate spatial gradient calculation of the level-set (LS) method [21]. Defining the fluid vapor as the secondary phase for tracking, its volume fraction equation is

$$\frac{\partial}{\partial t}(\alpha_{\mathrm{v}}\rho_{\mathrm{v}}) + \nabla \cdot (\alpha_{\mathrm{v}}\rho_{\mathrm{v}}\vec{v}) = \dot{m}_{\mathrm{lv}} \tag{1}$$

where $\dot{m}_{\mathrm{lv}}$ is a source term accounting for the volumetric mass transfer rate between phases (described in Section 2.2). For the analyzed cases, which correspond to flow boiling of a pure substance, the liquid and vapor volume fractions complement a value of 1 in each computational cell; therefore, the liquid (primary phase) volume fraction is obtained from the condition

$$\alpha_l = 1 - \alpha_{\mathrm{v}} \tag{2}$$

The level-set function ϕ is a signed distance that denotes the interface as those points where $\phi = 0$; the primary phase (liquid) is then defined as the set of points where $\phi > 0$, and the secondary phase (vapor) those points where $\phi < 0$. For the CLSVOF method, the advection equation for the LS function ϕ is also included for phase tracking as

$$\frac{\partial \phi}{\partial t} + \nabla \cdot (\vec{v}_{\mathrm{int}}\phi) = 0 \tag{3}$$

where $\vec{v}_{\mathrm{int}}$ is the characteristic velocity of the interface, calculated as $\vec{v}_{\mathrm{int}} = \vec{v} + m''_{\mathrm{lv}}/\rho$, and the term m''_{lv} denotes the phase-change mass flux due to vaporization (see Section 2.2).

The momentum equation is shared by the liquid and vapor phases and given as

$$\frac{\partial}{\partial t}(\rho\vec{v}) + \nabla \cdot (\rho\vec{v}\vec{v}) = -\nabla P + \nabla \cdot [\mu(\nabla\vec{v} + \nabla\vec{v}^{\mathrm{T}})] + \rho\vec{g} + \vec{F}_{\mathrm{st}} \tag{4}$$

where ρ and μ are the density and viscosity of the mixture, respectively, computed using the LS function as

$$\rho = \rho_{\mathrm{v}}(1 - H(\phi)) + \rho_{\mathrm{l}}H(\phi) \tag{5}$$

$$\mu = \mu_{\mathrm{v}}(1 - H(\phi)) + \mu_{\mathrm{l}}H(\phi) \tag{6}$$

where $H(\phi)$ is a Heaviside function given by

$$H(\phi) = \begin{cases} 1 & \text{if } \phi > 0 \\ 0 & \text{otherwise} \end{cases} \quad (7)$$

This formulation makes the LS function continuous across the interface, which enables it to compute sharper and more accurate interfaces than the VOF model, which has the disadvantage of being discontinuous across the interface [22]. The unit normal vector to the interface and the interface curvature are calculated using the LS function as

$$\vec{n} = \left. \frac{\nabla\phi}{|\nabla\phi|} \right|_{\phi=0} \quad (8)$$

$$\kappa = \nabla \cdot \left. \frac{\nabla\phi}{|\nabla\phi|} \right|_{\phi=0} \quad (9)$$

The volumetric surface tension force ($\vec{F}_{\mathrm{st}}$) is calculated using the continuum surface force (CSF) model by Brackbill *et al.* [23], with the LS function ϕ given by

$$\vec{F}_{\mathrm{st}} = \sigma \frac{\rho\kappa\delta(\phi)\vec{n}}{\frac{1}{2}(\rho_l + \rho_{\mathrm{v}})} \quad (10)$$

$$\delta(\phi) = \begin{cases} 0 & |\phi| \geq \alpha \\ \frac{1+\cos(\pi\phi/\alpha)}{2\alpha} & |\phi| < \alpha \end{cases} \quad (11)$$

where h is the grid spacing and $\alpha = 1.5h$. This approach can lead to a more accurate surface tension calculation than the VOF method, which is desirable for microscale applications where such effects are important.

Despite the surface representation advantages of the LS method over the standard VOF technique, the former has the disadvantage of not being conservative in its mass formulation, leading to calculation errors of up to 20% in this variable. In contrast, with the CLSVOF method developed by Sussman and Puckett [19], mass is conserved within a fraction of a percent. With this hybrid approach, both LS and VOF functions are employed for the interface reconstruction, using the volume fraction equation to get the cut size in the cell where the interface moves, while the LS function is used to get the direction of such an interface [22]. The piecewise linear interface calculation (PLIC) algorithm [24] is used for the interface reconstruction, in which linear slopes are used within each cell for computing the fluid advection through the faces, then the distances from a given point to the front-cut segments are minimized, and finally the LS function is reinitialized [22].

The energy conservation equation is also included in this model in the following form:

$$\frac{\partial}{\partial t}(\rho E) + \nabla \cdot [\vec{v}(\rho E + P)] = \nabla \cdot (k \nabla T) + S_{\mathrm{e}} \tag{12}$$

where E is the mixture energy, computed as a mass-averaged variable:

$$E = \frac{\alpha_{\mathrm{l}} \rho_{\mathrm{l}} E_{l} + \alpha_{\mathrm{v}} \rho_{\mathrm{v}} E_{\mathrm{v}}}{\alpha_{\mathrm{l}} \rho_{\mathrm{l}} + \alpha_{\mathrm{v}} \rho_{\mathrm{v}}} \tag{13}$$

The energy for each phase (E_{l} and E_{v}) is obtained using the corresponding specific heat and temperature, and the mixture thermal conductivity k is computed similarly to Eqs. (5) and (6).

2.2. *Phase-change model*

The modeling procedure for calculating the volumetric mass transfer due to phase change is described in this section. In the kinetic theory of gases, the interfacial mass flux for a vaporization process of a given liquid can be represented by the Hertz–Knudsen equation as

$$m''_{\mathrm{lv}} = \frac{2\sigma_{\mathrm{e}}}{2 - \sigma_{\mathrm{e}}} \left(\frac{M}{2\pi R_{\mathrm{u}} T_{\mathrm{sat}}} \right)^{1/2} (P_{\mathrm{v}} - P_{\mathrm{sat}}) \tag{14}$$

where P_{v} is the partial vapor pressure at the interface and σ_{e} is the accommodation coefficient for evaporation, representing the portion of vapor molecules absorbed at the interface, with a suggested value of $\sigma_{\mathrm{e}} = 1$ in the experimental studies by Cammenga *et al.* [25] and Maa [26]. For the phase-change process at the saturation condition, the pressure and temperature can be related through the integrated form of the Clausius–Clapeyron equation:

$$(P_{\mathrm{v}} - P_{\mathrm{sat}}) = -\frac{h_{\mathrm{lv}}}{T_{\mathrm{sat}}(1/\rho_{\mathrm{v}} - 1/\rho_{\mathrm{l}})}(T - T_{\mathrm{sat}}) \tag{15}$$

Combining Eqs. (14) and (15), the interfacial mass flux may be calculated as

$$m''_{\mathrm{lv}} = \frac{2\sigma_{\mathrm{e}}}{2-\sigma_{\mathrm{e}}}\left(\frac{M}{2\pi R_{\mathrm{u}} T_{\mathrm{sat}}}\right)^{1/2} h_{l\mathrm{v}}\left(\frac{\rho_{\mathrm{v}}\rho_{l}}{\rho_{l}-\rho_{\mathrm{v}}}\right)\left(\frac{T - T_{\mathrm{sat}}}{T_{\mathrm{sat}}}\right) \tag{16}$$

The volumetric mass transfer rate $\dot{m}_{\mathrm{lv}}$ required for the volume fraction (Eq. (1)) can be obtained as the product of the phase-change mass flux and the interfacial area density. As suggested in the model by Lee [13], and adapted for the CFD studies in Refs. [27–29], the constant quantities in Eq. (16) may be grouped to cast a simplified expression as

$$\dot{m}_{\mathrm{lv}} = \lambda_{\mathrm{l}}\alpha_{\mathrm{l}}\rho_{\mathrm{l}}\frac{T_{\mathrm{l}} - T_{\mathrm{sat}}}{T_{\mathrm{sat}}^{3/2}} \quad \text{if } T_{\mathrm{l}} > T_{\mathrm{sat}} \tag{17}$$

where λ_{l} is a relaxation parameter that may be adjusted, as discussed in Ref. [29]. The formulation of Eq. (17) assures that phase change only occurs at computational cells that meet the specified criteria and are in liquid phase (i.e., the liquid volume fraction α_{l} weights the vaporization rate, including cells with both liquid and vapor phases). It is also important to note that any amount of liquid vaporized is accounted for in the corresponding vapor volume fraction equation, thus making this approach mass conservative, along with the other considerations associated with the level-set and VOF coupling. Energy is transported with mass during the boiling process, and such an effect is accounted for in Eq. (12) through the energy source term S_{e}, computed by multiplying the latent enthalpy of vaporization by the mass transfer rate due to phase change as

$$S_{\mathrm{e}} = h_{\mathrm{lv}}\dot{m}_{\mathrm{lv}} \tag{18}$$

3. Computational Domain and Boundary Conditions

Modeling transient multiphase problems is a computationally intensive task, especially at the microscale, where the mesh size plays a crucial role in determining the maximum time step through the

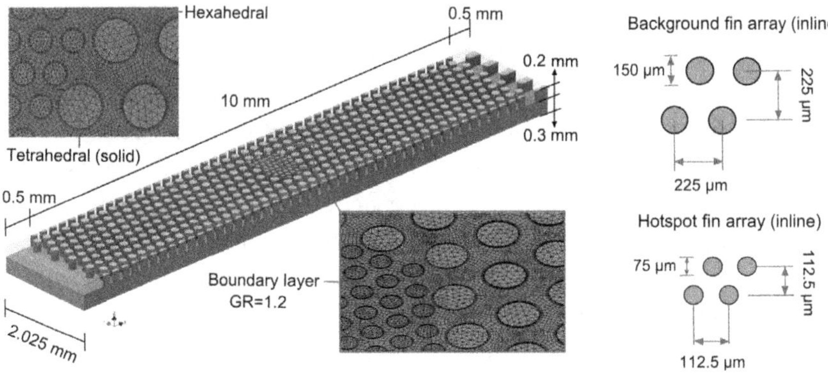

Figure 1. Schematic diagram of the computational domain, with the relevant dimensions and meshing details.

Courant number relation (suggested to be <1 for numerical stability). Because of the large simulation times that are known to result in the analysis of such flow boiling cases at the microscale, a symmetric portion of a silicon microgap proposed in Ref. [30] is analyzed as the computational domain; such a domain has enough width in the transversal direction to capture the relevant physics of flow boiling within the microgap (as validated in Section 5). Figure 1 shows a schematic diagram of the computational domain and its features, consisting of a portion of the silicon microgap with cylindrical pin fins. Relevant dimensions as well as meshing details are indicated in the different subsets of the figure; the heated length is the indicated central section of 10 mm width, and the background and hot spot pin fin arrays have an inline configuration with the indicated center-to-center spacing (pitch).

The boundary conditions for the numerical model are as follows. Heating is defined as a constant heat flux type at both the background and hotspot zones at the bottom surface of the silicon device. The numerical formulation consists of a conjugate heat transfer setup with the corresponding continuity and non-slip conditions at the solid–fluid interfaces, while also including the effect of phase change and latent heat of vaporization through the appropriate source term given in Eq. (12). The transversal cut surfaces of the computational domain are defined with periodic boundary conditions, and the remnant exterior surfaces can be properly modeled as adiabatic surfaces.

The background fin array zone was divided into four equal parts in the experimental study in Ref. [30] since the device has four independent heaters; in the current investigation, a heat flux value of $q''_B = 32$ W/cm^2 is chosen in order to compare with the available experimental data in Ref. [30]. The central hotspot fin array also has an independent heater, and the heat flux input (q''_H) is varied in the range from 0 to 150 W/cm^2. For the depicted computational domain in Figure 1 and its inlet dimensions, the corresponding mass flow rate input was calculated as 2.66×10^{-4} kg/s for a volume flow rate of 60 ml/min in the complete device, which also uses the dielectric refrigerant HFE-7200 as the two-phase coolant. Referencing the experimental conditions reported in Ref. [30], the inlet fluid temperature is kept at the saturation condition of 81°C, with an inlet fluid pressure of 365 kPa.

The solid domain (silicon) has relatively constant thermophysical properties, except for the thermal conductivity, which shows a significant temperature dependence; therefore, the values of this parameter reported in Ref. [31] were used and curve-fitted ($R^2 = 0.99$) to obtain the following polynomial relation:

$$k_{\mathrm{Si}} = 2122.1 - 16.765\,T + 4.8183 \times 10^{-2}\,T^2 - 4.7442 \times 10^{-5}\,T^3 \quad (19)$$

The values for temperature in Eq. (19) must be input in Kelvin (K), and the range of applicability for the above relation is 280–373 K, which is well within the temperatures of interest for the study of microelectronics cooling. Table 1 lists the details of the

Table 1. Thermophysical properties of silicon, liquid HFE-7200 and its vapor at $T_{\mathrm{sat}} = 81$°C.

Property	Symbol	Silicon	HFE-7200 (liquid)	HFE-7200 (vapor)
Density (kg/m^3)	$\rho_{\mathrm{si}}, \rho_{\mathrm{l}}, \rho_{\mathrm{v}}$	2,330	1,325	10.13
Thermal conductivity (W/m-K)	$k_{\mathrm{si}}, k_{\mathrm{l}}, k_{\mathrm{v}}$	Eq. (19)	0.07966	0.01436
Specific heat (J/kg-K)	$c, c_{\mathrm{pl}}, c_{\mathrm{pv}}$	712	1,333	972.2
Dynamic viscosity (Pa-s)	$-\mu_{\mathrm{l}}, \mu_{\mathrm{v}}$	—	3.79×10^{-4}	1.035×10^{-5}
Surface tension coefficient (N/m)	σ	—	0.01299	

thermophysical properties used to model the solid (silicon) and liquid (HFE-7200) domains.

4. Numerical Procedure and Grid-Independence

One of the main goals of proposing such a modeling scheme is to provide a replicable method that can be used by other researchers who employ commercial CFD–HT software. Although there are many approaches and options when it comes to numerical simulations, one of the main disadvantages and reasons why this field has not advanced as needed is the fact that the problem formulation is complex. While direct numerical simulations can offer desirable customization for thin-film phenomena, they are often very limited when it comes to predicting device-level behavior in practical applications. In the authors' opinion, advances aimed at the development of practical applications (such as microelectronics cooling) could benefit more from the implementation of less complex methods, which can be run in a high-performance computing (HPC) setup to better design effective cooling solutions in microdevices with thousands of geometric features. Hence, the widely used commercial software ANSYS® FLUENT® (R15) was used as the solver in the current study (and the authors' previous works) for the numerical solution of the presented governing equations using the finite volume method. With regard to multiphase formulation, the built-in CLSVOF within FLUENT was configured, and the phase-change modeling was included through a user-defined function (UDF) that accounts for the mass and energy source terms introduced in Section 2.2. The numerical simulations were set up as transient, laminar flow cases using a first-order explicit temporal discretization, whereas the spatial discretization used a second-order upwind scheme.

Although the semi-implicit method for pressure-linked equations (SIMPLE) algorithm can also be used for this type of formulation, the pressure implicit with splitting of operators (PISO) algorithm is suggested for the velocity and pressure coupling, as it provides a more stable calculation. The convergence criteria are considered when the residual values fall below 10^{-4} for the mass, momentum, and level-set equations and below 10^{-7} for the energy equation. Since time step adjustment is a key parameter to consider in such types of calculations, it is recommended to define an automated time step adjustment in the range from 10^{-8} to 10^{-6} s, with a maximum change factor of 2, with the objective of maintaining the value of the global Courant number below 2, with 1 being the most common choice since limiting the time step will, of course, reduce the risk of divergence. In the authors' experience, when running such simulations, the global Courant number can be increased up to a value of 5 once the amount of vapor has reached a quasi-steady state. More practical tips for the implementation of this method can be consulted in Ref. [30].

The numerical simulations using the described method were run on a HPC cluster due to the considerable number of geometric features, even though it is a reduced domain. The cluster consisted of 80 computing cores of Intel® Xeon® E5-2680 v2 processors with a base frequency of 2.8 GHz, interconnected in a parallelized configuration with a $4 \times$ QRD InfiniBand® Interconnection of 40 GB/s; the average solution time per case was 500 h due to the large amount of cells in the computational domain resulting from the many pin fin features. With even further reduced computational domains and sped up formulations using bolder choices of the Courant number, successful simulation times were reduced to less than 100 h.

The use of non-conformal meshing schemes is recommended for the analysis of conjugate heat transfer problems, where the fluid domain discretization benefits from the use of hexahedral elements and the coupled 3D heat conduction problem in the solid domain can save a significant number of elements and easily adapt to most geometries by employing tetrahedral elements. Figure 1 shows the details of such a meshing scheme in the analyzed computational domain, in which the boundary layers (or inflation layers) are defined around each pin fin structure. The mean element sizing in the fluid domain is 5 μm, and the boundary layer (growth) ration has a value of 1.2.

The mesh independence analysis was carried out by inspecting the effect of a range of element sizes. The complete details regarding such parametrization can be found in Ref. [30] and are not repeated here for the sake of brevity. The selected grid density (mean fluid element sizing of 5 μm) provides a cost-effective balance in terms of solution accuracy and total HPC time. With the simulation framework set up, the time-marching simulation of two-phase cooling requires initialization by loading a data solution file, which contains a steady-state point with an isothermal condition (at the saturation temperature) and an adiabatic flow field. Such a procedure is similar to preparing for a two-phase experiment by reaching steady-state conditions with no heating before switching the heaters on. More details on such implementation can be found in Ref. [30].

5. Results

A key aspect of any modeling scheme is its capability to predict the relevant features of the analyzed problem; therefore, the results section starts with the temperature validation plot shown in Figure 2. While the authors clearly acknowledge that the current model has its own idealizations and an extensive discussion of trade-offs and multiple trials of such methods is presented in detail in Ref. [30], the adopted modeling approach has been refined to a point where a relatively complex problem with practical application features can be accurately predicted and therefore constitutes a valuable tool in the design of two-phase cooling systems.

The dotted curves in Figure 2 indicate the numerical model results for the prediction of temperature variation along a central line in the flow direction (x coordinate), which passes through the hotspot, as indicated by the central stripe in Figure 2. The experimental average temperature is represented by the red squares, which is calculated through the measurement of an electrical resistance and then correlated to a temperature; more details and calibration curves can be found in Ref. [30]. Owing to the type of platinum heaters and measurement procedure used, only five temperature values (four background heaters and the hotspot heater) could be obtained for those experiments; hence, the numerical simulation constitutes a very valuable tool since it provides a fully resolved temperature distribution in all directions. As shown in Figure 2, the numerical model results are in good agreement with the reported experimental values in Ref. [30] and therefore indicate an outstanding capability

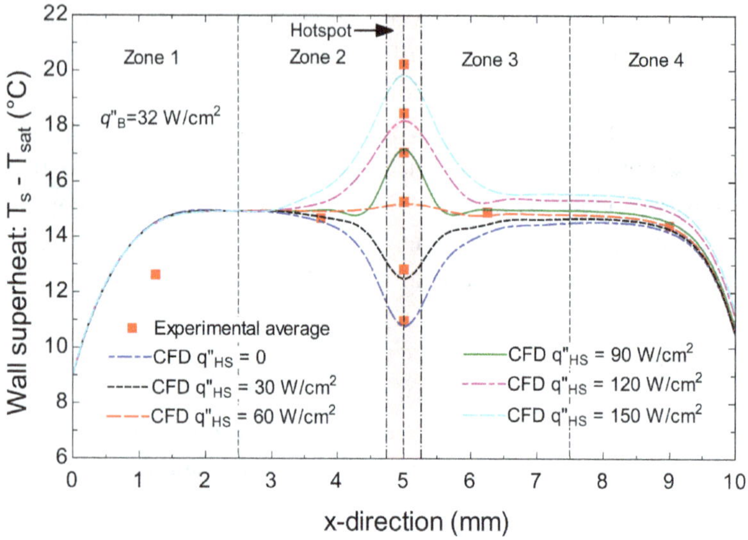

Figure 2. Temperature validation plot of the numerical model with experimental data [30] for a background heat flux of $q''_{\mathrm{B}} = 32\ \mathrm{W/cm^2}$, hotspot heat flux q''_{HP} in the range of 0–150 $\mathrm{W/cm^2}$, and a flow rate of 60 ml/min.

Source: Reprinted from Ref. [32], Copyright (2019), with permission from Elsevier.

to predict temperature distribution in non-trivial cases operating under two-phase conditions. A remarkable feature is the capability of the model to accurately predict the temperature jumps around the hotspot for all the heat flux inputs, and an even more interesting observation is how the temperature profile remains almost flat for the case in which the hotspot heat flux is nearly double the background heat flux ($60/32\ \mathrm{W/cm^2}$); such evidence indicates the important fact that the heat transfer coefficient could be approximated as constant throughout the whole device (even at the hotspot, which has smaller flow passages), and hence, the temperature response could be easily estimated by knowing the heat transfer surface area. Since it is known that the hotspot area density is double that of the background zones, locally doubling the heat flux results in such behavior. Another key finding is that the effect of flow bypass near the hotspots (due to a higher flow resistance) in two-phase cooling conditions does not significantly affect heat transfer performance, which is proven by the remarkable ability of the localized pin fin array to handle heat fluxes up to five times higher than the background zones. Such an observation is also relevant because it indicates that the surface area may be specifically tuned only in the vicinity of hotspots and not the entire device, which allows a significant pressure drop reduction while alleviating excessive temperatures in these areas. This is an important statement which is relevant for the design and optimization of hybrid cooling layers in practical applications where hotspots are present, such as microelectronics.

An important advantage of two-phase cooling is that, unlike single-phase cooling, where the mean fluid temperature increases in the flow direction, the resulting temperatures remain relatively constant at the so-called wall superheat, which is around 15°C for all of the analyzed cases. The sharp drops and spikes observed in Figure 2 are due to the effects of heat spreading and the fact that the observed behavior is along a centerline. Hence, the advantage of obtaining a fully resolved temperature field through the numerical simulation is described as follows. Figure 3 shows the quasi-steady solution ($t > 20$ ms), calculated using the numerical model, for the case in which there is no heat input to the hotspot, which consequently turns into a sink. Readers interested in more details of the transient solution are referred to Refs. [30, 32] for the intermediate steps that show the onset of boiling and its evolution before reaching the described state. Figure 3(a) depicts the temperature distribution on a cross-sectional plane ($x - y$), in which the scale is indicated in terms of the wall superheat. As expected, there exists heat spreading at the inlet/outlet zones as well as the central fin array, which is due to the lack of power inputs near these zones. However, the maximum temperature gradient in the indicated scale is lower than 6°C in the shown plane, which confirms the capability of the analyzed device to handle heterogeneous power inputs. Keeping a relatively uniform temperature is a very desirable aspect for improved performance and device reliability since high temperature gradients are associated with induced stresses in the silicon material, which can eventually be a cause of failure. Figure 3(b) depicts the quasi-steady-state two-phase flow regime, in which the vapor interface is clearly indicated by the blue surface. In accordance with experimental observations,

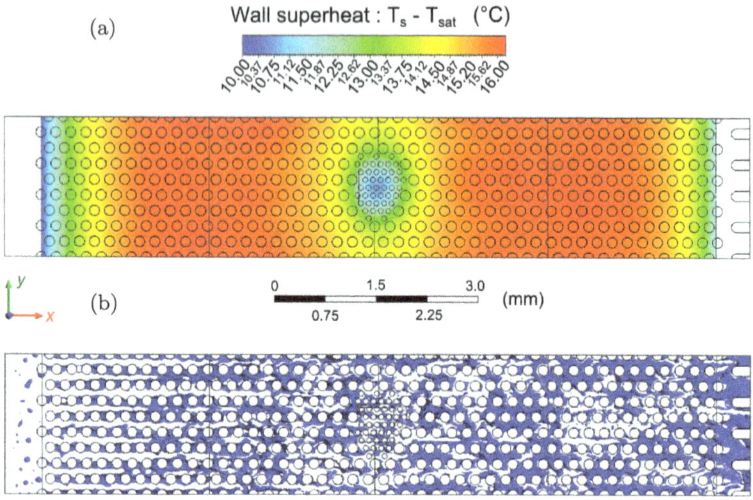

Figure 3. Quasi-steady results with a background heat flux of $q''_{\mathrm{B}} = 32\ \mathrm{W/cm^2}$, hotspot turned off ($q''_{\mathrm{HP}} = 0$), and a flow rate of 60 ml/min: (a) temperature field, (b) two-phase flow regime.

Source: Reprinted from Ref. [32], Copyright (2019), with permission from Elsevier.

the nucleation mechanism occurs mainly at the back side of the pin fins due to the flow recirculation condition that is generated in that zone, whereas the inline array helps in the evacuation of such vapor through the straight flow lanes, while also allowing the transversal two-phase flow interaction as the vapor flows and merges into bigger churns downstream.

Figure 4 shows the temperature and vapor phase distributions for an alternative case in which the hotspot is now powered at 90 W/cm^2. As expected, there is a localized temperature increase at the hotspot, as shown in Figure 4(a); however, the larger heat transfer area in this zone helps maintain a wall superheat below 7°C under such operating conditions. Slight temperature increases are also observed in the peripheral strips downstream and spanwise, which explain the small offsets in Figure 2. As expected, these offsets increase with the magnitude of the hotspot heat flux. The quasi-steady-state two-phase flow regime, as shown in Figure 4(b), is practically of the same shape as that of the previous case. As explained in more detail in Ref. [30], the effect of the hotspot heat flux on the two-phase flow regime is

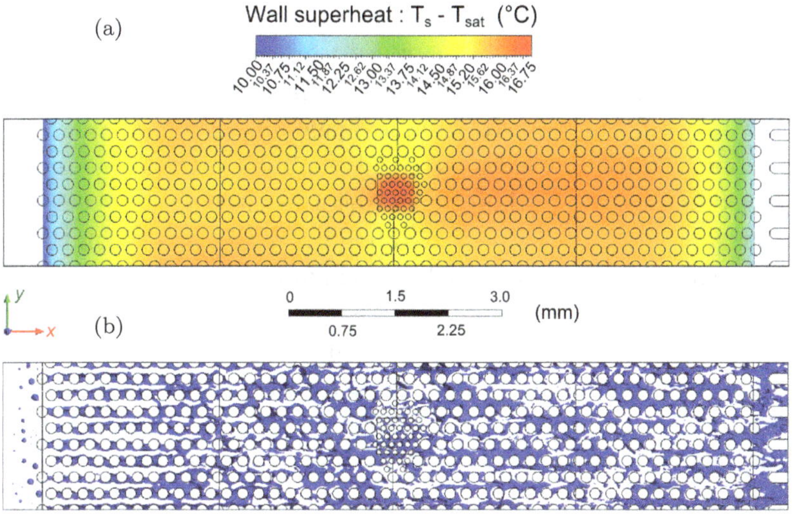

Figure 4. Quasi-steady results with a background heat flux of $q''_{\mathrm{B}} = 32$ W/cm^2, hotspot heat flux of $q''_{\mathrm{HP}} = 90$ W/cm^2, and a flow rate of 60 ml/min: (a) temperature field, (b) two-phase flow regime.

Source: Reprinted from Ref. [32], Copyright (2019), with permission from Elsevier.

negligible due to its relatively small size and the significant amount of vapor that is advected from upstream regions.

The other key flow variable to analyze is the two-phase pressure drop since it is known that multiphase flow requires higher pressure gradients to remove the generated vapor due to boiling, and keeping it to a minimum is one of the objectives of the analysis and optimization of such systems. Although many empirical correlations exist for the prediction of two-phase pressure drops in boiling flows (macro- and microsystems), their intrinsic empiricism limits their range of applications. For the analyzed cases, none of the correlations available in the literature are suitable due to the relatively complex geometry with a hotspot and variable geometrical features, hence the need for more robust models that can predict a wider range of conditions. As a complement to the model validation, Figure 5 shows a plot that compares experimental measurements [30] and model predictions for both single- and two-phase pressure drops at a system flow rate of 60 ml/min. The shown single-phase measurements and model predictions are under adiabatic conditions at the saturation point, while for the two-phase comparison, the background heat flux is 32 W/cm^2 and the hotspot heat flux ranges from 0 to 150 W/cm^2. The reason to include single-phase measurements and model predictions in that plot is the availability of such results in both experiments and simulations before switching the heaters on (initialization point), while also adding further confidence in the prediction capabilities of the modeling technique. The horizontal lines at the bottom of the plot in Figure 5 indicate the steady-state values for the single-phase comparison between the experimental measurement and the numerical model, which are in very good agreement at 23.41 and 22.88 kPa, respectively. Regarding the two-phase results, a significant jump can be observed due to the aforementioned reasons, where the red dotted horizontal line represents the experimentally measured pressure drop at the indicated flow rate [30], and the solid blue line corresponds to the numerical model results. One of the advantages

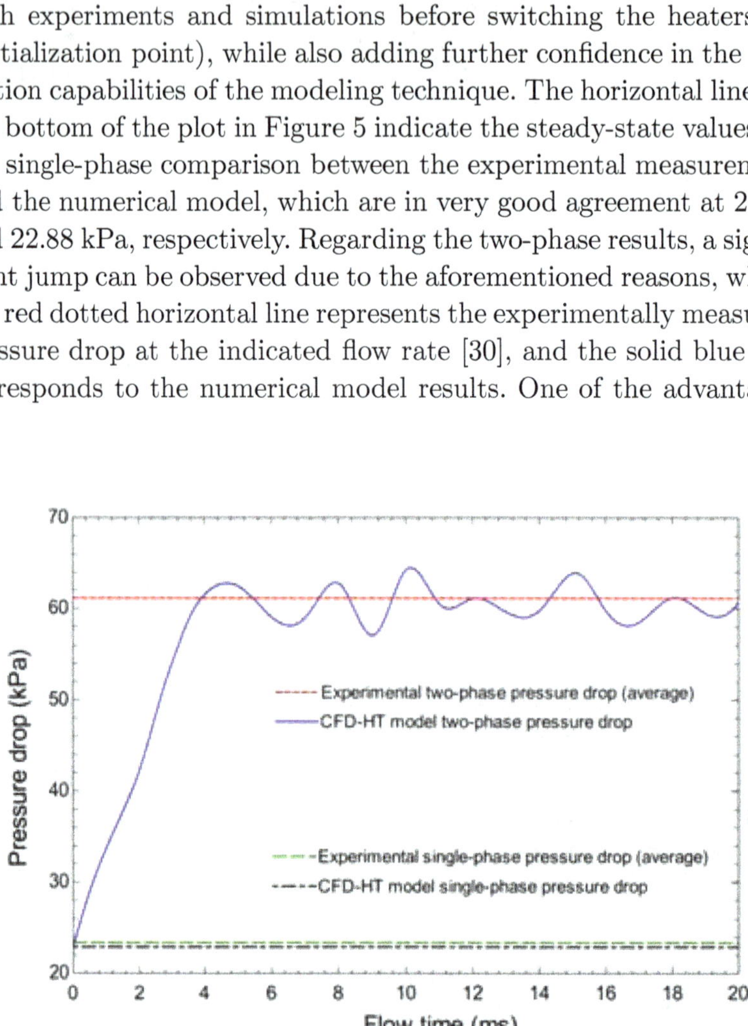

Figure 5. Comparison of the single- and two-phase pressure drop measurements in experiments and simulations with HFE-7200 at 60 ml/min.

Source: Reprinted from Ref. [32], Copyright (2019), with permission from Elsevier.

of the numerical model is that it can provide insights into transient behavior in great detail, allowing us to observe how the model starts at the single-phase pressure drop and then increases sharply during the first milliseconds of flow due to the boiling effect, until reaching a periodic behavior, which is attributed to the constant removal and generation of vapor within the computational domain (quasi-steady state). In order to provide an objective comparison between the CFD model predictions for two-phase pressure drop and the experimental value, the fluctuating values after 8 ms were averaged at 60.63 kPa; with the experimental average at 61.15 kPa, it can be stated that the current numerical model does also provide a very accurate estimation of this important parameter. More results that further validate the prediction capabilities of this model for a number of cases and operating conditions can be consulted in Ref. [30].

An important observation is that the effect of hotspot heat flux on the two-phase pressure drop of the system is also negligible due to its relatively small dimensions compared to the background zones, and this is supported by both experimental measurements [30] and numerical results. In order to provide more quantification of the results discussed so far, Table 2 lists a number of comparisons between the experimental and CFD–HT results for the silicon microgap at different heating combinations. Recalling the negligible effect of hotspot input on the background zones, only the representative values are listed while leaving the localized hotspot temperature in a

Table 2. Two-phase CFD–HT model validation with the experimental data [32].

		Average zone temperature				Hotspot temperature	Pressure drop
Background heat flux, q''_B(W/cm^2)	Result type	ΔT_{wz1} (°C)	ΔT_{wz2} (°C)	ΔT_{wz3} (°C)	ΔT_{wz4} (°C)	ΔT_{HP} (°C)	ΔP (kPa)
13	Exp.	4.15	4.91	5.29	5.98	3.06–8.05	55.35
	Model	3.95	4.12	4.95	5.36	2.79–7.85	54.64
32	Exp.	12.63	14.68	14.89	14.41	10.98–20.23	61.15
	Model	13.82	14.86	14.93	14.72	10.63–19.85	60.63

range; there is an excellent agreement between the numerical model and experiments for other operating conditions as well, indicating the great potential in using such a tool for parametric studies to predict the wall superheats of the background and hotspot zones (ΔT_{wz} and ΔT_{HP}, respectively) and the two-phase pressure drop (ΔP).

6. Two-Phase Flow Comparison with Experimental Results

Figure 6 shows a comparison of the different flow boiling behaviors with the experimental results reported in Ref. [30] at three selected locations in the silicon microgap: the inlet, hotspot, and outlet zones. In accordance with multiple observations (experimental and numerical) in the authors' previous works [28–32], the presented results show once again how the recirculation zones between pin fins result in being the dominant mechanism for bubble nucleation and growth. As observed in the subsets on the left of Figure 6, once the bubbles generate, they keep growing until they are detached by the shear forces of the liquid flowing on the adjacent lanes; this is identified as the main boiling mechanism in both the background and hotspot zones. As noted in the experimental observations in Ref. [30], two-phase flow visualization at the microscale is a significant challenge due to a number of factors, and although the obtained high-speed visualizations are of a very good quality, as seen in Figure 6, there are intrinsic limitations that affect full image tracking, such as the lighting, plane focus, and reflections, which make it extremely difficult to focus on the thousands of vapor structures that are flowing. Although many bubbles (presumably on the same x–y plane) are clearly resolved by the camera, many of the vapor films and churns show a lower resolution due to variances in height and also the high speed at which they are flowing. Hence, the availability of a reliable CFD–HT model is highly valuable due to the fact that the vapor interface can be tracked across the entire domain and observed fully resolved, helping with a better insight into how the vapor structures merge and interact as they leave the computational domain.

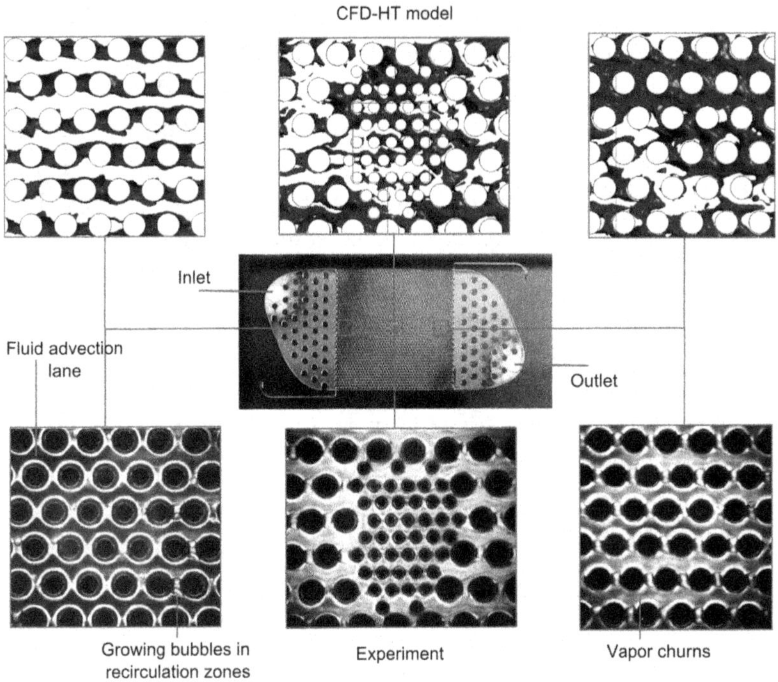

Figure 6. Comparison of the flow boiling regimes from experimental visualization and numerical results of the CFD–HT model.

Source: Reprinted from Ref. [32], Copyright (2019), with permission from Elsevier.

Although exploring the effect of pin fin spacing and array (staggered vs. inline) is beyond the scope of this study, a remark worth mentioning is that, as has been found in the author's previous investigations [28–32], inline pin fin configurations are preferred for boiling conditions not only because of a reduced pressure drop but also because such an array results in a more effective vapor removal through the formed straight lanes and therefore constituting a design suggestion that is backed by both experiments and simulations.

7. Further Testing of the Numerical Model: Effect of Multiple Hotspots

Once sufficient proof has been presented and discussed to validate the numerical model, it is of interest to perform parametric variation analysis for more complex systems in relevant applications. The remaining part of this work is dedicated to assessing the effect of multiple hotspots on heat transfer, pressure drop, and two-phase flow regimes for silicon microgaps with relevant features in cooling microelectronic systems. One of the hypotheses that motivated such variation is that selectively tuning the heat transfer area in the vicinity of hotspots can also efficiently manage the presence of multiple, heterogenous heat sources. An important advantage is that the present modeling scheme allows a relatively easy adjustment of the geometrical parameters and operating conditions, allowing the generation of data matrixes for optimization purposes and also saving a significant amount of resources if such variations were to be experimentally proven. For such iterations, the number of hotspots has been increased to three zones located at the inlet, center, and outlet of the analyzed geometry, as shown in Figure 7.

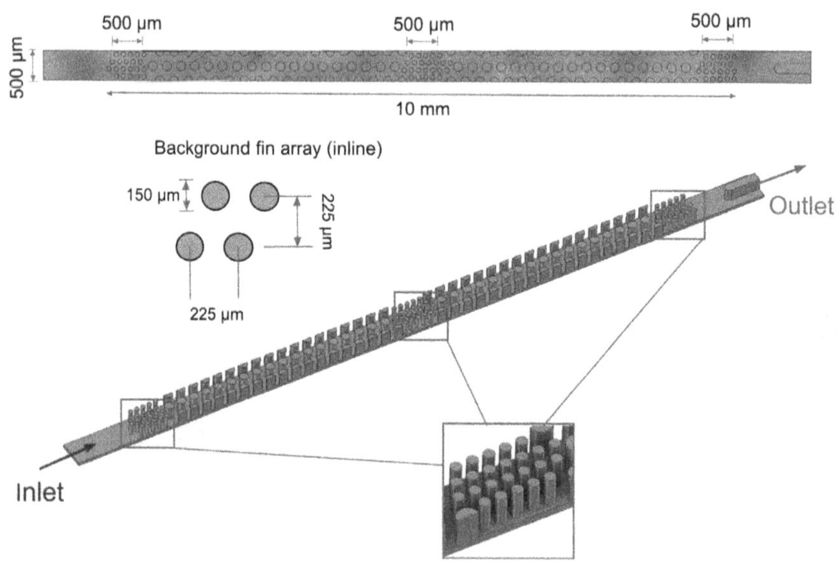

Figure 7. Schematic diagram of the analyzed silicon microgap portion with multiple hotspots.

Source: Reprinted from Ref. [33], Copyright (2019), with permission from AIP Publishing.

For the current case, the computational domain has been reduced in the y-direction, considering symmetry boundary conditions to simplify the numerical solution. In such cases also, the hotspots extend in the transversal direction, and the effect of the hotspots will therefore be more pronounced in the upstream regions. Regarding dimensions, the pin fin diameters for the background and hotspot zones are the same as those in the previous cases, at 50 and 25 μm, respectively. The silicon microgap height is also set at 200 μm, and the other relevant dimensions are clearly denoted in Figure 7. The active heating section of the domain is the indicated 10 mm in extension, with a width of 500 μm.

Regarding the boundary conditions, the background heat flux was defined for all cases with a value of $q''_{\mathrm{B}} = 30$ W/cm^2, and the range of hotspot heat flux (q''_{HS}) was varied with values of 60, 90, and 120 W/cm^2, i.e., a ratio of 2–4 times the background heat flux. Two cases were analyzed, in which the magnitude of the heat flux was changed according to the location in order to assess the effect of such an action on the numerical results. The solid and fluid domains are kept as silicon and HFE-7200, respectively; hence, the thermophysical properties are the same as those already introduced in the previous sections. Although the selected flow rate is still 60 ml/min for the current alternative cases, the flow inlet boundary condition is updated due to the narrower channel, resulting in a mass flow rate of 1.099×10^{-7} kg/s. The inlet fluid temperature is kept at the saturation condition of 81°C from the previous cases.

In order to demonstrate the useful capabilities of the numerical model to analyze the time evolution of field variables before reaching the quasi-steady state, the transient response of the temperature contours is obtained, as shown in Figure 8, for the analyzed cases with a background heat flux of 30 W/cm^2 and variable hotspot heat flux magnitudes/locations. The plotted temperature contours correspond to the wall superheat, and it can be noted how this variable starts increasing first at the hotspot locations as a function of its magnitude. Figure 8 depicts the two analyzed cases, in which the hotspot heat flux magnitudes and their positions are clearly indicated, with the flow direction being from left to right. The hotspot areas with

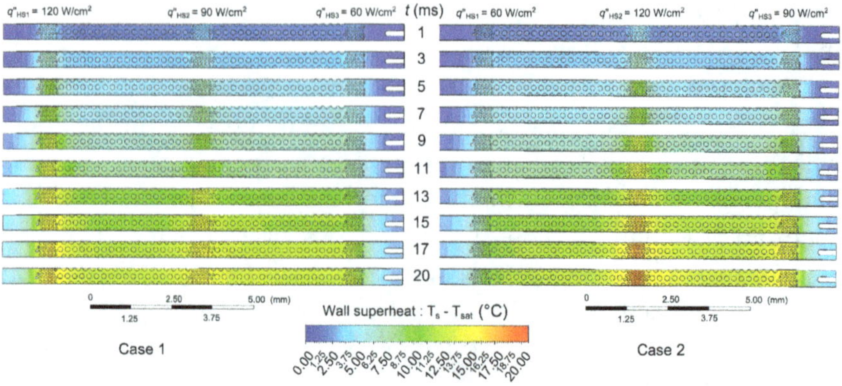

Figure 8. Transient response of the temperature contours for the analyzed cases with a background heat flux of 30 W/cm^2 and variable hotspot heat flux magnitudes/locations.

Source: Reprinted from Ref. [33], Copyright (2019), with permission from AIP Publishing.

the heat flux of 120 W/cm^2 start spreading faster to the adjacent areas, which is expected; however, an interesting observation is that the localized temperatures for the hotspots are of similar magnitudes in their different locations. In agreement with previously discussed results, the background zones remain at a relatively uniform temperature during the entire process, until reaching quasi-steady-state conditions, which is one of the sought-after advantages of using two-phase cooling methods.

As can be noted in Figure 8, the results at times 17 and 20 ms are practically the same and therefore denote the peak values of operation under the defined settings (quasi-steady conditions). Comparing both cases under such conditions, it is observed that the hotspot magnitude/location does have a visible effect on the resulting zone temperature. For the highest heat flux of 120 W/cm^2, it can be stated that the location will affect its terminal temperature; for Case 1 (inlet zone at 120 W/cm^2), the average hotspot superheat is ~16°C, while changing the hotspot location of that magnitude to the center (Case 2) results in a slightly higher hotspot superheat of ~18°C. As the magnitude of the heat flux decreases, the temperature offset resulting from such location changes also decreases, as observed for

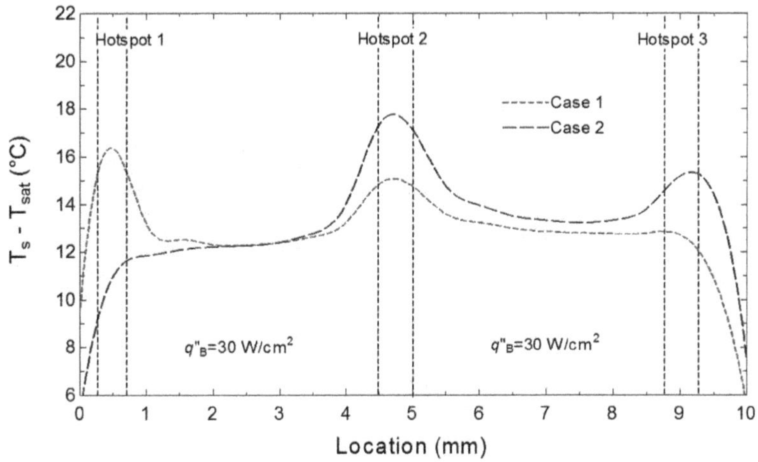

Figure 9. Average wall superheat along the active heated length for the analyzed cases with a background heat flux of 30 W/cm^2 and variable hotspot heat flux magnitudes/locations.

Source: Reprinted from Ref. [33], Copyright (2019), with permission from AIP Publishing.

the magnitude of 90 W/cm^2, in which, for Case 1 (central hotspot at 90 W/cm^2), the wall superheat is ~15°C and only ~0.5°C higher than that for Case 2, in which the 90 W/cm^2 heat flux is located at the outlet zone. These effects are mainly associated with the local heat spreading to the adjacent adiabatic zones at both the inlet and outlet plena.

Figure 9 shows a plot of the average wall superheat variation along the heated length, where the dashed vertical lines indicate the positions of the multiple hotspots, labeled as 1, 2, and 3, and corresponding to the inlet, central, and outlet clusters. The behaviors that were described earlier can be alternatively described with this plot, noting the peaks in temperatures across the multiple hotspots and their variation into higher values as they are moved toward the outlet. Although the difference in peak values between hotspot superheat is lower than 2°C for all of the cases, the trend indicates that slightly lower temperatures may be achieved at the hotspot areas with the highest heat fluxes if they are located near the inlet, where the combined effects of spreading can further act as an additional sink.

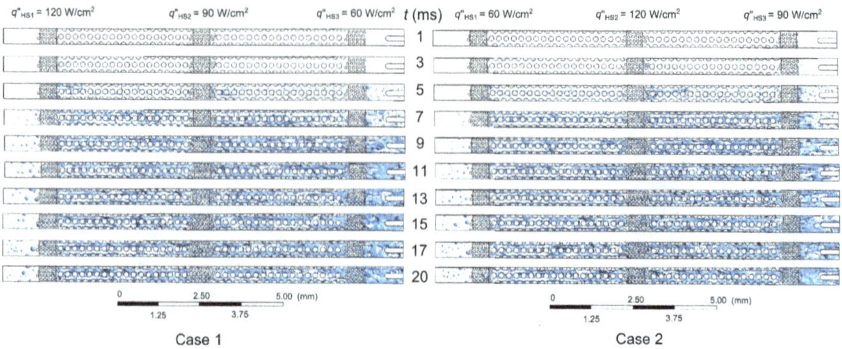

Figure 10. Transient response of the two-phase flow regime for the analyzed cases with a background heat flux of 30 W/cm^2 and variable hotspot heat flux magnitudes/locations.

Source: Reprinted from Ref. [33], Copyright (2019), with permission from AIP Publishing.

Figure 10 depicts the transient two-phase flow regime evolution during 20 ms of flow, in which the vapor interface is clearly indicated by the blue surface, the flow direction is from left to right, and the location and magnitude of the heat fluxes are indicated in the upper part for both cases. Since the numerical simulations are initialized with the steady-state, single-phase solution at the saturation point, there are no distinguishable vapor structures during the first millisecond of flow. As the temperature starts increasing, especially in those hotspot areas with the highest heat flux, the adjacent fluid zones that meet the phase-change criteria begin to generate vapor bubbles, as can be noted at t = 3 ms. For the next shown time step of t = 5 ms, a considerably higher amount of vapor has been generated, and the nucleation of larger bubbles in the recirculation zones can be observed, which again is in agreement with all of the numerical and experimental observations on such types of configurations [30]. At t = 7 ms, the grown and detached vapor structures from recirculation zones can be observed as they move toward the adjacent flow lanes, merging in the process with other vapor structures and forming the so-called vapor churns in the downstream zones. An interesting observation is how the two-phase flow regime achieves the quasi-steady state faster than the temperature field; as can be noted,

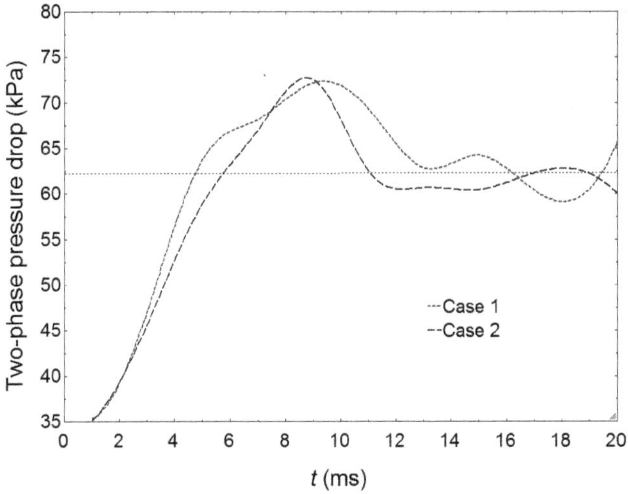

Figure 11. Transient response of the two-phase pressure drop for the analyzed cases with a background heat flux of 30 W/cm^2 and variable hotspot heat flux magnitudes/locations.

Source: Reprinted from Ref. [33], Copyright (2019), with permission from AIP Publishing.

after 9 ms of flow, the two-phase flow regimes appear to be practically the same for the remnant time steps. The comparable amounts of vapor and their distribution for flow times >10 ms point to a balance between the evacuated and generated vapors in the cooling process, which is the criterion for reaching the quasi-steady state of this variable.

Finally, the transient response of the two-phase pressure drop for both cases is plotted in Figure 11; this plot starts at a minimum value of 35 kPa, which corresponds to the single-phase pressure drop for such geometry at a flow rate of 60 ml/min. Similarly to the behavior described in Figure 5, a steep increase in pressure can be observed at the early stages of two-phase flow ($t < 7$ ms), which is clearly due to the relatively fast appearance and buildup of vapor structures within the computational domain. The curves in Figure 11 provide a valuable quantification of the transient behavior shown in Figure 10. Notably, the peak value at $t \sim 9$ ms indicates the point when the vapor buildup has reached a maximum within the domain right before its removal and a periodic behavior of generation/evacuation from the domain, hence generating the sinusoidal behavior in Figure 11, which has a mean value of ~62 kPa for both cases and is denoted by the horizontal dashed line.

8. Conclusion

The numerical modeling of flow boiling at the microscale is a key element for the advancement of cooling technologies in microelectronics. A comprehensive CFD–HT model for two-phase cooling analysis, which is relatively easy to implement in a commercial software, has been described, and a comparison of its results with the non-trivial thermal-hydraulic behavior observed in the experimental measurements of two-phase cooling in silicon microgaps showed its remarkable capability for predicting these results with a high level of detail and accuracy. The temperature jumps at the hotspot area were closely predicted, in addition to the relatively uniform temperature across the background zones due to the two-phase mode of operation. On the hydraulic side, the model is capable of predicting results with excellent accuracy under both the initial saturation conditions for pressure drop (single phase) and under evolving conditions as a function of time, until reaching a quasi-steady-state equilibrium at which the dynamic two-phase pressure closely matched the average experimental values and range of dynamic variation.

The prediction capabilities of the numerical model provided useful insight into the analysis of the flow boiling mechanism in a silicon microgap, which was characterized by variable pin fin clustering and heterogeneous power inputs that are relevant to the study of microelectronics. The flow boiling of the dielectric fluid HFE-7200 was explored under saturated conditions, leading to the following main observations:

1. Under flow boiling operation, the background zones with constant heat flux conditions can be kept at a relatively uniform temperature due to the fluid's latent heat of vaporization, regardless of the location of the multiple hotspots and their magnitudes in the analyzed cases.

2. The central hotspot heat flux may be considerably amplified for conditions where the local heat transfer area has been selectively increased; for the present cases, it was enlarged up to five times the background heat flux values, with the temperature difference being 10°C less than the temperature of the other zones.
3. The wall superheats present a linear behavior for the hotspot cases, with twice the heat flux input than the background zones; such a pattern confirmed that the heat transfer coefficient can be safely assumed to remain constant at the hotspot zone since both the heat flux and surface area had been doubled.
4. The location of the hotspot and its heat flux magnitude have an effect on the local wall superheat; such results were associated with the heat spreading effect, as the inlet and outlet zones are more likely to experience a sink effect due to the adjacent solid zones (fluid plena) in adiabatic conditions, unlike the central hotspot zones, which are only adjacent to the background heat flux. For such scenarios, it is then suggested to relocate the hotspots with higher power inputs near the inlet or outlet zones.
5. The effects of the hotspot location and magnitude may be considered negligible on the two-phase pressure drop response, given that the power input and geometries are maintained.

The numerical modeling of such phenomena in these cases had remained relatively unexplored due to a number of challenges; therefore, the direct validation of this model constitutes a key advance for the detailed thermal design and optimization of complex microgap cooling layers for two-phase cooling systems.

References

[1] S.V. Garimella, C.B. Sobhan, Transport in microchannels: A critical review, *Annu. Rev. Heat Transfer*, 13, 2003, 1–50.

[2] J.R. Thome, Boiling in microchannels, *Int. J. Heat Fluid Flow*, 255, 2004, 128–139.

[3] S.S. Bertsch, E.A. Groll, S.V. Garimella, Review and comparative analysis of studies on saturated flow boiling in small channels, *Nanoscale Microscale Thermophys. Eng.*, 12, 2008, 187–227.

[4] S.M. Ghiaasiaan, *Two-Phase Flow, Boiling and Condensation: In Conventional and Miniature Systems*, Cambridge University Press, New York, 2008.
[5] S.G. Kandlikar, Scale effects on flow boiling heat transfer in microchannels: A fundamental perspective, *Int. J. Thermal Sci.*, 49, 2010, 1073–1085.
[6] S. Krishnamurthy, Y. Peles, Flow boiling of water in a circular staggered micro-pin fin heat sink, *Int. J. Heat Mass Transfer*, 51, 2008, 1349–1364.
[7] J.-M. Koo, S. Im, L. Jiang, K.E. Goodson, Integrated microchannel cooling for three-dimensional electronic circuit architectures, *ASME J. Heat Transfer*, 127, 2005, 49–58.
[8] J.-M. Kim, Y.K. Joshi, A.G. Fedorov, Y.-J. Lee, S.-K. Lim, Thermal characterization of interlayer microfluidic cooling of three-dimensional integrated circuits with non-uniform heat flux, *ASME J. Heat Transfer*, 132, 2010, 041009.
[9] X. Han, A. Fedorov, Y. Joshi, Flow boiling in microgaps for thermal management of high heat flux microsystems, *ASME J. Electron. Packaging*, 138 (4), 2016, 040801.
[10] P. Asrar, X. Zhang, C.E. Green, M. Bakir, Y. Joshi, Flow boiling of R245fa in a microgap with integrated staggered circular cylindrical pin fins, *Int. J. Heat Mass Transfer*, 121, 2018, 329–342.
[11] H.L. Wu, X.F. Peng, P. Ye, Y.E. Gong, Simulation of refrigerant flow boiling in serpentine tubes, *Int. J. Heat Mass Transfer*, 50, 2007, 1186–1195.
[12] Z. Yang, X.F. Peng, P. Ye, Numerical and experimental investigations of two-phase flow during boiling in a coiled tube, *Int. J. Heat Mass Transfer*, 51, 2008, 1003–1016.
[13] W.H. Lee, A pressure iteration scheme for two-phase flow modelling, in: *Multiphase Transport Fundamentals, Reactor Safety Applications*, T.M. Verizoglu (ed.), Hemisphere Publishing, Washington, DC, 1980, pp. 407–432.
[14] S.C.K. De Schepper, G.J. Heyderickx, G.B. Marin, Modeling of the evaporation of a hydrocarbon feedstock in the convection section of a steam cracker, *Comput. Chem. Eng.*, 33, 2009, 122–132.
[15] C. Fang, D. Milnes, A. Rogacs, K. Goodson, Volume of fluid simulation of boiling two-phase flow in a vapor-venting microchannel, *Front. Heat Mass Transfer*, 1, 2010, 013002.
[16] R. Zhuan, W. Wang, Simulation of subcooled flow boiling in a microchannel, *Int. J. Refrig.*, 34, 2011, 781–795.
[17] M.S. Plesset, S.A. Zwick, The growth of vapour bubble in superheated liquid, *J. Appl. Phys.*, 25, 1954, 493–500.
[18] M. Magnini, J.R. Thome, Computational study of saturated flow boiling within a microchannel in the slug flow regime, *ASME J. Heat Transfer*, 138 (2), 2015, 021502.
[19] M. Sussman, E.G. Puckett, A coupled level set and volume-of-fluid method for computing 3D and axisymmetric incompressible two-phase flows, *J. Comput. Phys.*, 162 (2), 2000, 301–337.

[20] C.W. Hirt, B.D. Nichols, Volume of fluid (VOF) method for the dynamics of free boundaries, *J. Comput. Phys.*, 39, 1981, 201–225.
[21] S. Osher, J.A. Sethian, Fronts propagating with curvature-dependent speed: Algorithms based on Hamilton–Jacobi formulations, *J. Comput. Phys.*, 79, 1988, 12–49.
[22] *ANSYS® FLUENT® Theory Guide*, Release 14.0, November 2011, Canonsburg, PA, USA: ANSYS, Inc.
[23] J.U. Brackbill, D.B. Kothe, C. Zemach, A continuum method for modeling surface tension, *J. Comput. Phys.*, 100, 1992, 335–354.
[24] D.L. Youngs, Time-dependent multi-material flow with large fluid distortion, in: *Numerical Methods for Fluid Dynamics*, K.W. Morton and M.J. Baines (eds.), Cambridge, Massachusetts: Academic Press, 1982, pp. 273–285.
[25] H.K. Cammenga, F.W. Schulze, W. Theuerl, Vapor pressure and evaporation coefficient of water, *J. Chem. Eng. Data*, 22, 1977, 131–134.
[26] J.R. Maa, Evaporation coefficient of liquids, *Ind. Eng. Chem. Fundam.*, 6, 1967, 504–518.
[27] D. Lorenzini, Y. Joshi, Computational fluid dynamics modeling of flow boiling in microchannels with non-uniform heat flux, *J. Heat Transfer*, 140 (1), 2018, 011501.
[28] D. Lorenzini, Y. Joshi, Comparison of the volume of fluid and CLSVOF methods for the assessment of flow boiling in silicon microgaps, *J. Heat Transfer*, 139 (11), 2018, 111506.
[29] D. Lorenzini, Y. Joshi, CFD Study of flow boiling in silicon microgaps with staggered pin fins for the 3D-stacking of ICs, *Proceedings of the 15th IEEE ITHERM*, 2016, pp. 766–773.
[30] D. Lorenzini, Computational modeling and experimental validation of single phase and boiling flows in microgap cooling layers, PhD Thesis, Georgia Institute of Technology, 2019.
[31] Engineering Equation Solver (EES), Academic Professional Version 9.911, Madison, WI, USA: F-Chart Software, LLC. (2015).
[32] D. Lorenzini, Y. Joshi, Numerical modeling and experimental validation of two-phase microfluidic cooling in silicon devices for vertical integration of microelectronics, *Int. J. Heat Mass Transfer*, 138 (1), 2019, 194–207.
[33] D. Lorenzini, Y. Joshi, Flow boiling heat transfer in silicon microgaps with multiple hotspots and variable pin fin clustering, *Phys. Fluids*, 31, 2019, 102002.

https://doi.org/10.1142/9789811279379_0004

Chapter 4

Chip-Embedded Two-Phase Cooling

Timothy Chainer*, Pritish Parida†, and Mark Schultz‡

IBM T. J. Watson Research Center, NY, USA

**tchainer@us.ibm.com*

†prparida@us.ibm.com

‡markds@us.ibm.com

Abstract

This chapter describes the experimental and modeling studies of two-phase chip-embedded cooling. The experimental investigation included chip-embedded cooling of a thermal test vehicle (TTV) and an embedded-cooled microprocessor (ECM). The modeling study included the development of both full- and reduced-physics two-phase flow boiling models, which were validated against the experimental data. The TTV was designed with chip-embedded radially expanding microchannels and a power map to mimic an eight-core microprocessor. The TTV enabled an experimental investigation of the limits of two-phase cooling, demonstrating cooling capability well beyond that of current microprocessor heat dissipation. Thermal modeling of the TTV chip-embedded cooling showed good agreement with the experimental data. The ECM module provided a test vehicle to directly compare chip-embedded cooling to a commercial air-cooled system. The air-cooled microprocessors in an IBM server were fully characterized in an environmental chamber and then modified by creating chip-embedded microchannels on the backside of the microprocessor die. A set of workloads was used to compare the performance of air- and embedded-cooled microprocessors. The ECM showed a significant reduction in chip junction temperature compared to air cooling, with the model in good agreement with experiments. The results showed that two-phase dielectric flow boiling provided highly effective cooling in microchannels with dimensions compatible with 3D chip stack cooling.

Keywords: Electronics cooling, two-phase cooling, two-phase flow, chip-embedded cooling

1. Introduction

1.1. *Thermal challenges of advanced packaging*

The electronics packaging industry is deploying highly dense 2.xD and 3D packaging, which provides system performance improvement but also creates thermal challenges. In 3D chip stack structures, as compared to those in 2D, dies are stacked rather than placed side by side, as shown in Figure 1. As the heat dissipated by multiple chips in a stack is confined to the footprint of a single die, the heat density (W/cm^2) of the top die in the stack is increased. In addition, stacked dies are thinned to 50 μm or less to include through-silicon vias (TSVs) and occupy a volume that is less than that of a full-thickness 750 μm die, which increases the volumetric heat density (W/cm^3).

When utilizing traditional cooling methods by placing a heat sink or cold plate on the backside of the top die in the stack, stacking requires heat generated by the Nth die in the stack to conduct through $(N-1)$ dies in the stack to reach the heat sink. As the heat generated is conducted through interconnects and back-end-of-line (BEOL) structures, as shown in Figure 2, which include multiple

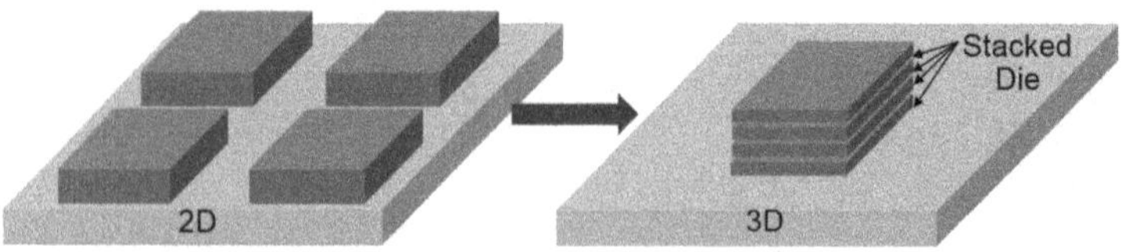

Figure 1. 2D vs. 3D chip stack.

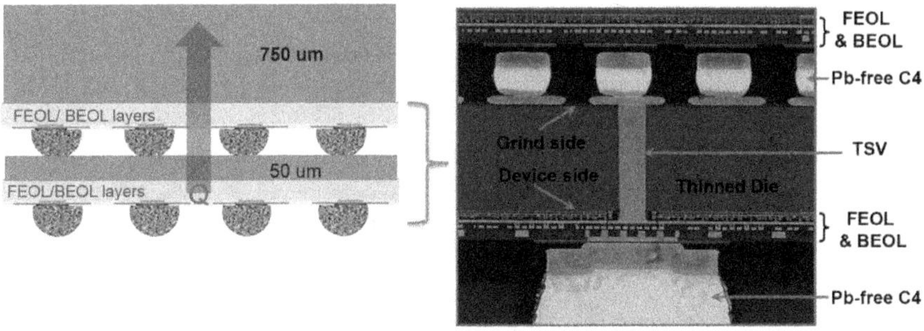

Figure 2. 3D chip stack FEOL and BEOL structures.

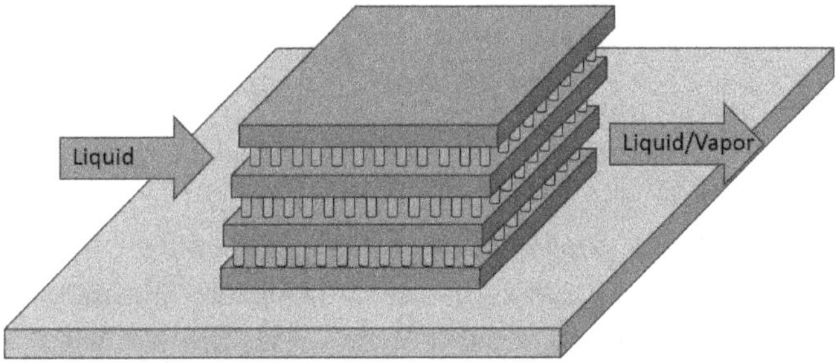

Figure 3. Embedded cooling.

wiring levels and dielectric materials, this creates a significant thermal resistance within the stack.

1.2. *Two-phase chip-embedded cooling*

A solution to the thermal challenges of 3D chip stack cooling is to flow liquid between the dies in the stack, as shown conceptually in Figure 3. This provides cooling directly within the stack, rather than relying on thermal conduction through the stack. This intra-chip cooling approach requires that the fluidic structures be co-designed with the electrical interconnects within the stack, which provides both power and signal I/O. The co-design of the structure must also be consistent with the thinned die in the stack, which is limited by the aspect ratio of TSVs to a maximum thickness of roughly 50 μm.

The cooling approach we utilized is dielectric two-phase cooling, in which a dielectric fluid is pumped into microchannels and extracts heat by changing phase from liquid to vapor. As the fluid

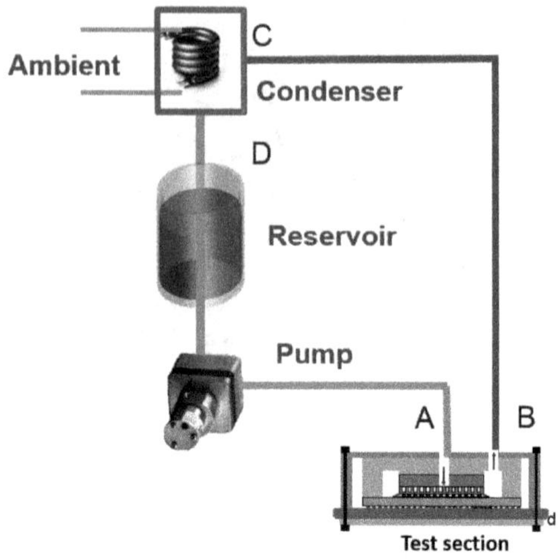

Figure 4. Two-phase cooling loop.

is a dielectric, it can contact power and signal interconnects in the stack, which greatly simplifies the 3D co-design requirements. In comparison, conductive fluids, such as water, require electrical isolation of the interconnects from the fluid.

As shown in Figure 4, the two-phase cooling loop operates in a fully closed system. The coolant is pumped into a module as a single-phase liquid (location A). The flow rate is determined by controlling the pump speed. The coolant absorbs heat as it flows through the device and changes phase from liquid to vapor. The liquid/vapor mixture exits the device (location B), and the downstream liquid/vapor mixture enters the condenser (location C), which is cooled by a secondary loop to extract heat and condense the vapor back into liquid. The condensed coolant exits the condenser and enters the reservoir (location D). The output of the reservoir is fed into the pump.

1.3. *Modeling two-phase cooling*

The development of a two-phase cooling solution using a pumped dielectric coolant requires high fidelity and computationally manageable conjugate thermal models, both at the chip-package level as well as at the system level. Chip-level models include microscale modeling of *in situ* thermal and hydrodynamic phenomena (e.g., boiling and bubble dynamics, two-phase heat transfer rates, and pressure drop) of two-phase flow boiling with good accuracy.

At the chip level, cooling is achieved by utilizing the latent heat of vaporization of a chip-to-chip interconnect-compatible dielectric fluid. During the cooling process, the coolant is converted from liquid to vapor phase as it flows through the chip-embedded micron-scale cavities. By utilizing two-phase cooling, large amounts of heat can be removed with low coolant flow rates and small changes in coolant and chip temperatures across the coolant cavity/channel [1–6]. The design of micron-scale cavities for the coolant flow requires high-fidelity thermal models of the chip(s) with spatially varying heat sources, together with a two-phase microfluidic convection network. The fundamental challenge is to integrate the variations in coolant saturation temperature, local heat transfer rates, friction coefficients, and vapor quality with the complex conduction in the microprocessor package [7, 8]. The combined two-phase flow and heat transfer behavior at these scales is itself an extremely complex phenomenon to model accurately [9].

The chip-level thermal models can be broadly categorized into the following two types: (i) full-physics models and (ii) reduced-physics/order models. In the full-physics model, a set of continuity, momentum, and energy equations is solved for each phase [9–12]. The coupling between liquid and vapor phases is captured through their common pressure field and through momentum and energy exchange sub-models, such as the interphase drag, lift, continuum surface force, heat and mass transfer models, across the fluid–solid interface [10]. In addition, turbulence modeling is needed to capture the turbulent interaction between the liquid and vapor phases [10]. These interactions include the turbulent dispersion and homogenization of the distribution of the dispersed phase (i.e., the vapor phase in the sub-cooled boiling and bubbly flow regimes and the liquid phase in the post dry-out/misty flow regime). Capturing turbulent interaction is important for flow boiling in a micropin field where recirculation zones exist immediately downstream of inlet restrictors as well as at the trailing edge of each pin. Further, at the fluid–solid interfaces/walls, the mechanistic wall heat flux partitioning model is used to capture the two-phase heat transfer rates [10].

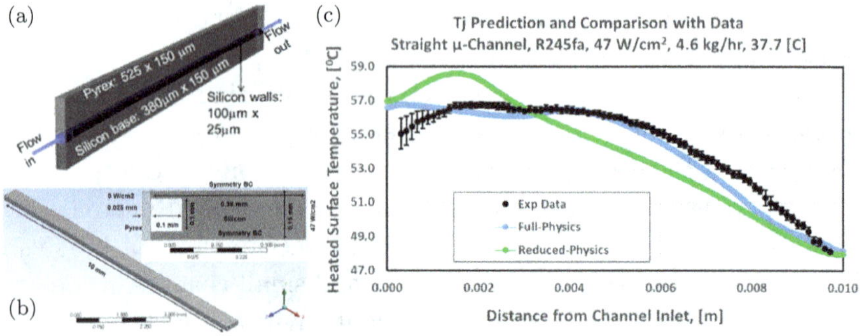

Figure 5. (a) Geometry used for full-physics simulation [11]. (b) Geometry used for the reduced-physics simulation [14]. (c) Width-wise average junction temperature profile prediction and comparison against experimental data [17] for uniform cross-section microchannel case.

Reduced-physics/order models [13, 14] are cross-sectional averaged models, where the fluid domain is approximated in either one dimension (1D) or two dimensions (2D) and the solid domain is modeled in three dimensions (3D). The 1D/2D flow behavior is enforced by using orthotropic thermal properties for the fluid domain and directional momentum source terms [14]. These models provide a good first-order estimate of the temperature and pressure profiles, with an orders-of-magnitude speedup in computational time over full-physics models [15, 16]. These reduced-order models, however, require correlations for pressure drop and heat transfer rate predictions, which can be derived experimentally, from detailed physics simulations of representative sub-domains, or from both [16].

Figure 5 shows a one-to-one comparison of full- and reduced-physics thermal models for a two-phase flow through a uniform cross-section microchannel, where an applied heat flux of 47 W/cm^2 was cooled with R245fa flowing at a mass flow rate of 4.6 kg/h. Figure 5(a) and 5(b) show the unit-channel structure modeled by the full- and reduced-physics models, respectively. Note that the unit-channel structure for the two cases was slightly different. In the case of the reduced-physics model, the dimensionality reduction to 1D or 2D requires geometric simplifications. Thus, only the uniform cross-section channel downstream of the inlet orifice was modeled. Also, a thinner Pyrex top cover plate was considered, as its impact on

Table 1. Comparison of peak junction temperature, pressure drop, and solver time for full- and reduced-physics models for straight microchannels.

Straight μ-channel	$T_{j\ \max}$ [°C]	Δp [kPa]	Solver time[a]
Experimental data [17]	56.8 ± 0.2	68.2 ± 4.2	N/A
Full physics	56.8	140	~48 h on 12 cores
Reduced physics	58.6	62.2	100 s

Note: [a]Excludes serial processes, such as geometry creation, meshing, and post processing.

heat transfer was negligible. Figure 5(c) shows the width-wise average junction temperature profile prediction by the two models and the corresponding experimental data available in the literature [17]. Table 1 compares the maximum junction temperature, pressure drop across the pin field, and solver time for both models. A speedup of more than four orders of magnitude was achieved by the reduced-physics model due to the simplified formulation of the conservation equations, low-complexity computational domain, and simpler mesh with a low number of nodes. The pressure drop was observed to be underpredicted by the reduced-physics model, while the full-physics model overpredicted the pressure drop. Note that the pressure drop predictions by the full-physics model were significantly higher than the experimental data due to two key factors. First, the inlet and outlet plena were not modeled, resulting in the absence of lateral heat spreading and forcing higher than actual heat flux and heat input into the coolant near the inlet. Second, the higher amount of heat results in higher amounts of vapor that traverse the entire length of the cooling channel, effectively reducing the cross-section for coolant flow. Also, more vapor near the inlet or in the inlet orifice results in an increased pressure drop across the orifice. The peak junction temperature predictions by the reduced-physics models were a couple of degrees higher than the full-physics model predictions.

Figure 6 shows a one-to-one comparison of the full- and reduced-physics thermal models for two-phase flow through a field of micropin fins arranged in an inline manner (0° relative to the flow direction), where an applied heat flux of ~39 W/cm^2 was cooled with R1234ze

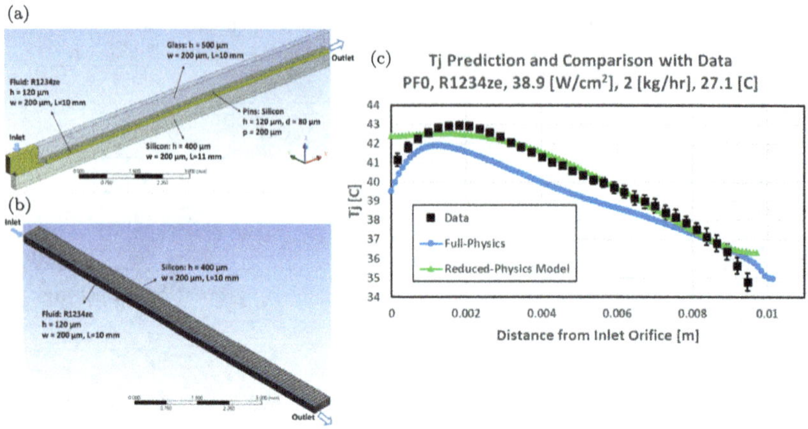

Figure 6. (a) Geometry used for full-physics simulation [12]. (b) Geometry used for the reduced-physics simulation [16]. (c) Width-wise average junction temperature profile prediction and comparison against experimental data for two-phase flow through a micropin field arranged in an inline manner.

Source: (a–b) Copyright (2016) IEEE.

flowing at a mass flow rate of ~2 kg/h. Figure 6(a) and 6(b) show the unit-channel structure modeled by the full- and reduced-physics models, respectively. Note that in the reduced-physics simulation, instead of resolving the mesh for features in the cooling cavity, the silicon micropin fins and the fluidic regions were approximated as a porous domain, allowing for a coarse mesh and reduced geometrical complexities. Figure 6(c) shows the width-wise average junction temperature profile predicted by the full- and reduced-physics models and the corresponding experimental data for the inline pin-fin arrangement.

The assembled ECM modules were mounted onto the server and connected to a two-phase cooling system, as shown in Figure 22(a). A schematic diagram of the system is shown in Figure 22(b), which shows the basic elements of the two-phase cooling loop, including a pump, a condenser, and a reservoir. Additional details of this test setup can be found in Ref. [25].

For both models, the junction temperature profile prediction shows excellent agreement with the data. Further, in this case, the

Table 2. Comparison of peak junction temperatures, pressure drops and solver times for full- and reduced-physics models for inline micropin fins case.

Micropin fin	$T_{j\,\max}$ [°C]	Δp [kPa]	Solver time[a]
Experimental data	43 ± 0.2	37.2 ± 1	N/A
Full physics	41.9	31	~48 h on 12 cores
Reduced physics	42.6	37.3	60 s

Note: [a]Excludes serial processes, such as geometry creation, meshing, and post processing.

junction temperature prediction by the reduced-physics model is slightly better than the full-physics model in the bulk of the cooling cavity with the micropin field. The full-physics simulation included the inlet and outlet plena, where significant lateral heat spreading resulted in relatively lower junction temperatures near the beginning and toward the end of the cavity. Thus, near the inlet and outlet regions of the cooling cavity, the full-physics model was able to better capture the increasing and sharply decreasing temperature trends, respectively [12]. Note that neither the inlet orifice nor the inlet and outlet plena were included in the reduced-physics simulations, as they require special treatment and are not suitable for dimensionality reduction. The focus of the reduced-physics model was faster estimation of thermal performance in the bulk of the cooling structure/cavity and not at the edges. Table 2 compares the peak junction temperatures, pressure drops across the pin field, and solver times for both the models. Here again, the reduced-physics model showed a speedup of over four orders of magnitude compared to the full-physics model. The pressure drop was well predicted by the reduced-physics model, while the full-physics model underpredicted the pressure drop by roughly 17%. The peak junction temperatures predicted by the reduced-physics models were only slightly higher than the full-physics model predictions. A similar model performance was observed for pin fins arranged at different angles of 27° and 45° (staggered), relative to the flow direction [15, 16].

These validated thermal models were used for the design and thermal performance prediction of the two experimental demonstrations described in Sections 2 and 3.

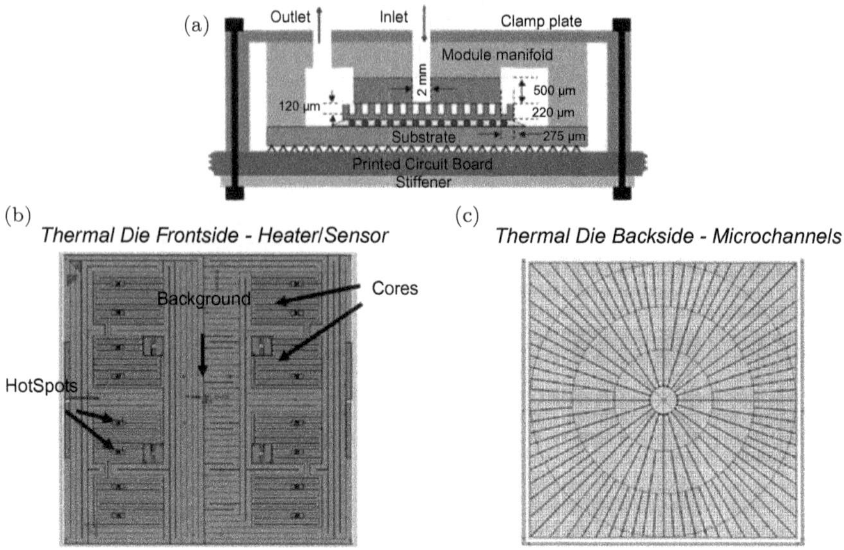

Figure 7. (a) Thermal test vehicle. (b) Heater layout [19]. (c) Radial channel design.

Source: (b) Copyright (2016) IEEE.

2. Thermal Test Vehicle

2.1. *Design and fabrication*

An experimental demonstration of chip-embedded cooling using two-phase flow was performed by creating a thermal test vehicle (TTV) chip-package module to mimic the heat generation capability of microprocessors without requiring actual transistor-based circuitry [18–20]. This allowed for operation beyond the heat density of today's microprocessors to understand the limits of chip-embedded cooling technology.

The TTV was constructed using a 20 × 20 mm silicon die pair, which included a top manifold die and a bottom thermal die mounted onto a ceramic substrate with a lid, as shown in Figure 7. The lid was bonded to both the top manifold die and the ceramic substrate to provide a leak-tight module with fluid inlet and outlet ports. The top manifold die included a central 2 mm diameter through

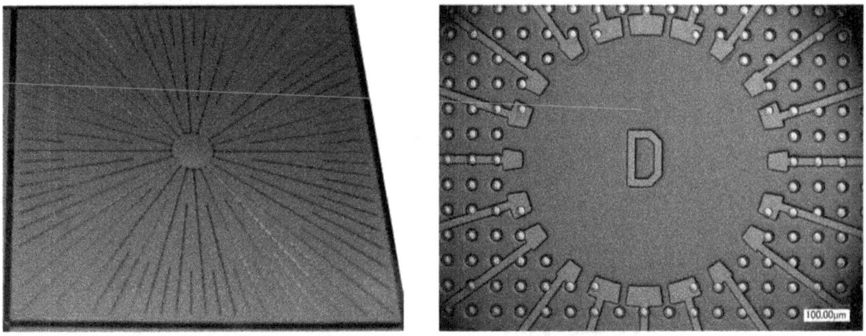

(a) (b)

Figure 8. (a) Thermal die backside radial microchannels. (b) Magnified image of the inlet region showing the manifold inlet, orifices, micropin field, and radial microchannel walls.

hole to provide a fluidic inlet to the microchannels and was bonded to the bottom die to form the top surface of the microchannels. The frontside of the thermal die was patterned with resistive wiring heaters to mimic the heat density map of an eight-core microprocessor, as illustrated in Figure 7(b), including separate heaters for areas which represented the microprocessor cores, background, and hotspots. The core heaters measured 3.6 mm $\times$ 4.8 mm, and the hotspots measured 200 μm $\times$ 200 μm. The frontside of the thermal die also included 37 resistance temperature detectors (RTDs) to facilitate thermal measurements. The backside of the thermal die was etched to provide microchannels for two-phase flow boiling using an expanding radial channel design, as shown in Figure 7(c) for one of the quadrants of the thermal die.

An optical image of the thermal die showing the radial microchannels etched into the backside to a depth of 120 μm is shown in Figure 8. The central inlet fluid was divided into five segments per quadrant, for a total of 20 segments. A magnified image of the inlet region in Figure 8(b) shows the manifold inlet, orifices, micropin field, and radial expanding channel walls.

The radial channels cover the entire die area, which is populated with silicon pins of 80 μm diameter at a 200 μm spacing to mimic structures for TSVs, which would be present in the flow field for 3D die stacking. The expanding radial channel approach for

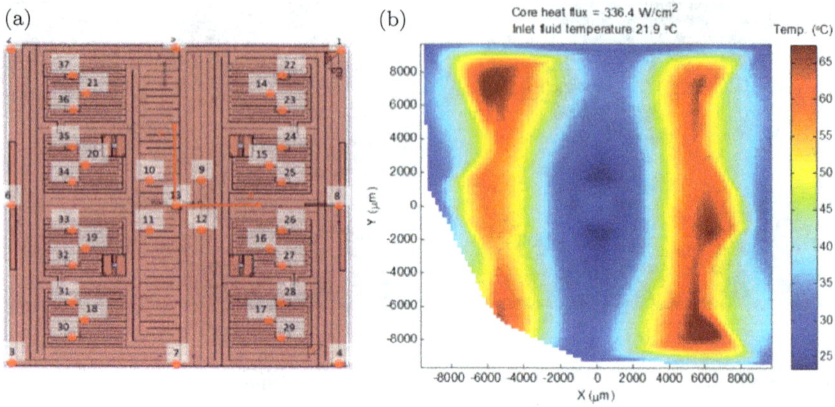

Figure 9. (a) Heater layout with 37 temperature sensors [23]. (b) Interpolated thermal map for a TTV with core heat flux of 336 W/cm^2 and cooled using R1234ze flowing at a rate of 15 kg/h.

Source: (a) Copyright (2018) IEEE.

chip-embedded cooling offers several advantages over parallel channels, which have been widely investigated [2, 3, 21] and have faced challenges of flow stability exacerbated by non-uniform power maps. The expanding radial design increases the channel cross-sectional area along the channel flow path, which reduces the pressure drop as liquid is converted to vapor during the flow, compared to parallel channels. The implementation of the inlet port at the center of the die also reduces the flow path and pressure drop compared to those in parallel channels extending across the entire die. The expanding cross-sectional area also contributes to stabilizing the two-phase flow. This approach provides a path to higher energy efficiency and maximum critical heat flux [6].

2.2. *Experimental results*

The TTV was operated using the coolant R1234ze over a range of mass flow rates. The 37 RTD sensors shown in Figure 9(a) were measured, and a "junction temperature" was calculated, which accounted for the ~5% additional thermal resistance due to the 500 nm oxide layer between the sensors and silicon. The result shown in Figure 9(b) is an interpolated graphic of the RTD sensors, with all cores powered

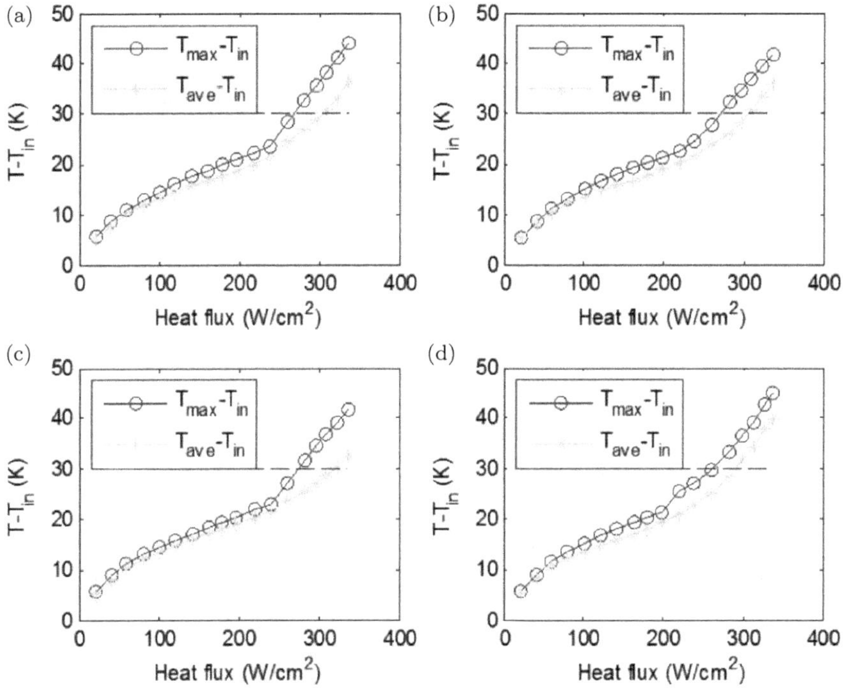

Figure 10. TTV temperature vs. heat flux for quadrants: (a) Q1, (b) Q3, (c) Q2, and (d) Q4.

at 336 W/cm^2 and an overall background heat flux of 20 W/cm^2, resulting in a total power of 450 W. The missing data in the lower-left quadrant were due to non-functional RTDs. The total pressure drop across the module was 17.3 psi (119.3 kPa). The temperature rise above the inlet temperature for the individual quadrants Q1, Q2, Q3, and Q4 versus heat flux is shown in Figure 10, which showed similar behavior across all quadrants.

The results demonstrate effective cooling of core power densities of 336 W/cm^2, which is more than 3× those of modern microprocessors, while utilizing 120 μm height channels with a volumetric efficiency of more than 10× that of conventional backside liquid cooling solutions. The volumetric efficiency is the ratio of the volume of a conventional cold plate to the volume occupied by embedded cooling.

An experimental investigation of two-phase cooling of hotspots was also conducted, where the core and background heat fluxes were

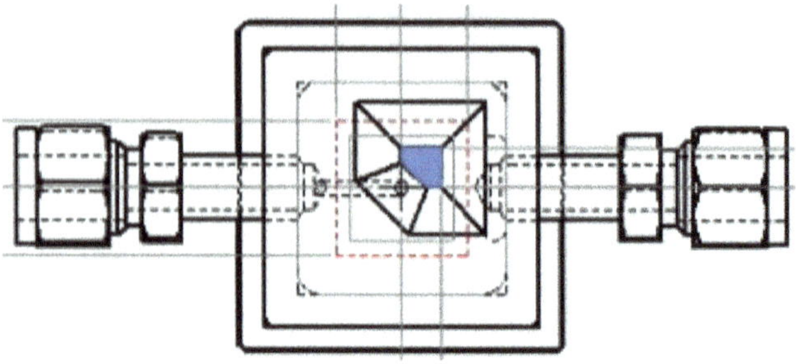

Figure 11. Schematic diagram of TTV module with optical viewing port.

kept constant while varying the hotspot heat flux. It was demonstrated that the embedded radial microchannels could manage a heat flux of 2.1 kW/cm^2 with a hotspot temperature below 63°C for 200 μm × 200 μm hotspots at different locations on the thermal die [20]. Two different mass flow rates, 10 and 15 kg/h, were tested, which showed similar trends [20]. The local thermal resistance gradually reduced with increasing hotspot heat flux, likely as a result of heat spreading and increased local heat transfer rates. At the highest heat flux condition, the overall local heat transfer coefficient, including heat spreading, was observed to be as high as 636,000 W/m^2K [20]. Overall, the TTV results demonstrated that radial microchannels with pin-field structures could reliably cool "core" power densities of 350 W/cm^2 and hotspot densities up to at least 2.1 kW/cm^2 at reasonable pressure drops using R1234ze as the coolant.

A special TTV module which includes an optical port to allow for viewing two-phase flow boiling phenomena utilizing a high-speed camera to observe a portion of one TTV quadrant is shown in Figure 11. In this unique TTV module, the process step of silicon manifold die attachment to the thermal die with solder, as previously described, was replaced with a glass manifold die attachment to the thermal die with the PKHC™ phenoxy thermoplastic. The copper manifold was modified to include a viewing port through a portion of this glass. The dotted red outline is where the operational TTV die is located. The blue area is where the window for optical examination is located.

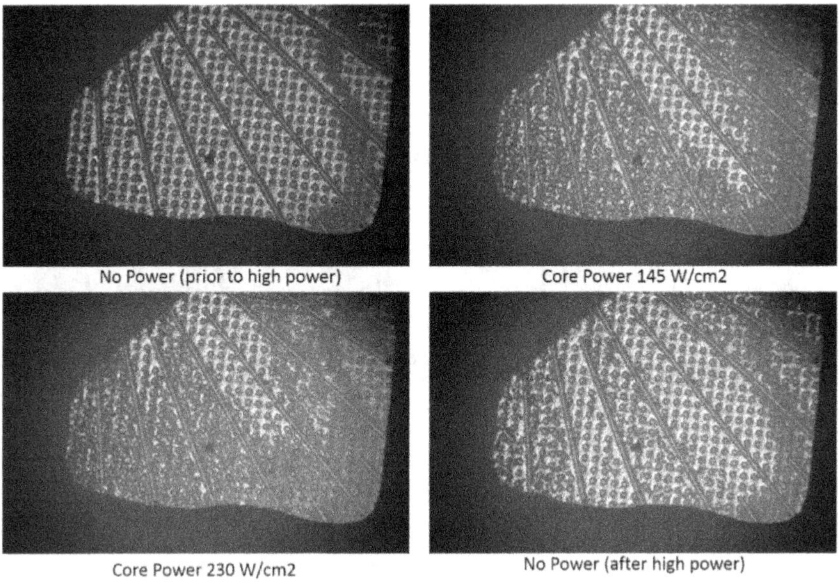

Figure 12. Snapshot images of the entire view window at varying core powers.

Figure 12 shows the full-window images of the bubble field. On the upper left is the pattern observed initially without power applied. Bubbles are seen pretty much only in the left edge segment, where there is orifice-induced flashing. In the upper right is the pattern observed with 145 W/cm^2 of core power applied. The central area of the image corresponds to the location of the unpowered notch in one half of a pair of core heaters. The lower-right corner of the image is near the location of the sensors 14, 17, 18, and 21 for each of the four quadrants, respectively. In the central segment, without background heating, flashing does not occur until after the flow crosses the core heater edge. Shown in the lower left is the pattern observed with 230 W/cm^2 of core power applied. The flashing "front" moves closer to the center of the chip (upstream) in the central segment. Shown in the lower right is the pattern observed a few minutes after power is removed. Certain bubbles remain attached to the pins in the central and neighboring segments, serving as generation points for adiabatic detaching bubbles for quite some time (hours) after power has been removed. This could be the source of hysteretic behavior during

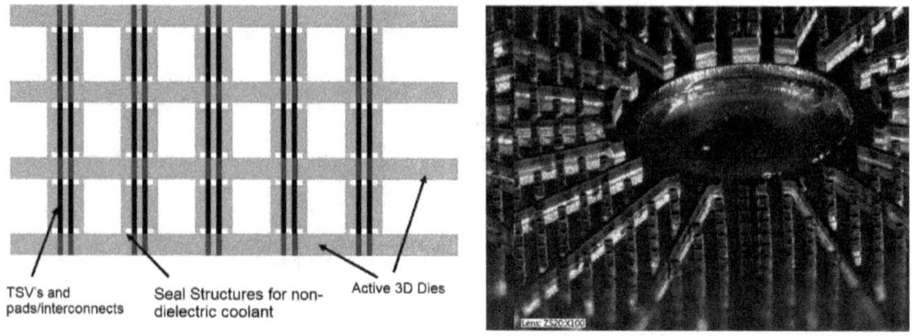

(a) (b)

Figure 13. (a) Schematic diagram of stackable pin and die structures. (b) Pins and guiding walls fabricated on a glass handler and bonded to a die [23].
Source: Copyright (2018) IEEE.

the application and removal of power. Additional flow visualization results can be found in Ref. [22].

To apply the experimental results to the thermal management of 3D chip stacks, which require dies to be thinned below a thickness of 50 μm for the inclusion of TSVs, a new method of fabricating channels and TSV structures for fluid flow between the dies in the stack was developed. This process includes forming pins for TSVs and wall structures on a glass handler wafer and then transferring and bonding the structures to the backside of a die, as shown in Figure 13.

This process provides a path forward for scalable intra-chip embedded cooling. Further details of this process can be found in Ref. [23].

2.3. *TTV modeling*

Due to its symmetry, only a quarter of the TTV was modeled. Figure 14 shows the different views of the TTV's radial quadrant geometry that was modeled along with the boundary conditions applied to the model. Table 3 summarizes the boundary conditions applied in a couple of exemplary cases. The geometry modeled consists of three main simulation domains: the solid (silicon) domain, the fluid domain (from the inlet manifold to just after the orifice), and five radial channels (S1, S2, S3, S4, and S5) modeled as porous domains for reduced-physics modeling [7]. The solid domains consist

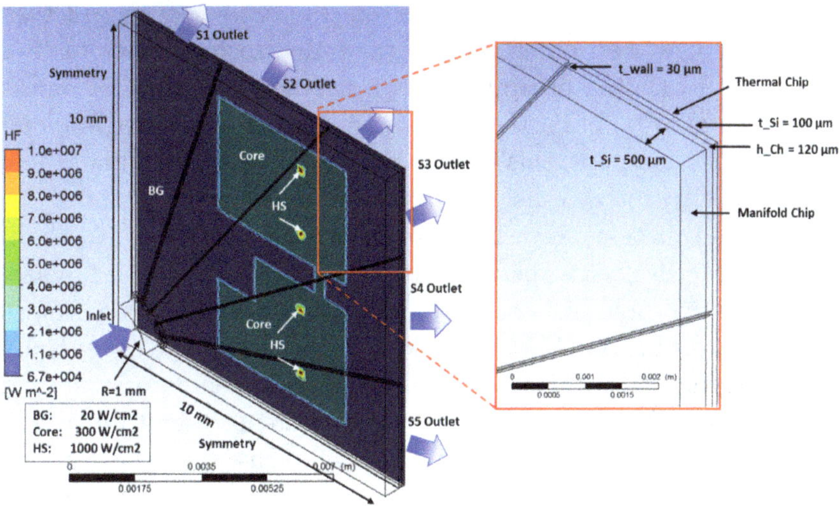

Figure 14. Simulation model for one quadrant of the TTV showing the geometrical dimensions and a representative input power map [7].
Source: Copyright (2017) IEEE.

Table 3. Boundary conditions applied to the TTV simulation.

	Case 1	Case 2
Coolant	R1234ze	R1234ze
Mass flow rate	15 kg/h (full TTV); 3.75 (1 Qtr)	20 kg/h (full TTV); 5 kg/h (1 Qtr)
Inlet temperature	21.4 [°C]	21.2 [°C]
Outlet pressure	448.5 kPa	448.5 kPa
Power map	BG = 20, Core = 300, HS = 1,000 W/cm^2	BG = 20, Core = 300, HS = 1,000 W/cm^2
Total power	113 W (1 Qtr)	113 W(1 Qtr)

of the TTV thermal chip, which has a total thickness of 220 μm, with a 100 μm thick base and 120 μm high channel walls, and the TTV manifold chip, which has a total thickness of 500 μm [7].

By design, the pressure drop associated with inlet orifices is due to the area contraction ratio, which is strongly dependent on the width of the orifices. Additionally, depending on the operating conditions, the two-phase pressure drop in the radial sectors can be comparable to the pressure drop across the orifices, which in turn affects the flow distribution among the radial sectors. Therefore, in order to capture the pressure losses from the inlet manifold across the orifice and to obtain the flow distribution among the five radial channels per quadrant, this fluid domain was modeled in detail using the continuity, momentum (all three directions), energy, and $k - \varepsilon$ turbulence equations for a single-phase flow [7]. In the radial channels, a reduced-physics thermal modeling approach was used, where the silicon micropin fins and the fluidic regions are approximated as a porous domain [7]. A detailed description of the governing equations and their implementation in a commercial computational fluid dynamics (CFD) solver can be found in reference literature [7, 14]. During the simulation, information, such as mass flow rate through each orifice, coolant temperature, and wall heat flux, are transferred as input from the detailed simulation to the reduced-physics simulation. Next, information such as pressure drop across the radial channels, wall heat transfer coefficient profile, and local coolant temperature are computed by the reduced-physics simulation and transferred as input to the detailed simulation. This entire iterative simulation process, along with appropriate convergence monitors, were integrated within a commercial CFD software using custom scripts [7].

Table 4 summaries the flow distribution among the five radial channels for the simulated non-uniform power map cooled using the two different mass flow rates of 15 and 20 kg/h. The table also shows the orifice widths for each radial channel. Figures 15 and 16 show the thermal and hydrodynamic predictions for the 15 and 20 kg/h cases, respectively. Figures 15(a) and 16(a) show the velocity streamlines in the inlet manifold domain, i.e., from the entrance to the exit of the orifices for the two cases. As designed, the radial channel S3 receives the highest flow in both cases.

Figures 15(b) and 16(b) show the pressure contours at the plane passing through the middle of the radial channels. The pressure drop across each radial channel is different, with the highest pressure drop occurring across the radial channel S3. Since the overall pressure drop across all the radial channels was the same, the difference in

Table 4. Inlet orifices width and flow distribution in each radial channel for the two different TTV simulations.

Channel #	Orifice width [rm]	Case 1: Predicted flow [%]	Case 2: Predicted flow [%]
S1	80	13.2	12.6
S2	170	22.0	21.6
S3	264	25.5	26.4
S4	194	21.3	21.9
S4	126	18.0	17.5

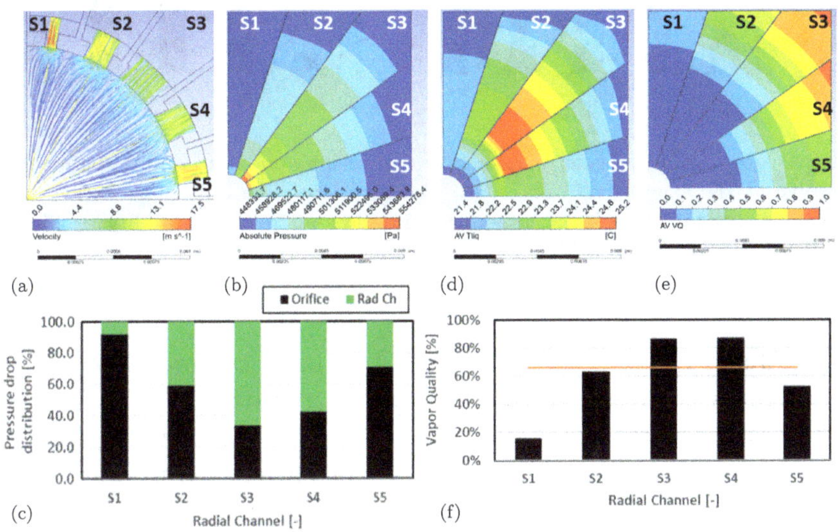

Figure 15. (a) Velocity streamlines in the inlet manifold section [7]. (b) Pressure profile at the channel mid-plane [7]. (c) Pressure-drop distribution for each radial channel [7]. (d) Local coolant temperature profile [7]. (e) Vapor quality profile at the plane passing the middle of the radial channels. (f) Average vapor quality at the exit of each radial channel for the 15 kg/h case.

Source: (a–d): Copyright (2017) IEEE.

pressure came from the pressure drop across the orifices. Note that there is a strong coupling between the pressure drop across the orifices, the two-phase pressure drop across the radial channels, and the flow distribution among the radial channels [7]. The total pressure drop across the inlet and outlet ports of the TTV was measured to

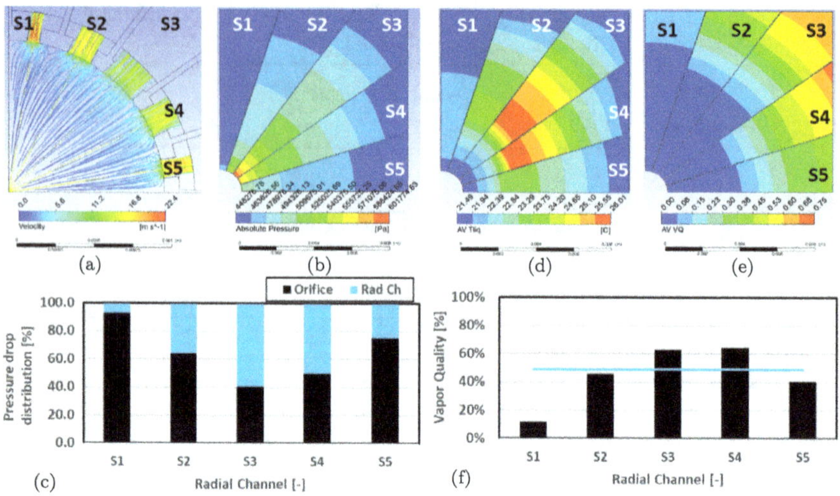

Figure 16. (a) Velocity streamlines in the inlet manifold section. (b) Pressure profile at the channel mid-plane. (c) Pressure-drop distribution for each radial channel. (d) Local coolant temperature profile and (e) vapor quality profile at the plane passing the middle of the radial channels. (f) Average vapor quality at the exit of each radial channel for the 20 kg/h case.

be 179 and 271 kPa for the 15 and 20 kg/h cases, respectively. The total pressure drop from the inlet to the outlet plenum predicted by the model was 158 and 255 kPa, respectively. These predictions are lower than the data because they do not include the pressure drop from the outlet plenum to the outlet port.

Figures 15(c) and 16(c) show the pressure drop breakdown for each of the five radial channels in both simulation cases, respectively. The pressure drop across the orifices (black bars) is single phase, whereas the pressure drop in the radial channels (green/blue bars) is a mix of single-phase and two-phase pressure drops. The pressure drop across the orifices was observed to vary from ~34% (S3) to ~93% (S1) of the total pressure drop across the TTV. Such high pressure drops across the orifices (especially for S1 and S5) helped balance the flow passively among the five radial channels. Figures 15(d) and 16(d) show the liquid temperature profile in the radial channels. Since the liquid at the inlet of the radial channels is sub-cooled, the coolant temperature first increases along the radial direction until the flow conditions reach local saturation conditions, and then, the coolant temperature decreases with the local fluid pressure [7]. Note that in some high-flow-rate cases, the pressure drop across the orifices could be significantly high to cause the flow conditions to reach local saturation conditions right across the orifice, forcing the coolant to flash (instantaneous boiling) at the orifice exit, which causes the coolant temperature inside the radial channels to drop along the flow direction [7].

Figures 15(e) and 16(e) show the vapor quality (= mass of vapor/total mass of coolant) profile in the radial channels. A value of 0 for the vapor quality corresponds to sub-cooled coolant. Owing to the highly non-uniform nature of the chip power map and a preliminary orifice design, a large variability in the vapor quality profile among the five radial channels was observed. Figures 15(f) and 16(f) show the average vapor quality at the exit of each radial channel relative to the TTV exit vapor quality for the two simulation cases. A large variability in exit vapor quality among the radial channels can result in the onset of dry-out conditions in one or more radial channels at relatively low TTV exit vapor qualities (~50–60%). In the subsequent radial channel designs, e.g., for the ECM module, both single-phase pressure drop across the orifices and two-phase pressure drop across the radial channels were considered, resulting in a tighter distribution of exit vapor quality among the radial channels [10].

Figure 17 shows the junction temperature prediction for the TTV with a non-uniform power map, where the background heat flux is 20 W/cm^2, core heat flux is 300 W/cm^2, and the hotspot heat flux is 1,000 W/cm^2. This core power density is ~3× the core power densities of today's microprocessors [7]. The TTV was cooled by R1234ze flowing at a rate of 15 kg/h in case 1 (Figure 17(a)) and 20 kg/h in case 2 (Figure 17(b)). The yellow cross-hairs in the figure represent the locations of the RTD sensors relative to the power map in the simulated geometry. Figure 17(c) and 17(d) show a comparison of the junction temperature predictions at these 11 different sensor

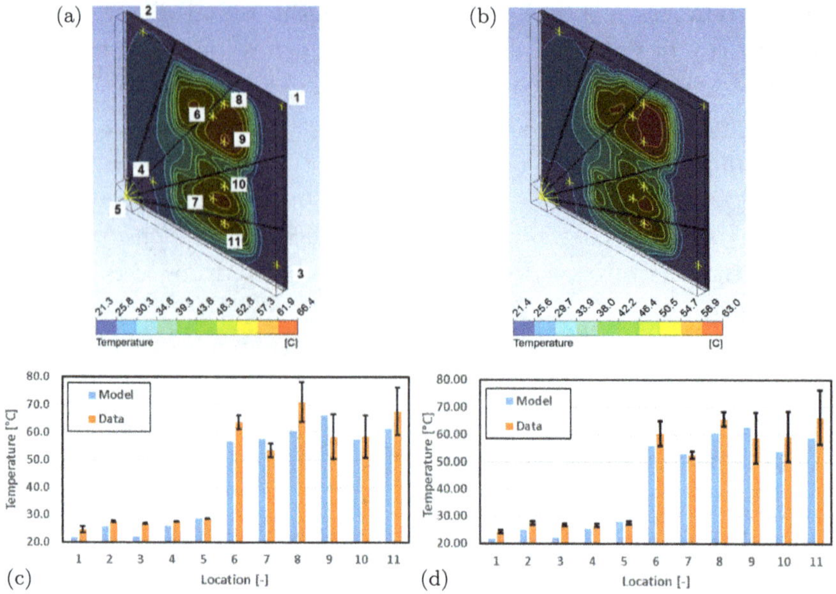

Figure 17. Junction temperature prediction for (a) case 1 [7] and (b) case 2. Cross-hairs show the thermal sensor locations used for model validation. Sensor-by-sensor comparison of junction temperature prediction against experimental data for (c) case 1 [7] (d) case 2.

Source: (a) and (c): Copyright (2017) IEEE.

locations with the corresponding experimental data (averaged across the four quadrants) for both cases. The error bars show the maximum of experimental uncertainty and standard deviation (σ) in the measured value across the four quadrants. A small variability indicates even flow distribution and uniform power and thermal behavior across the four TTV quadrants, whereas a large variability indicates either an uneven flow distribution across quadrants, partial bypass within a quadrant, or variation in the electrical insulation layer thickness near the heater, which affects both the power density and reported equivalent junction temperature [7]. Overall, the reduced-physics model predicted a maximum junction temperature of 66.3°C (data: 70.9°C) and an average hotspot temperature of 61.5°C (data: 63.9°C) for the 15 kg/h case. For the 20 kg/h case, the model predicted a maximum junction temperature of 63°C (data: 66.4°C) and an average hotspot temperature of 59.1°C (data: 62.6°C).

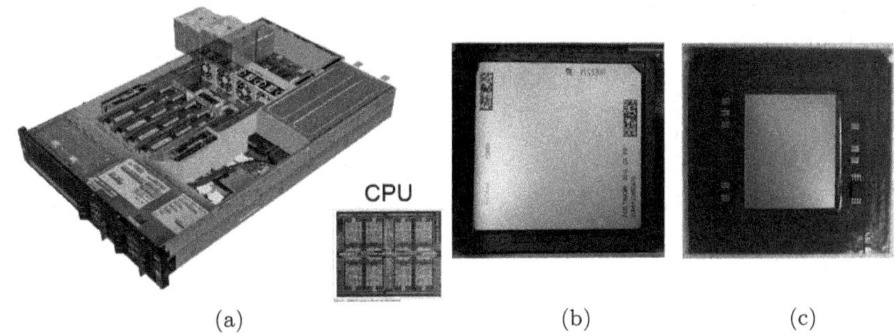

Figure 18. (a) IBM POWER® 730 server. (b) Microprocessor module. (c) Microprocessor module with the lid removed.

3. Embedded Two-Phase Liquid Cooled Microprocessor

3.1. *Design/fabrication*

To demonstrate the compatibility of two-phase chip-embedded cooling with the electronic circuitry of a functional device, a commercial microprocessor was modified to include embedded channels. An IBM POWER® server was chosen, which had two sockets each with an eight-core microprocessor which were originally air cooled, as shown in Figure 18(a). The processor module shown in Figure 18(b) comprises a single die on an organic substrate, which is mounted onto a land grid array socket for easy removability. Figure 18(c) shows the processor with the top lid removed.

The demonstration of chip-embedded cooling was initiated with a detailed characterization of the air-cooled microprocessor modules, which would later be modified for two-phase cooling. This was done by placing the servers in a controlled temperature environment, executing preselected workloads, and measuring the corresponding processor power and temperature. The processor temperature was measured using 40 digital thermal sensors (DTSs) available on the processor die. The processor power was characterized over a range, from roughly 90 W to nearly 250 W, by changing the processor frequency and operating temperature.

The processor modules were then modified for embedded two-phase liquid cooling. The modules were first disassembled by cutting through the seal band of the organic substrate and removing the module lid. The modules were cleaned to remove any thermal interface material and seal band and placed in a deep reactive-ion etching (DRIE) tool, which etched radial patterns into the backside of the die to a depth of 120 μm. An image of the microprocessor die on the organic substrate after DRIE is shown in Figure 19. The fluid network consists of a fluid cavity populated with micropins and guiding walls extending radially from the central inlet region. An optical microscopy image of the same fluid network is shown in Figure 19(b). Figure 19(c)–(e) show scanning electron microscopy (SEM) images of the inlet region, inlet orifices, micropin fins, and wall structures. To form the top of the channels, a manifold die was bonded to the etched silicon die. As the microprocessor die on the organic module was curved due to a mismatch in the coefficient of thermal expansion (CTE) between the silicon die (4 ppm) and the organic substrate (17 ppm), a flexible manifold die of silicon CTE-matched borosilicate glass of 200 μm thickness was chosen. To form the bond, a thin layer of approximately 3 μm of thermoplastic was deposited onto the glass manifold die. The manifold die was aligned to the silicon die and placed in a vacuum bag to provide a uniform force on the manifold die to match the curvature of the module. The bag was then placed in an oven to heat the thermoplastic to form the bond and then slowly cooled to room temperature, as illustrated in Figure 20.

Next, a lid is attached to the microprocessor module with inlet and outlet ports, as illustrated in Figure 21. To provide a leak-tight seal, a structural adhesive was dispensed between the periphery of the lid and the organic substrate. To form the inlet seal, either a structural adhesive or a neoprene gasket was placed between the glass manifold and the lid. An optical image of a module with a cutaway is shown in Figure 21(b). Further process details can be found in Ref. [24].

3.2. *Experimental results*

The coolant chosen for this study was R1234ze. As shown schematically in Figure 22, the coolant is pumped into the module as a single-phase liquid. The flow rate is determined by controlling the

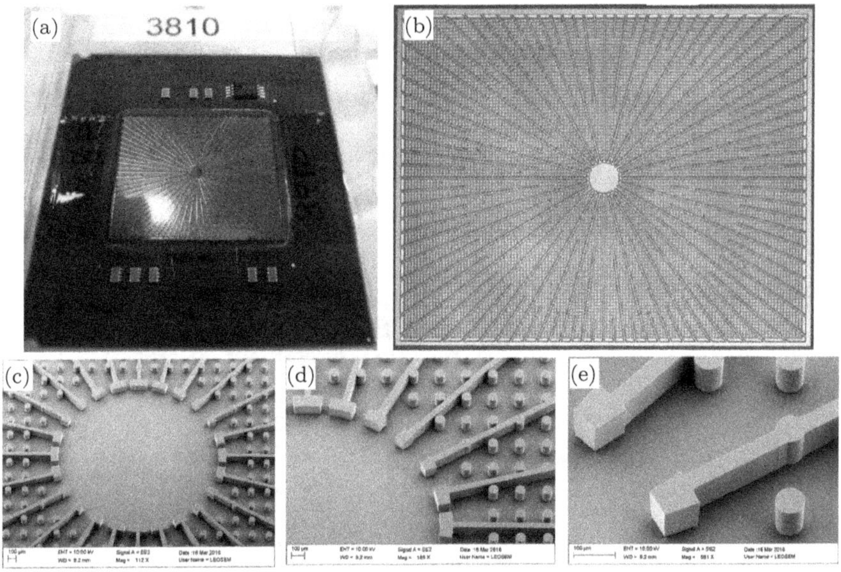

Figure 19. Channels etched into microprocessor [24]. © 2017 IEEE.

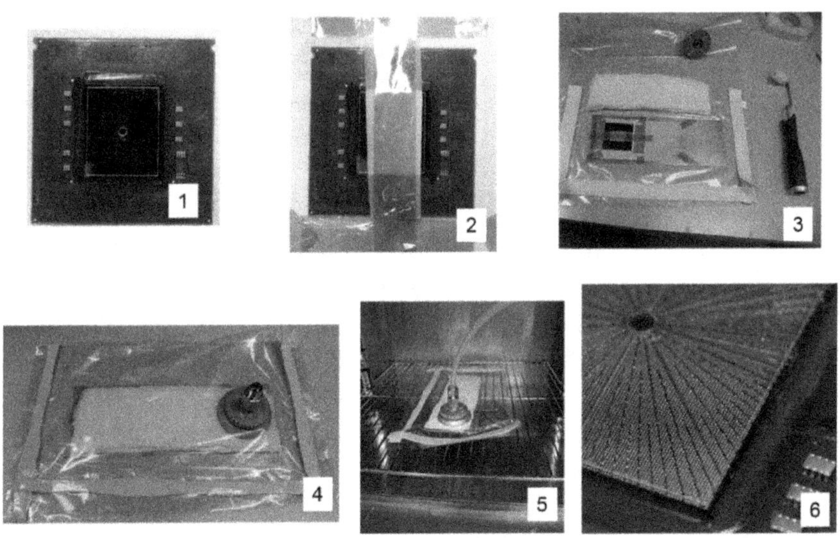

Figure 20. Process of attaching the manifold die.

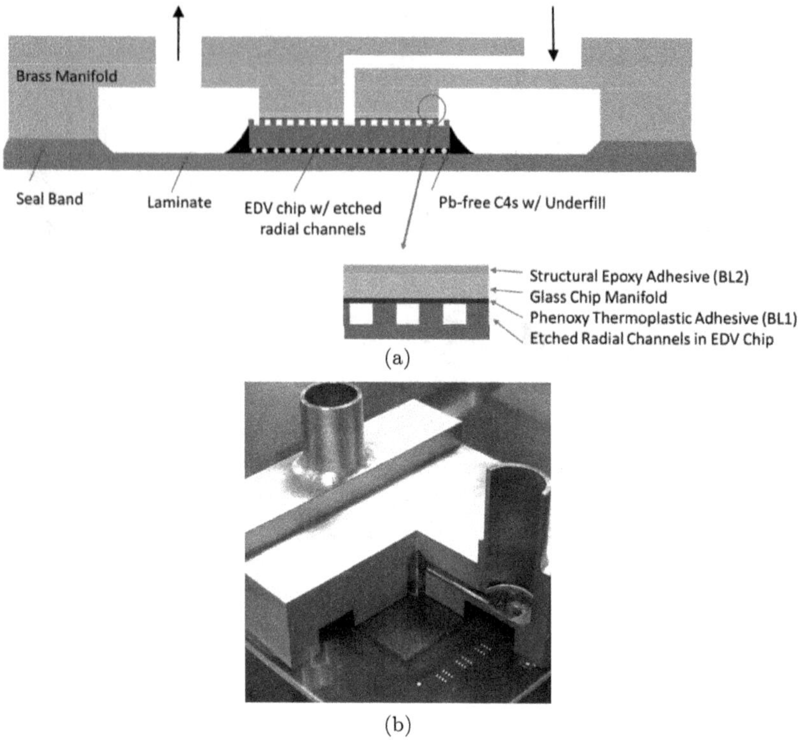

Figure 21. (a) Cross-sectional schematic diagram and (b) cut-away section of an assembled ECM module [8].
Source: Copyright (2017) IEEE.

pump speed. The fluid flow is distributed in the module by the 24 inlet orifices into their corresponding radial expanding channels, which were symmetrically designed — six per quadrant. The cooling structures, including the inlet, orifices, pins, and radial expanding channels, were optimized using both the detailed full-physics and reduced-physics models summarized in Section 1.3. The coolant absorbs heat as it flows through the channels and undergoes a phase change from liquid to vapor. The liquid/vapor mixture exits the microprocessor module on all four sides of the die into the plenum formed between the copper lid and the substrate, where it exits to the module outlet. The downstream liquid/vapor mixture enters the condenser, which is cooled by a secondary loop to extract heat from

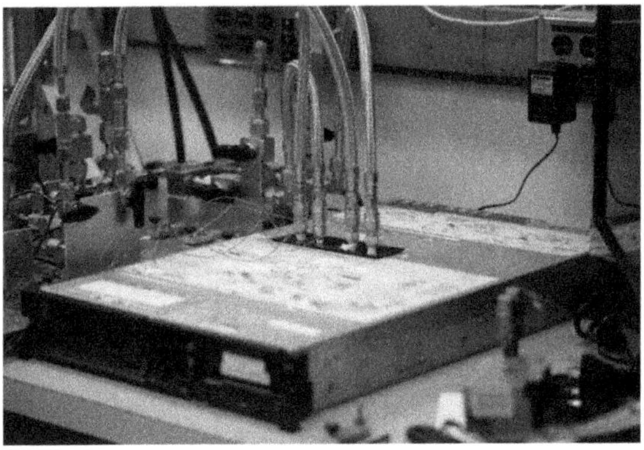

(a)

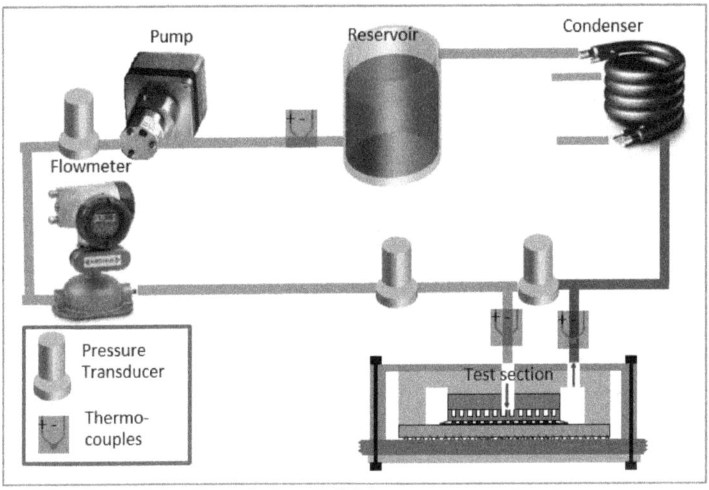

(b)

Figure 22. (a) ECM modules installed in a server [25]. (b) Schematic diagram of the cooling loop [26].

Source: (a) Copyright (2017) IEEE. (b) Copyright (2016) IEEE.

the mixture and condense the vapor back into liquid. The single-phase coolant exits the condenser and enters the reservoir, from where it is fed back into the microprocessor module through the pump.

The initial characterization of the performance of the ECM modules was obtained as a function of the coolant mass flow rate using the 40 on-chip DTSs. As shown in Figure 23, the results were not

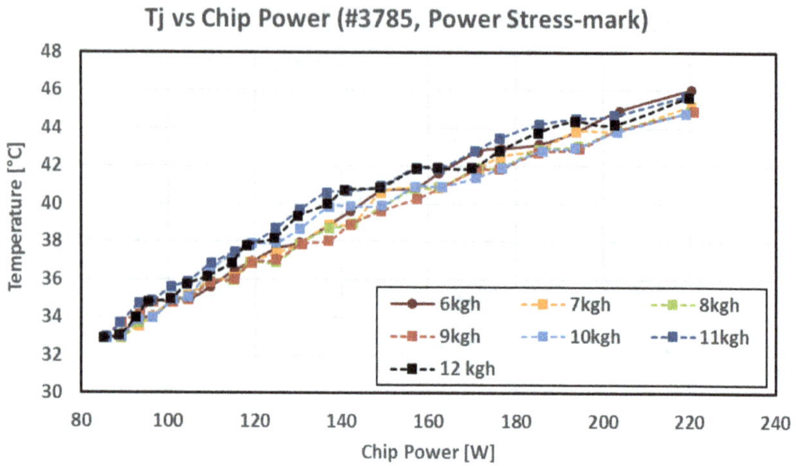

Figure 23. Chip temperature vs. chip power at different coolant mass flow rates [25].

Source: Copyright (2017) IEEE.

strongly dependent on the flow rate, which suggests that the system can operate over a wide range.

Figure 24 shows the pressure-drop study for the ECM module at full power, idle power, and power off over a range of mass flow rates. The pressure drop variation between the three power states was relatively small. Thus, a coolant supply system utilizing a simple "on/off" pump control without flow or pressure control would produce only small changes in flow and/or pressure drop with changes in operating power level. This suggests that modules may operate in parallel without any active flow balancing between modules. The pressure drop was stable in time and did not show evidence of flow instability.

A detailed comparison of the air-cooled and chip-embedded two-phase liquid cooled microprocessors was performed. The coolant inlet temperature in both cases was 25°C. The dielectric coolant mass flow rate was 9 kg/h, which produced a pressure drop of 75 kPa (~11 psi) across the module. The average core temperatures of the air-cooled modules and those of the modules after modification for chip-embedded liquid cooling are shown in Figure 25, which shows a significant improvement in the core temperature for the

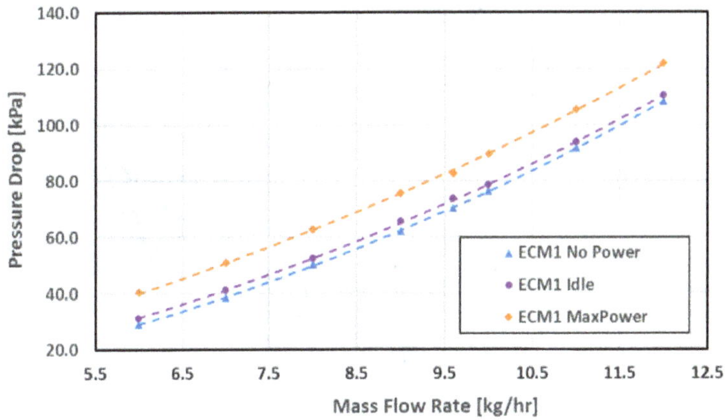

Figure 24. Pressure drop vs. mass flow rate for ECM module at three different power levels [25].
Source: Copyright (2017) IEEE.

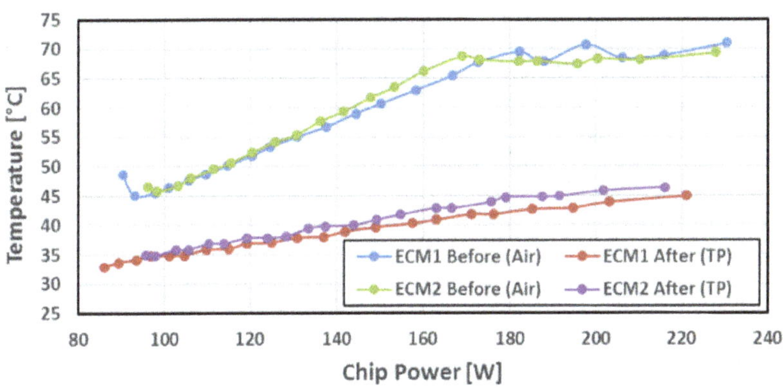

Figure 25. Average core temperature vs. chip power for air-cooled and 9 kg/h embedded-cooled cases for two modules [25].
Source: Copyright (2017) IEEE.

chip-embedded two-phase liquid cooling, achieving roughly 25°C at the highest power. It is important to observe that the core temperature of the air-cooled modules plateaus at roughly 70°C due to the ramping up of the system fans' speeds (∼65%) to prevent overheating [26]. The power and thermal information of the power systems, including the microprocessor, memory, and fan power/speed, were

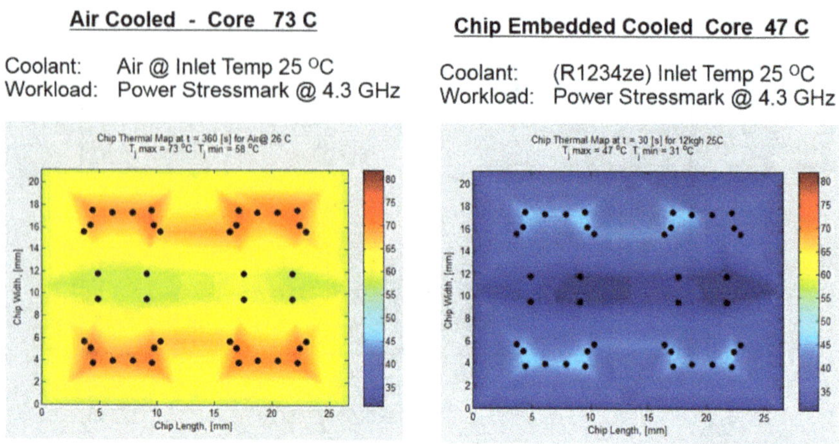

Figure 26. Air cooled vs. chip-embedded two-phase liquid cooled microprocessor.

measured during the execution of the chosen workloads. At the highest power operation (4.3 GHz) the reduced operating temperature resulted in an over 10 W decrease in the power consumed by the microprocessor, along with a significant (~15 W) reduction in fan power. The two-phase cooling loop pump utilized about 1 W. An interpolated thermal map of the air-cooled and chip-embedded two-phase liquid-cooled microprocessors is shown in Figure 26 for the maximum power case of 4.3 GHz.

3.3. *ECM modeling*

The model of the ECM consists of two major components: (i) an electrical model of the ECM to simulate the spatially varying temperature-dependent heat dissipation from the chip at different operating frequencies [8], and (ii) a conjugate thermal model to simulate the flow and heat transfer behavior [8, 10].

Figure 27 shows a quarter-symmetry model of the ECM module along with the boundary conditions (including an example of a converged power map) that were applied to the model. The geometry modeled consists of six radial segments (S1, S2, S3, S4, S5, and S6) and two solid domains: silicon chip and glass cover [8]. Table 5 summarizes the boundary conditions applied. Note that the electrical and thermal behavior of the ECM module are interdependent, requiring

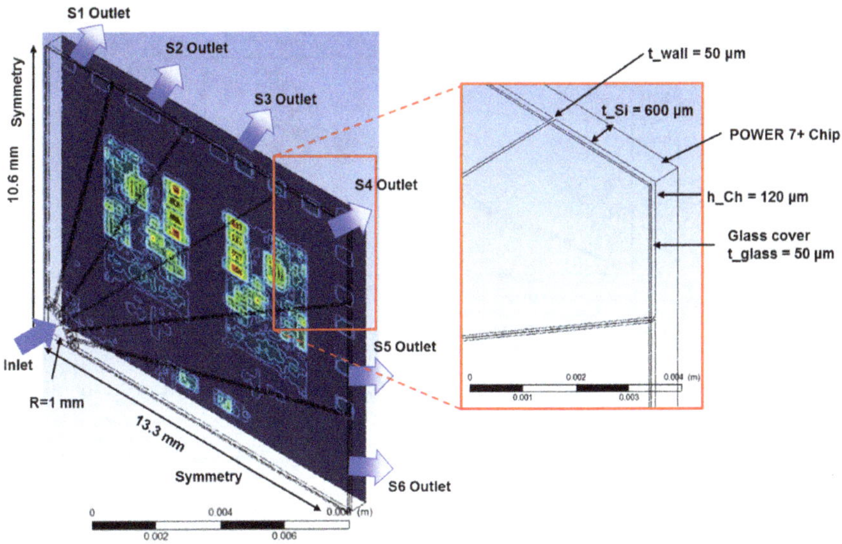

Figure 27. Simulation model for one quadrant of the ECM showing the geometrical dimensions and an example of the converged power map [8].
Source: Copyright (2017) IEEE.

Table 5. Boundary conditions applied to the TTV simulation [8].

Coolant	R1234ze
Mass flow rate	12 kg/h (full ECM); 3 kg/h (1 Qtr)
Inlet temperature	24.8 [°C]
Outlet pressure	495.5 kPa (P_{sat} 24.8 [°C])
Power map/total power	Simulation dependent

Source: Copyright (2017) IEEE.

the power map and temperature field in the simulation to converge. As a result, the converged chip power map is also an output of the simulation.

Figure 28 shows the ECM chip power prediction and comparison against data for five different ECM modules collected over a wide range of operating frequencies (2.2–4.3 GHz) and coolant mass flow rates ($6 - 12 \pm 0.06$ kg/h). For all the validation cases, the model predictions were within $\pm 10\%$, with higher errors observed in the lower chip power cases. At higher chip powers, the model predictions

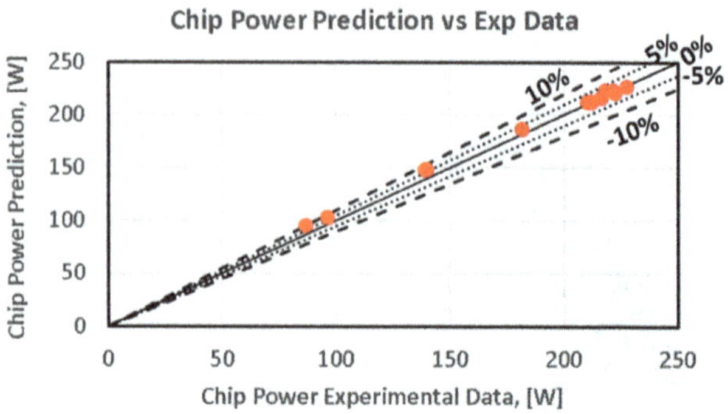

Figure 28. Chip power prediction and its comparison with ECM data for different validation cases [8].
Source: Copyright (2017) IEEE.

were within ±5% of the data. Figure 29 shows the prediction of the average junction temperature rise above the coolant inlet temperature and its comparison with the ECM data for the same validation cases. The blue error bars represent the variability in the average junction temperature across quadrants (i.e., the minimum and maximum average quadrant temperature values). Larger error bars signify higher variability across quadrants, which could be possibly due to fabrication variations resulting in non-uniformity in quadrant power dissipation or flow imbalances between quadrants [8]. Smaller error bars signify good uniformity across all four quadrants. For all the cases, the model predictions were within ±30%, with higher errors observed for the cases where non-uniformity across quadrants was present, while errors less than ±10% were observed for well-behaved ECM modules. The green arrow in Figure 29 highlights the ΔT_j comparison for a well-behaved case, for which a detailed sensor-by-sensor temperature comparison is shown in Figure 30 and the thermal and hydrodynamic predictions are shown in Figure 31.

Figure 30 shows the comparison of junction temperature predictions against the processor's DTS measurements. Figure 30(a) shows the location of the thermal sensors P1–P10 relative to the converged power map for the ECM quadrant [8]. Figure 30(b) shows the comparison of the model predictions with the experimental data

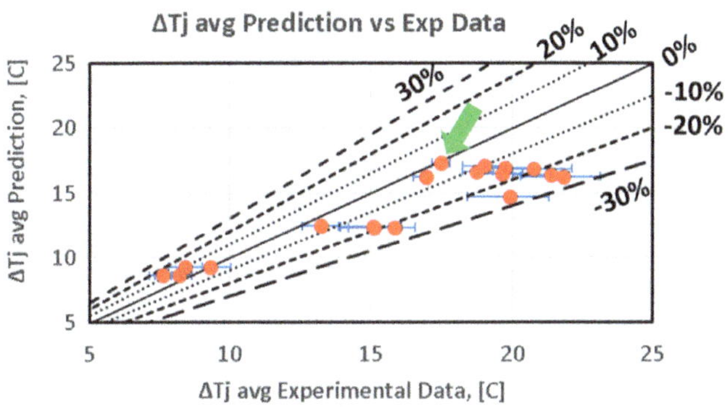

Figure 29. Average T_j relative to coolant inlet temperature and comparison with ECM data for different validation cases. Error bars represent the variability across quadrants [8].
Source: Copyright (2017) IEEE.

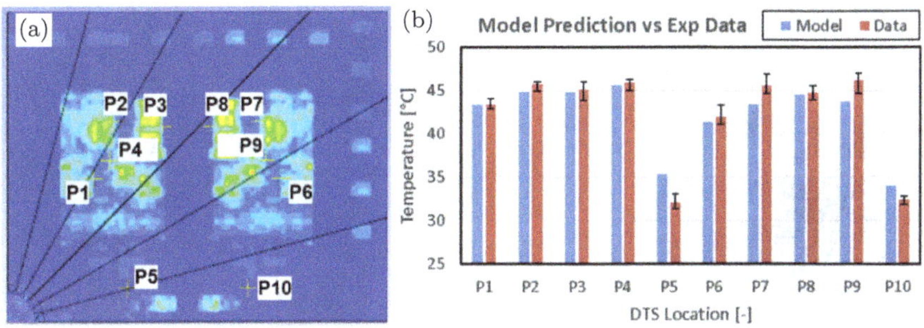

Figure 30. (a) DTS locations (P1–P10) overlaid on converged power map and (b) T_j prediction and comparison with ECM data [8].
Source: Copyright (2017) IEEE.

(averaged across all four quadrants) for each of these 10 sensors for one exemplary case, in which the ECM was operated at 4.3 GHz and operated under the maximum-power workload. The maximum-power workload is a static workload and provides a stable power map [8]. The ECM was cooled by R1234ze pumped at a rate of 12 kg/h (174.5 mL/min). The model prediction closely matches the ECM experimental data.

Figure 31(a) shows the velocity streamlines in the inlet manifold domain, i.e., from the inlet to just after the orifices. Figure 31(b)

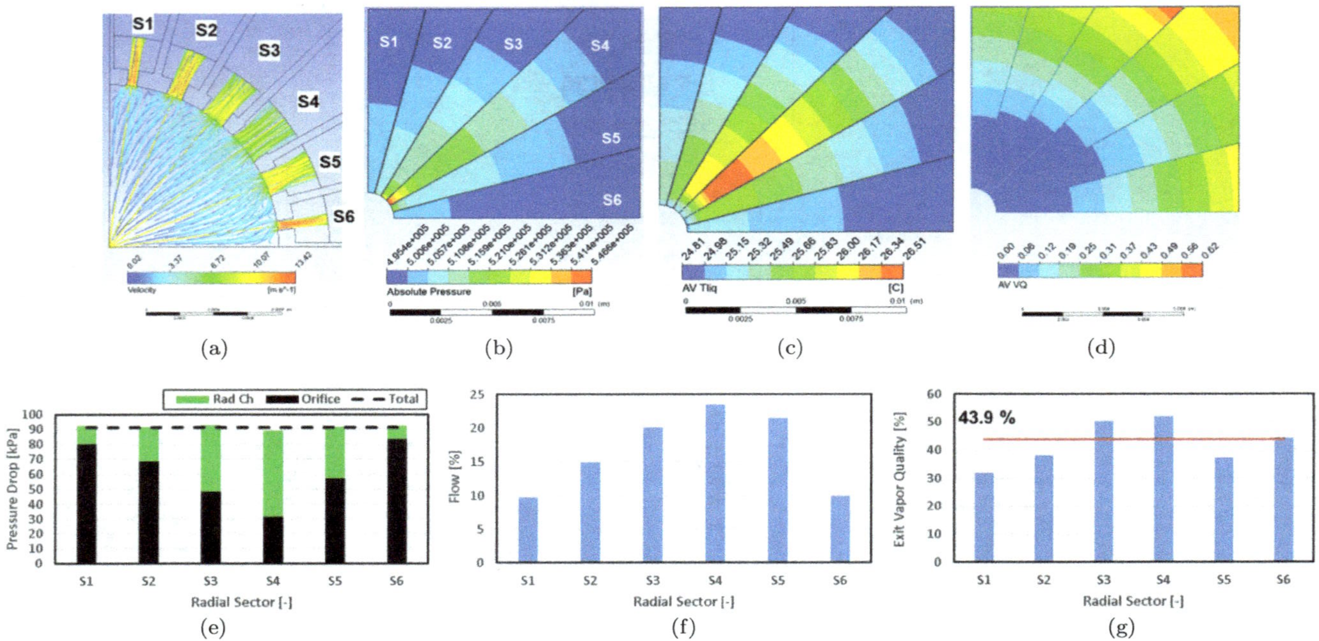

Figure 31. (a) Velocity streamlines in the inlet manifold section. (b) Pressure profile, (c) local coolant temperature profile, and (d) local vapor quality profile at mid-plane of the radial channels. (e) Pressure drop breakdown, (f) flow rate distribution across, and (g) exit vapor quality for all six radial channels [8].

Source: Copyright (2017) IEEE.

shows the pressure contours at the plane passing through the middle of the radial channels. The contribution to the overall pressure drop from each radial channel is different, with the highest pressure drop occurring across the channel S4, as expected, and a difference in pressure occurring across the orifices. Figure 31(c) and 31(d) show the liquid temperature and vapor quality profiles in the radial channels, respectively. Since the liquid at the inlet of the radial channels is sub-cooled, the coolant temperature first increases along the radial direction until the flow conditions reach local saturation conditions, and then, the coolant temperature decreases with the local fluid pressure and the vapor quality increases along the flow direction. The vapor quality in the sub-cooled region is reported as 0. Owing to a good channel design in the ECM, the rate of change of vapor quality along the radial direction is very similar across all six radial channels.

The total pressure drop from the inlet to the outlet plenum of the ECM was predicted to be 91.1 kPa (13.2 psi). Figure 31(e) shows that the pressure drop across the orifices ranges from ~35% (S4) to ~90% (S1 and S6) of the total pressure drop across the ECM. Such high pressure drops across the orifices (especially for S1 and S6) helped balance the flow in all six radial channels under different chip operating conditions. Figure 31(f) shows the flow rate distribution across the six radial channels, and Figure 31(g) shows the average vapor quality at the exit of each of the six radial channels. Owing to a good channel design, all six radial channels received enough coolant flow to achieve an exit vapor quality that is within ±10% of the ECM exit vapor quality. This tight distribution of exit vapor quality, which is far better than the distribution observed for the TTV, highlights the robustness of the radial microchannel design. For this simulation case, the average vapor quality at the ECM outlet was observed to be ~44%. Such a tight distribution of the exit vapor quality across the radial channels was also predicted for other flow rates and operating frequencies.

A detailed simulation of the ECM module was also performed. Table 6 summarizes the boundary conditions for the full-physics simulations of the ECM radial quadrant. Figure 32 shows the simulation

Table 6. Boundary conditions applied in the ECM simulation [10]. © 2018 IEEE.

Coolant	R1234ze
Mass flow rate	12 kg/h (Full ECM); kg/h (1 Qtr) S1-S6: 0.29, 0.45, 0.61, 0.71, 0.65, 0.3[kg/h]
Inlet temperature	24.9 [°C]
Outlet pressure	495.5 kPa(P_{sat} 24.8 [°C])
Power map	Converged power map from ECM reduced-physics model
Total power	56.6 [W]

predictions for an ECM module under a maximum-power workload at 4.3 GHz and cooled by R1234ze flowing at a mass flow rate (full ECM) of 12 kg/h (174.5 mL/min). In this full-physics simulation, to reduce simulation complexities, only a spatially varying power map was applied. This power map was not temperature dependent and was obtained as a converged solution from the reduced-physics simulation of the ECM [10]. Since each radial sector has its own inlet, the mass flow rate into each radial sector was applied explicitly. The value of mass flow rates into each radial microchannel was determined from the reduced-physics simulation. R1234ze was used as the coolant. To maintain solution stability, saturation temperature (T_{sat}) was set to a constant value of 26°C and not as a pressure-dependent function.

Figure 32(a) shows the pressure contours at a plane passing through the center of the microchannels. The predictions show that the pressure gradient across each sector is different and that sector S1 has the lowest pressure drop and sector S4 has the highest pressure drop. Figure 32(b) shows the thermal map of the active layer and the location of the thermal sensors P1–P10 relative to the chip temperature map. Figure 32(c) shows a comparison of both the full-physics and reduced-physics model predictions against the data for each sensor location. The average chip temperature (the average of the values at P1–P10) was predicted to be 43.3°C by the full-physics model and 42.1°C by the reduced-physics model, whereas that observed in the experiment was 42.3°C. This average chip temperature is over 25°C lower than that observed for typical air-cooled processors.

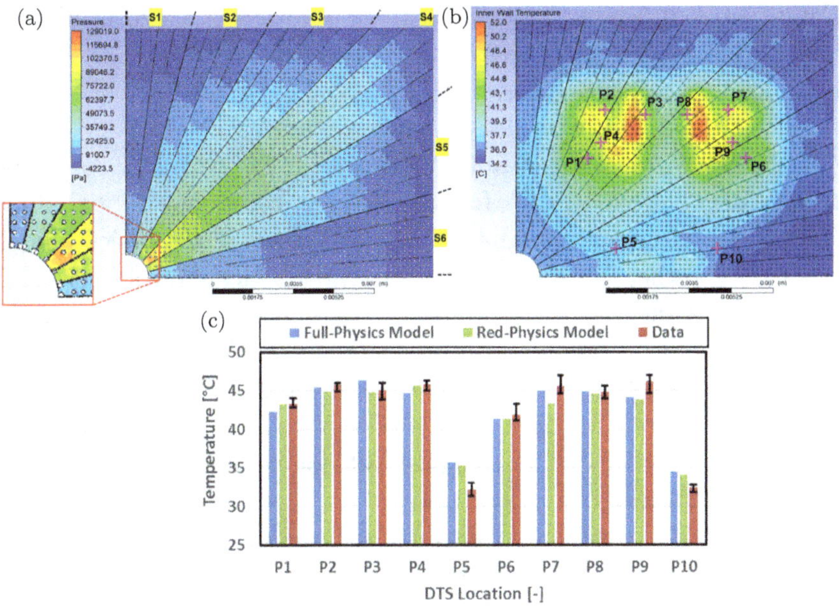

Figure 32. ECM model predictions: (a) pressure contours at the cavity mid-plane; (b) junction temperature contours and approximate locations of chip DTS; (c) comparison of ECM full- and reduced-physics model predictions against ECM experimental data [10].
Source: Copyright (2018) IEEE.

Overall, both the reduced- and full-physics models described here can predict the flow and heat transfer characteristics with very high fidelity. A speedup of four orders of magnitude above that of the full-physics model enables the use of reduced-physics models for parametric design space exploration.

4. Two-Phase System-Level Model

At the system level, a macroscopic model is required to understand the impact of several design choices and system operating parameters, such as the impact of condenser type and size, using large- versus small-diameter tubing, having a reservoir in the loop or not, having two dissimilar components (such as evaporators) in parallel, and heat dissipation and flow rate, on the system's coefficient of performance (COP). The two-phase cooling system (illustrated in

Figure 33) can be simulated using tools reported by Chen *et al.* [27] and Parida *et al.* [28]. The following paragraphs describe the use of one such tool called Sim2Cool [28] for system-level two-phase cooling simulation and design evaluation.

The Sim2Cool tool is built upon an enthalpy/energy balance-based formulation [27–29] for the system components, including the condenser, reservoir, and tubing, and a reduced-physics model (described earlier in this section) for chip-/device-level model. The basic goal of the system-level model is to calculate the macroscopic variables, namely the pressure (p), temperature (T) or enthalpy (h), and vapor quality (x), at any location in the system while considering the mass flow rate to be a known (user-selectable) parameter [27]. The tool uses REFPROP software [30] for computing the various intrinsic properties of the fluid.

The Sim2Cool system-level simulator was developed in Matlab® and has a graphical user interface (GUI), through which a user can interactively change the cooling system parameters (see Figure 33). Using the push-button tabs on the top-left side of the simulator GUI, a user can select a cooling system configuration from the four displayed options. For each system component, user inputs can be put in the corresponding purple-colored boxes, and the simulator-predicted values are presented in the corresponding green boxes. The COP values predicted by the simulator include the indoor and outdoor loop pumping powers [28].

Figure 34 shows the four different cooling system configurations. In configuration 1 (Figure 34(a)), a pump supplies liquid coolant to the device under test (DUT), which could be a TTV or an ECM module. The liquid coolant cools the DUT and exits as a liquid/vapor mix, which is condensed back to liquid in the condenser. In the condenser, the heat from the liquid/vapor mix coolant is transferred to the facility-side coolant loop. The performance of the condenser is determined by the type of condenser, the facility-side coolant and its flow rate, the refrigerant-side coolant and its flow rate and vapor quality, and the amount of heat exchanged across the condenser [28]. The performance of the condenser determines the operating conditions of the entire loop. The condensed liquid from the condenser

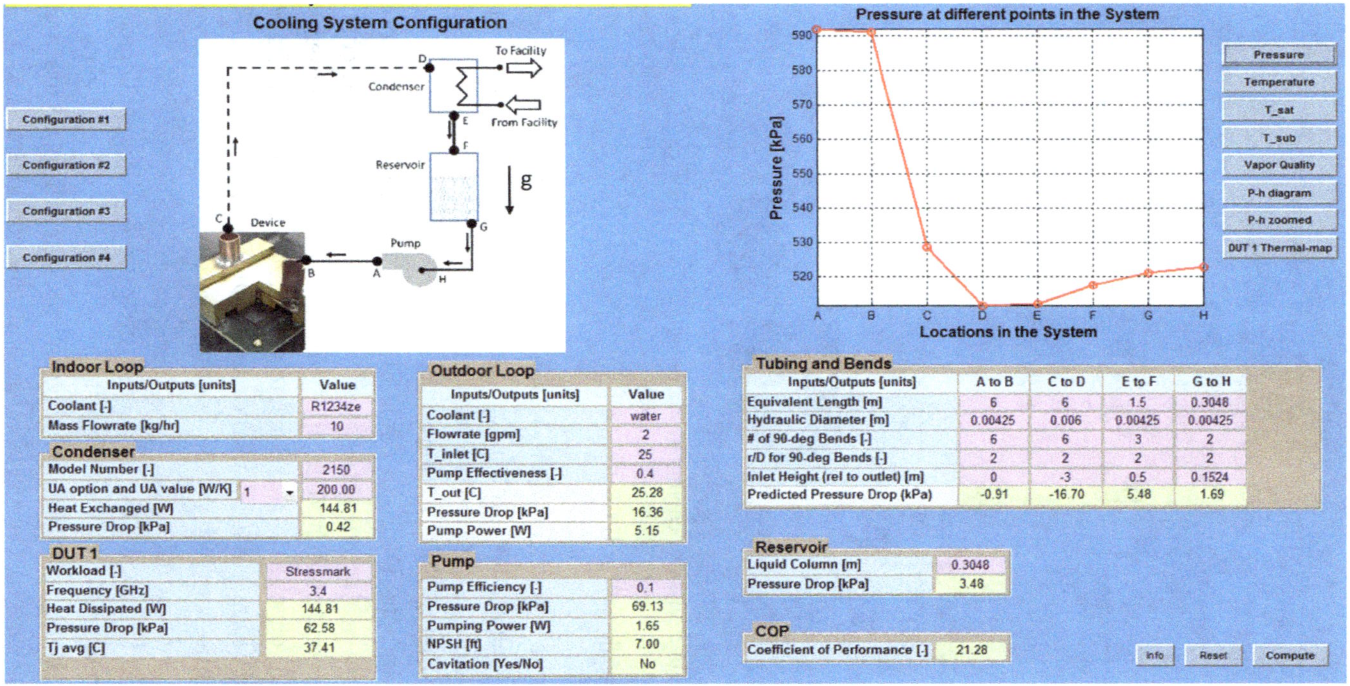

Figure 33. Graphical user interface (GUI) of the Sim2Cool system-level simulator [28].
Source: Copyright (2018) IEEE.

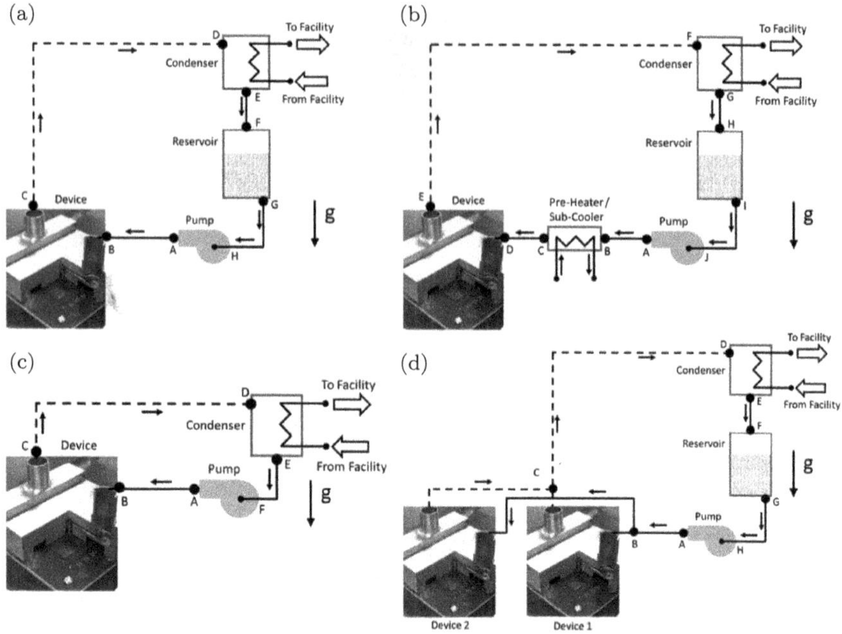

Figure 34. Four different cooling system configurations that can be selected in the Sim2Cool system-level simulator [28].
Source: Copyright (2018) IEEE.

flows into the reservoir, where any uncondensed vapor is separated from the liquid phase, and a certain liquid column is maintained to ensure there is no cavitation occurring in the pump. In configuration 2 (Figure 34(b)), there is a pre-heater/sub-cooler between the pump and the DUT. In configuration 3 (Figure 34(c)), there is neither a pre-heater nor a reservoir. In this case, the liquid column within the condenser and in the tube between the condenser and the pump is enough to avoid cavitation in the pump [28]. In configuration 4 (Figure 34(d)), two or more DUTs can be connected in parallel, while the rest of the cooling system components are the same as in configuration 1.

Figure 35(a) shows the expected pressure at different locations in the cooling system. The biggest pressure drop occurs across the DUT (i.e., B–C) and the next biggest pressure drop occurs across the tubing between the evaporator and the condenser (i.e., C–D). The pressure drop across the section A–B tubing is negligible, as

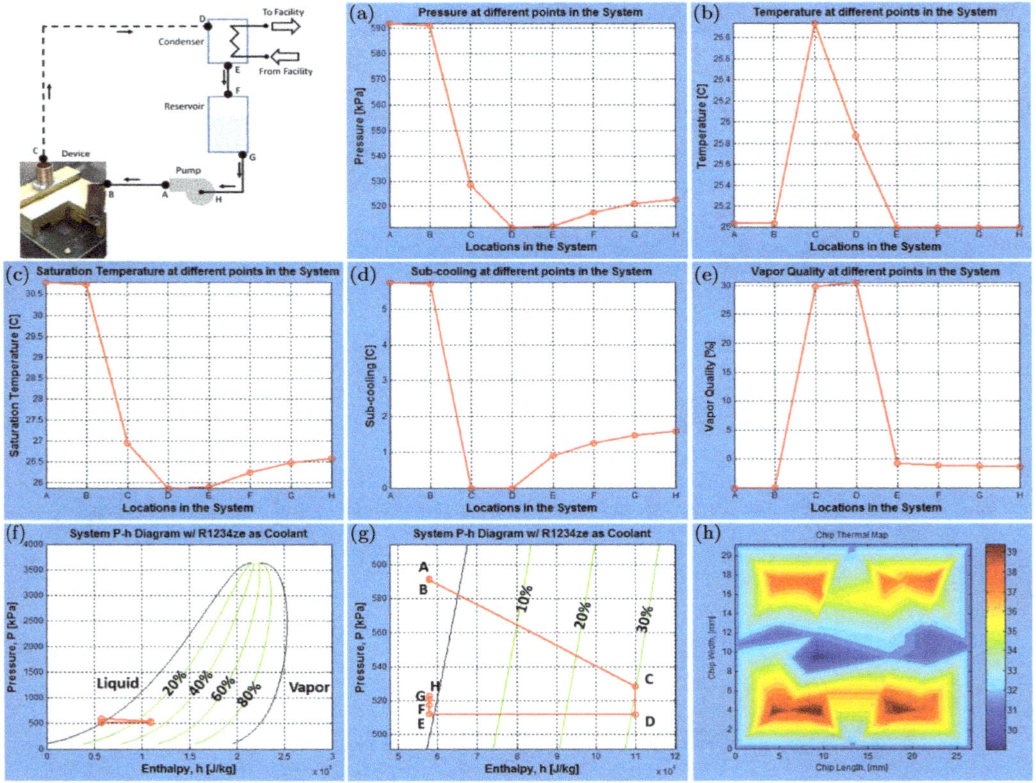

Figure 35. Different graphs generated by the Sim2Cool system-level simulator: (a) pressure profile, (b) coolant temperature, (c) saturation temperature, (d) sub-cooling, (e) vapor quality, (f) system $P-H$ diagram, (g) zoomed-in $P-H$ diagram, and (h) thermal map of the DUT [28].

Source: Copyright (2018) IEEE.

the fluid flowing in that section is in the liquid phase. The pressure drop across the condenser (i.e., D–E) is also negligible since the two-phase pressure drop within the condenser is compensated for by the hydrostatic pressure of the condensed liquid within the condenser. Across the sections E–F and G–H and across the reservoir (F–G), the coolant is in the liquid state, and the vertical height difference between the inlet and the outlet results in an increased hydrostatic pressure of the fluid. As there is no heat transfer in these sections, the increased fluid pressure results in increased sub-cooling at the pump inlet (i.e., at location H) [27].

The pressure difference (~69 kPa) between the points A and H is the head provided by the pump at the selected mass flow rate of 10 kg/h and a heat dissipation of ~145 W at the evaporator. Note that the heat dissipation at the evaporator (ECM module) is a function of the device temperature, operating frequency, and workload. The average junction temperature (T_j) reported here is the average temperature of the ECM cores. The pumping power required to operate the pumped two-phase loop at these settings is less than 1.7 W at a pump efficiency of 0.1. The outer loop requires ~5 W to remove the heat from the condenser, resulting in a COP of over 21.

As shown in Figure 35(a), the pressure recovery due to the vertical tubing E–F is very similar to that across the reservoir. Thus, in order to reduce the size and weight of the system, the reservoir and the tubing G–H can be removed from the system of configuration 1, resulting in a system of configuration 3 without significantly affecting the system or device performance [28].

Figure 35(b) shows the coolant temperature expected at different locations in the system. The temperature at each location is similar, except for locations C and D. The temperature at location D (condenser hot-side inlet) is roughly 0.9°C higher than the cold-side temperature. This temperature delta is determined by the thermal performance of the condenser and the amount of heat transferred across the condenser. For a given amount of heat exchanged across the condenser, a more effective condenser will result in a lower temperature delta. On the other hand, a less effective condenser will result in a higher temperature delta. The temperature at location C (evaporator outlet) is an additional 1°C higher. This temperature delta is directly proportional to the two-phase pressure drop across the tubing C–D. The higher the two-phase pressure drop, the higher the temperature delta. Note that, within the operating environment envelope of the dielectric coolant, the saturation temperature increases monotonically with the coolant pressure. Hence, the saturation (boiling) temperature of the coolant within the evaporator is equal to or greater than the temperature at location C. Further, for the current ECM modules' flow network, most of the pressure drop occurs in the inlet section and across the orifices, and the remaining minor portion of the pressure drop occurs in the radial channels. Thus, the temperature at location C can be used as an indicator of the saturation temperature within the evaporator [28].

The coolant state at the outlet of the evaporator is critical as it significantly influences the thermal performance of the evaporator. A condenser with poor thermal performance (i.e., lower effectiveness) will result in a higher saturation temperature at the evaporator, causing a higher device operating temperature. Alternatively, a long, thin tube connecting the evaporator and the condenser can result in a similar behavior by creating a large pressure drop between the evaporator outlet and the condenser inlet. The results show that reducing the diameter of tubing between locations C and D by ∼30% can increase the pressure drop in this section by nearly 5×, resulting in a ∼5°C increase in the DUT temperature. Hence, the proper selection of components downstream of the evaporator/DUT is very critical for optimum system and device performance [28]. Figure 35(c) and (d) show the saturation temperature and degree of sub-cooling expected at different locations in the cooling system, respectively. The coolant is 5.8°C sub-cooled at the evaporator inlet, which is enough to avoid any coolant flashing in the inlet section of the evaporator. Further, it can be observed that the coolant is roughly 1–2°C sub-cooled at the pump inlet, which reduces the potential for cavitation within the pump.

Figure 35(e) shows the vapor quality variation expected at different locations in the cooling system. A negative vapor quality means

Table 7. Key model predictions and comparison against data [28].

Location	Data	Sim2Cool
Chip power [W]	144	144.8
Chip T_j [°C]	37	37.4
Inlet pressure [kPa]	594	592
Pressure drop [kPa]	63.4	62.6
DUT inlet temperature [°C]	25	25
DUT inlet sub-cooling [°C]	6	5.8
DUT outlet temperature [°C]	27.1	26.9
Condenser inlet temp. [°C]	25.8	25.9
Condenser outlet temp. [°C]	25	25

Source: Copyright (2018) IEEE.

that the coolant is sub-cooled at that location. Figure 35(f) shows the pressure–enthalpy ($P - H$) diagram for this pumped two-phase cooling system. Figure 35(g) shows a zoomed-in image of the $P - H$ diagram, indicating the predicted state of the coolant at different locations of the system. The pressure is highest at location A. From location A to location B, there is a slight decrease in pressure at constant enthalpy due to a single-phase pressure drop without any heat addition/removal. From B to C (i.e., across the DUT), the pressure drops, along with heat addition into the coolant loop. From C to D, the pressure drops at constant enthalpy. From D to E (i.e., across the condenser), the pressure may drop or recover, along with heat removal from the coolant. From E to F to G to H, the pressure increases at constant enthalpy. Figure 35(h) shows the expected thermal map of the ECM module when operated at 3.4 GHz with a maximum-power workload and cooled by R1234ze flowing at 10 kg/h. The junction temperature can be expected to be less than 40°C for a facility-side coolant temperature of 25°C.

A comparison of the model predictions described in Figures 33 and 35 against the data showed excellent agreement. Table 7 shows the key model predictions and the corresponding experimental data. In this exemplary case, the chip power predictions were within 1 W ($< \pm 1\%$) of the experimental data. In general, the chip and cooling system temperature predictions were observed to be within 0.5°C

($\pm$4%) of the experimental data. The pressures at different locations in the system were also observed to be in excellent agreement ($\pm$2%).

Figure 36 shows the system-level predictions for two ECM modules operated in parallel at different power levels. Despite the significant difference in power levels, its impact on the flow distribution among the two ECMs is negligible. This highlights the robustness of the ECM channel design to passively balance the flow among multiple parallel components. Further, the parallel mode of operation helps improve the system COP to over 40 while maintaining a difference of less than 17°C between the outer-loop coolant temperature and the device operating temperature. Note that the tubing from the DUT outlet to the condenser is ~40% larger than that used in the single DUT case, which ensures a small pressure drop and temperature penalty across that section.

Figure 37(a) shows the thermal map for the two ECM modules operated at different power levels. Figure 37(b)–37(d) show the pressure, temperature, and sub-cooling predictions at different locations in the system, respectively. Figure 37(e) shows the $P - H$ diagram for this system configuration. The largest pressure drop occurs across the devices and again across the tubing connecting the devices to the condenser. The larger-diameter tubes between the devices and the condenser help keep the pressure drop below 15 kPa and the corresponding temperature penalty to less than 1°C. The model predictions suggest that the coolant would be 6°C sub-cooled at the evaporator inlet, which is enough to avoid any coolant flashing before the orifices in the evaporator. Further, it can be observed that the coolant would be roughly 3°C sub-cooled at the pump inlet, which reduces the potential for cavitation within the pump.

The key takeaways from the system-level modeling are summarized as follows:

i. Most of the pressure drop occurs across the evaporator and the tubing connecting the evaporator and the condenser.
ii. A higher pressure drop across the tubing between the evaporator and the condenser can result in a higher device operating temperature.

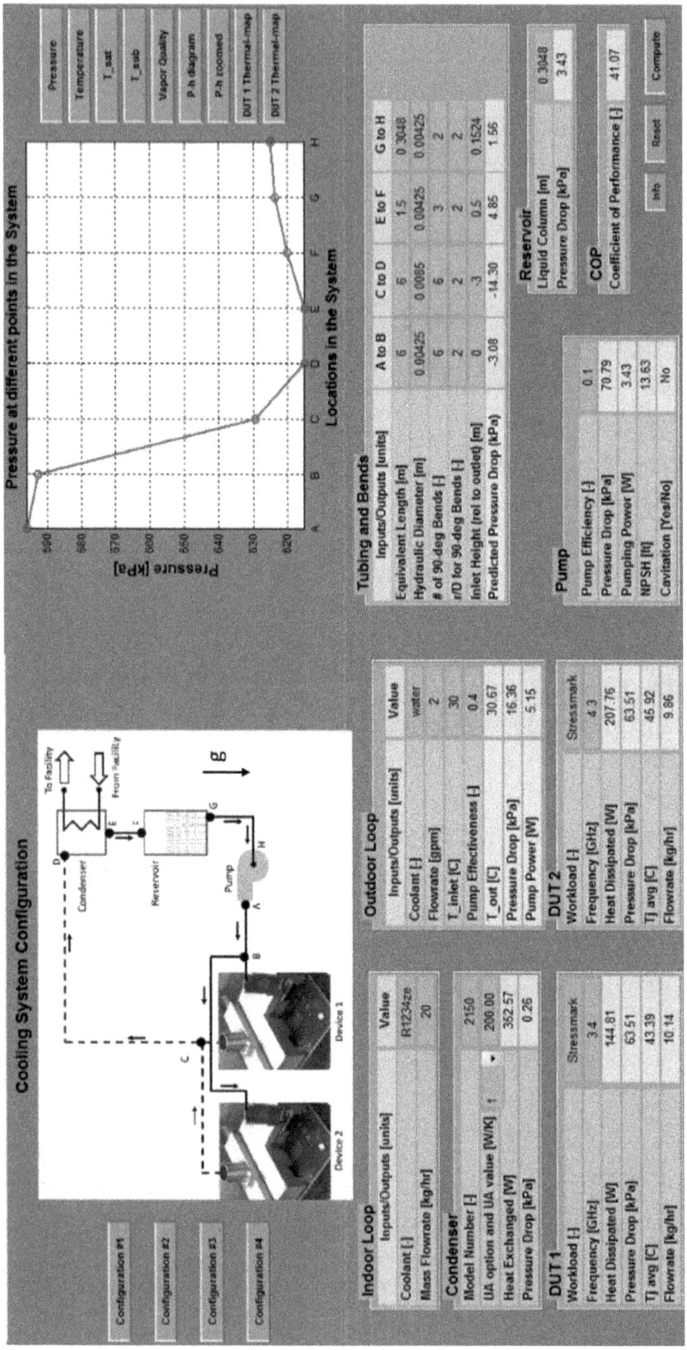

Figure 36. Model predictions for two ECM modules operated in parallel at different power levels.

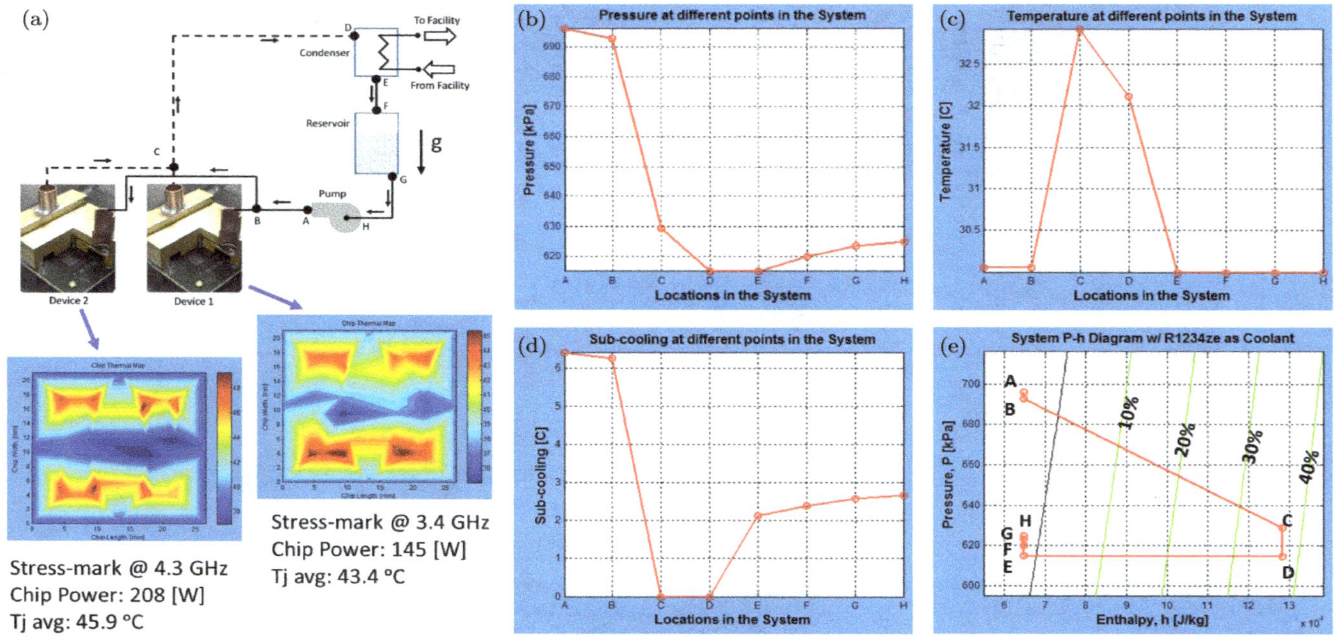

Figure 37. (a) Thermal map of the two ECM modules operating at different power levels. (b) Pressure profile, (c) coolant temperature, and (d) sub-cooling at different locations in the system. (e) System P–H diagram.

iii. The type and performance of the condenser affect the coolant state in the loop, thereby affecting the device(s) operating temperature.
iv. The reservoir, pre-heater, and sub-cooler can be removed from the cooling system without significantly affecting the device performance.

5. 3D Co-Design Summary

In summary, a 3D chip stack with embedded two-phase dielectric cooling significantly increases the interdependencies between thermal, mechanical, electrical, and computational performances, which in turn necessitates a detailed trade-off analysis.

Thermal performance: This requires an evaluation of trade-offs between fluid channel dimensions, critical heat flux criteria, and pressure drop to determine the range of channel dimensions and pin/interconnect spacing options for different heat flux criteria while also taking into account process fabrication, electrical and mechanical design, and computational performance.

Mechanical stability: Mechanical modeling of the 3D chip stack is needed to predict thermal stresses created during fabrication as well as stresses associated with operation. The optimization of interconnect size and pitch, consideration of CTE mismatch between chips and package, and thermal excursions during fabrication are important considerations for mechanical stability [26].

Electrical modeling/power: A power integrity analysis must be performed in 3D chip stack design. Placing the highest power chips closest to the package providing power delivery reduces IR drop and I^2R power losses. Such a configuration is generally not possible without embedded cooling.

Electrical modeling/signal I/O: Signal integrity analysis through the chip-to-chip connections in the presence of a coolant surrounding the connections needs to be performed for each type of coolant envisioned for the 3D chip stack cooling approach. Depending on the electrical properties of the coolant, the proximity of the coolant to the interconnects, and expected data transfer rates through the interconnects, the link losses could either be significant (e.g., water in proximity to interconnects) or marginal (e.g., a refrigerant flowing through a field of interconnects). Prior reported results indicate good signal integrity, with the link loss being only marginally (below 1%) impacted by having a dielectric coolant in direct contact with the inter-die connections for data rates of the order of 25 Gb/s [26].

Computational modeling/system performance simulations: Computational modeling focuses on the benefits of increased bandwidth and low latency in 3D chip stacks versus conventional 2D packaging. Prior results showed that the higher the utilization of the chip-to-chip bus, the larger the benefits of a 3D approach versus a 2D approach [26].

6. Acknowledgments

The research work reported in this chapter was developed with funding from the U.S. Defense Advanced Research Projects Agency (DARPA) Microsystems Technology Office ICECool Fundamentals Program under the award number HR0011-13-C-0035 and ICECool Applications Program under the award number FA8650-14-C-7466. Disclaimer: The views, opinions, and/or findings expressed are those of the author(s) and should not be interpreted as representing the official views or policies of the Department of Defense or the U.S. Government.

The authors would like to thank IBM ICECool team members for their valuable contributions and discussions, including IBM Zurich Research Lab members Thomas Brunschwiler, Arvind Sridhar, Chin Lee Ong, Ozgur Ozsun, Ute Drechsler, and Stephan Paredes, and IBM T. J. Watson Research Center team members Cyril Cabral, Michael Gaynes, Gerard McVicker, Julie Pope, Fanghao Yang, John Knickerbocker, Evan Colgan, Bing Dang, Yang Liu, Cornelia Tsang, Christy Tyberg, Pradip Bose, Alper Buyuktosunoglu, Hans Jacobson, and Augusto Vega.

The authors would also like to thank Avram Bar-Cohen, Ken Plaks, John Blevins, Kaiser Matin, Nicole DiLello, Joseph Maurer, Jim McLean, Abirami Sivananthan, John Blevins, and Glen "David" Via.

References

[1] S.G. Kandlikar, Review and projections of integrated cooling systems for three-dimensional integrated circuits, *Journal of Electronic Packaging*, 136 (2), 2014, 11.

[2] S.G. Kandlikar, History, advances, and challenges in liquid flow and flow boiling heat transfer in microchannels: A critical review, *J. Heat Transfer Trans. ASME*, 134 (3), 2012, 034001.

[3] J.R. Thome, Boiling in microchannels: a review of experiment and theory, *Int. J. Heat Fluid Flow*, 25 (2), 2004, 128–139.

[4] S. Krishnamurthy, Y. Peles, Flow boiling of water in circular staggered micro-pin fin heat sink, *Int. J. Heat Mass Transfer*, 51, 2008, 1349–1364.

[5] A. Reeser, A. Bar-Cohen, G. Hestroni, High vapor quality two phase heat transfer in staggered and inline micro pin fin arrays, in: *Back-Side and Interlayer Two-Phase Cooling, 14th IEEE ITHERM Conference*, Orlando, Florida, 2014, pp. 213–221.

[6] C.L. Ong, S. Paredes, A., Sridhar, B. Michel, T. Brunschwiler, Radial hierarchical microfluidic evaporative cooling for 3-d integrated microprocessors, in: *Proceedings of 4th European Conference on Microfluidics*, Limerick, 2014.

[7] P.R. Parida, A. Sridhar, M. Schultz, F. Yang, M. Gaynes, E. Colgan, B. Dang, G. McVicker, T. Brunschwiler, J. Knickerbocker, T. Chainer, Modeling embedded two-phase liquid cooled high power 3D compatible electronic devices, in: *Proceedings of 33rd IEEE Semi-Therm Symposium*, San Jose, CA, March 13–17, 2017.

[8] P.R. Parida, A. Sridhar, A. Vega, M. Schultz, M. Gaynes, O. Ozsun, G. McVicker, T. Brunschwiler, A. Buyuktosunoglu, T. Chainer, Thermal model for embedded two-phase liquid cooled microprocessor, in: *Proceedings of 16th IEEE ITherm Conference*, Orlando, FL, May 30–June 2, 2017.

[9] P.R. Parida, H. Tsuei, T.J. Chainer, Eulerian multiphase conjugate model for chip-embedded micro-channel flow boiling, in: *Proceeding of ASME InterPACK/ICNMM*, San Francisco, CA, July 6–9, 2015.

[10] P.R. Parida, M. Schultz, O. Ozsun, F. Yang, M. Gaynes, A. Sridhar, G. McVicker, T. Brunschwiler, T. Chainer, Eulerian multiphase conjugate model for embedded two-phase liquid cooled microprocessor, in: *IEEE Semi-Therm Symposium*, San Jose, California, 2018.

[11] C. Gorle, P.R. Parida, H. Lee, F. Houshmand, M. Asheghi, K. Goodson, Validation study for VOF simulations of boiling in a microchannel, in: *Proceeding of ASME InterPACK/ICNMM*, San Francisco, CA, July 6–9, 2015.

[12] P.R. Parida, T. Chainer, Eulerian multiphase conjugate model development and validation for flow boiling in micro-pin field, in: *Proceedings of 15th IEEE ITherm Conference*, Las Vegas, NV, May 31–June 3, 2016.

[13] A. Sridhar, P.R. Parida, C. Gorle, C. Lee-Ong, S. Paredes, T. Brunschwiler, B. Michel, E. Colgan, T. Chainer, K. Goodson, Thermal design of a hierarchical radially expanding cavity for two-phase cooling of integrated circuits, in: *Proceeding of ASME InterPACK/ICNMM*, San Francisco, CA, July 6–9, 2015.

[14] P.R. Parida, Reduce order modeling for chip-embedded micro-channel flow boiling, in: *Proceeding of ASME InterPACK/ICNMM*, San Francisco, CA, July 6–9, 2015.

[15] A. Sridhar, P.R. Parida, O. Ozsun, T. Brunschwiler, B. Michel, T. Chainer, Study of two-phase pressure drop and heat transfer in a micro-scale pin fin cavity: Part A, in: *IEEE ITherm 2016*.

[16] P.R. Parida, A. Sridhar, T. Brunschwiler, T. Chainer, Study of two-phase pressure drop and heat transfer in a micro-scale pin fin cavity: Part B, in: *IEEE ITherm 2016*.

[17] S. Szczukiewicz, Thermal and visual operational characteristics of multi-microchannel evaporators using refrigerants, EPFL Theses No. 5594, 2012.

[18] M. Schultz, F. Yang, E. Colgan, R. Polastre, B. Dang, C. Tsang, M. Gaynes, P.R. Parida, J. Knickerbocker, T. Chainer, Embedded two-phase cooling of large 3D compatible chips with radial channels, *J. Electron. Packag.*, 138 (2), 2016, 021005.

[19] F. Yang, M.D. Schultz, P.R. Parida, E. Colgan, B. Dang, G. McVicker, T. Chainer, Boiling instability in and sensitivity analysis of non-uniformly heated micro-scale devices, in: *Proceedings of 15th IEEE ITherm Conference 2016*, Las Vegas, NV, May 31–June 3 2016.

[20] F. Yang, M. Schultz, P. Parida, E. Colgan, R. Polastre, B. Dang, C. Tsang, M. Gaynes, J. Knickerbocker, T. Chainer, Local measurements of flow boiling heat transfer on hot spots in 3D compatible radial microchannels, in: *Proceedings of ASME InterPACK/ICNMM Conference 2015*, San Francisco, CA, July 6–9, 2015.

[21] S. Bhavnani, V. Narayanan, W.L. Qu, M. Jensen, S. Kandlikar, J. Kim, J. Thome, Boiling augmentation with micro/nanostructured surfaces: Current status and research outlook, *Nanoscale Microscale Thermophys. Eng.*, 18 (3), 2014, 197–222.

[22] M. Schultz, M. Gaynes, P.R. Parida, F. Yang, G. McVicker, O. Ozsun, T. Chainer, Two-phase flow characteristics in radial expanding channels with embedded pin fins, in: *Proceeding of IEEE ITherm*, San Diego, CA, May 29–June 1, 2018.

[23] M. Schultz, C. Cabral, C. Tsang, J. Maria, P. Andry, T. Chainer, Column interconnects: A path forward for embedded cooling of high power 3D chip stacks, in: *Proceeding of the 2018 68th ECTC Conference*, San Diego, CA.

[24] M. Schultz, P.R. Parida, M. Gaynes, O. Ozsun, G. McVicker, U. Drechsler, T. Chainer, Microfluidic two-phase cooling of a high power microprocessor Part A: Design and fabrication, in: *Proceedings of 16th IEEE ITherm Conference 2017*, Orlando, FL, May 30–June 2, 2017.

[25] M. Schultz, P.R. Parida, M. Gaynes, O. Ozsun, A. Vega, U. Drechsler, T. Chainer, Microfluidic two-phase cooling of a high power microprocessor Part B: Test and characterization, in: *Proceedings of 16th IEEE ITherm Conference 2017*, Orlando, FL, May 30–June 2, 2017.

[26] T. Chainer, M. Schultz, P.R. Parida, M. Gaynes, Improving data center energy efficiency with advanced thermal management, *IEEE Trans. Compon. Packag. Manuf Technol.*, 7 (8), 2017, 1228–1239.

[27] L. Chen, F. Yang, P.R. Parida, M. Schultz, T. Chainer, Enthalpy-based system-model for pumped two-phase cooling systems, in: *Proceeding of IEEE ITherm*, Las Vegas, NV, May 31–June 3, 2016.

[28] P.R. Parida, M. Schultz, T. Chainer, Sim2Cool: A two-phase cooling system simulator and design tool, in: *Proceeding of IEEE ITherm*, San Diego, CA, May 29–June 1, 2018.

[29] P.R. Parida, T. Chainer, M. Iyengar, M. David, M. Schultz, M. Gaynes, V. Kamath, B. Kochuparambil, R. Simons, R. Schmidt, System-Level design for liquid cooled chiller-less data center, in: *Proceedings of the ASME 2012 International Mechanical Engineering Congress & Exposition*, Houston, TX, November 9–15, 2012.

[30] E.W. Lemmon, M.L. Huber, M.O. McLinden, NIST standard reference database 23: Reference fluid thermodynamic and transport properties-REFPROP, Version 9.1, *National Institute of Standards and Technology, Standard Reference Data Program*, Gaithersburg, MD, 2013.

https://doi.org/10.1142/9789811279379_0005

Chapter 5

Embedded Cooling of High-Power RF Amplifiers

John Ditri

Lockheed Martin Rotary and Mission Systems, Moorestown, NJ, USA
john.ditri@lmco.com

Abstract

High-power radio-frequency (RF) amplifiers (HPAs) are widespread throughout commercial and military systems, from personal cell phones to large-scale communications networks, to the most advanced military radar and electronic warfare platforms. Ever-increasing performance demands, along with the desire for smaller sizes, have increased local heat fluxes beyond levels which can be effectively managed using conventional "remote" cooling paradigms, and the situation only worsens as developments in electronics fabrication outpace those in thermal management. This either leads to high operating temperatures, degraded system performance, and reduced life or to sub-optimal performance by "throttling back" the system to maintain suitable thermal control.

A very promising thermal management approach for reducing thermal resistance and restoring full system electrical potential is embedded cooling, in which the coolant is brought close to or in contact with the active heat source, effectively removing as many of the offending thermal interfaces as possible. This chapter reviews the embedded cooling concept and illustrates its benefits by comparing measured HPA thermal and RF performance using remote and embedded cooling approaches. Computational fluid dynamics and conjugate heat transfer simulations are shown to compare very favorably with the measured data, allowing accurate performance predictions to be made without the need for costly hardware iterations. Finally, some practical considerations for implementing embedded cooling in actual systems, such as the potential for leakage, clogging, and erosion, are discussed, and some recommendations are provided to mitigate such issues.

Keywords: high-power amplifiers (HPA), radio-frequency (RF) performance, thermal demonstration, power added efficiency (PAE), reliability of embedded cooling

1. Introduction

Thermal management is key to the reliable and efficient operation of electronic equipment. It is well documented that the operating life of electronic components is reduced when they are subjected to elevated temperatures. References such as MIL-HDBK-217F [1], which is an often-cited requirement for military system reliability predictions, include temperature correction factors to account for decreased life at elevated temperatures. Nearly all electronic components, including active semiconductors and passive resistors, capacitors, etc., exhibit reduced life with increased temperature to varying degrees, and high-power radio-frequency (RF) amplifiers (HPAs) are no exception.

Measurements show that the mean time to failure (MTTF) as a function of temperature takes the form indicated in Eq. (1):

$$\mathrm{MTTF}(T_\mathrm{J}) = \mathrm{MTTF}_\mathrm{RT} \times \exp\left[\frac{-E_\mathrm{a}}{k_\mathrm{B}}\left(\frac{1}{T_\mathrm{J}} - \frac{1}{T_\mathrm{RT}}\right)\right] \qquad (1)$$

where $\mathrm{MTTF}_\mathrm{RT}$ represents the MTTF at some reference temperature (T_RT), T_J represents the operating temperature (in absolute units), E_a represents the "activation energy" specific to the particular device in question, and k_B is Boltzmann's constant ($k_\mathrm{B} = 8.617\ldots \times 10^{-5}$ eV/K).

With regard to quantitative values, one relatively recent survey [2] of the reliability of gallium nitride (GaN) high electron mobility transistors (HEMTs) on silicon carbide (SiC) substrates shows that GaN/SiC HEMTs exhibit MTTFs of nearly 1×10^{12} h at 175°C, with an activation energy of $E_a = 2.5$ eV. While this can be considered a very robust thermal performance compared to previous semiconductor materials, such as high-voltage gallium arsenide [3] with an MTTF of 5×10^6 h at 150°C, Eq. (1) predicts that every roughly 7°C (or Kelvin) increase in junction temperature results in a *halving* of the MTTF and hence a reduction in reliability by a factor of two.

In addition to reducing the reliability of components and the systems which incorporate them, increased temperature also degrades the performance of many devices, including HPAs. The performance of digital integrated circuits, such as FPGAs and CPUs, is also negatively impacted by temperature, and they often include internal temperature sensing circuits to shut down operation if the temperature exceeds a certain limit. Given the relative thermal robustness of GaN/SiC HEMTs, performance degradation usually becomes the limiting thermal factor well before concerns about reliability set in.

As an example of the impact that temperature has on GaN HPA performance, Figure 1 shows measured data taken from Ditri *et al.* [4] on the RF output power and efficiency of a "typical" GaN/SiC amplifier, as the base temperature of the die is raised in 25°C increments from room temperature to 125°C. The particular HPA is an in-house designed, GaN/SiC, broadband millimeter wave (18–40 GHz) monolithic microwave integrated circuit (MMIC) with four HEMT stages along with on-MMIC impedance matching networks.

The plots on the left (referencing the left-hand vertical scale) correspond to (scaled) output power versus (scaled) input power, and the plots on the right (referencing the right-hand vertical scale) correspond to the scaled power-added efficiency (PAE). As can be seen from the data, both the output power and efficiency decrease substantially as the temperature increases. Quantitatively, the gain of the GaN HPA decreases at a rate of nearly 0.04 dB/°C, and the efficiency drops by ≈0.037 percentage points/°C. Practically, this means that if the base temperature of the HPA is raised by 100°C (which can easily occur in conventional thermal management schemes), its output power will drop by 60% compared to its value at room temperature, and its PAE will drop by nearly 4 percentage points. Similar gains

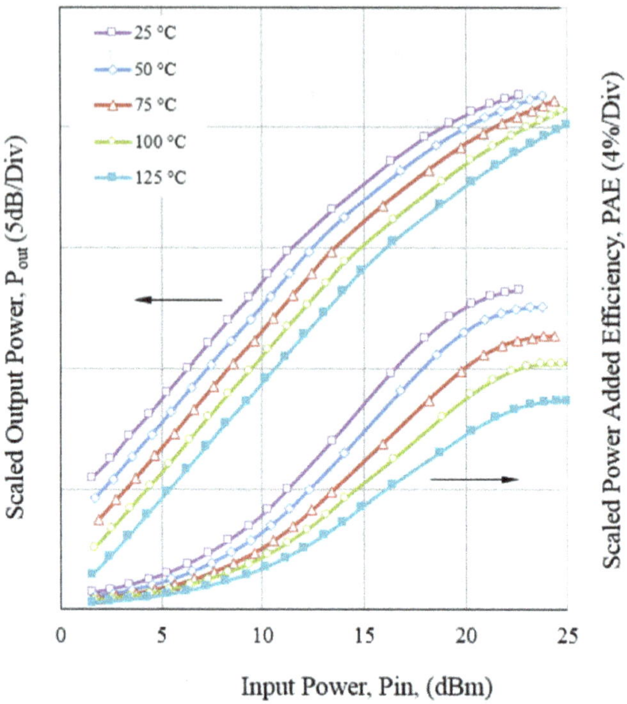

Figure 1. Measured impact of temperature on four-stage millimeter-wave GaN/SiC HPA output power and efficiency.

and efficiency sensitivities have been reported for other GaN/SiC HEMTs by other authors [5].

The deleterious effects of temperature on component reliability has been known for decades, and many thermal management techniques have been developed over the years to manage component operating temperatures. Textbooks [6] and countless review papers (see, for example, the reviews by Mudawar [7] and Sung and Mudawar [8]) have been written detailing emerging thermal management techniques to deal with the ever-increasing waste heat generated by electronic components. Many of these techniques, however, only deal with increasing the convective heat transfer at the fluid–wall interface(s) and do not address the issue of minimizing the thermal rise through multiple interfaces before the heat makes it to the heat transfer fluid.

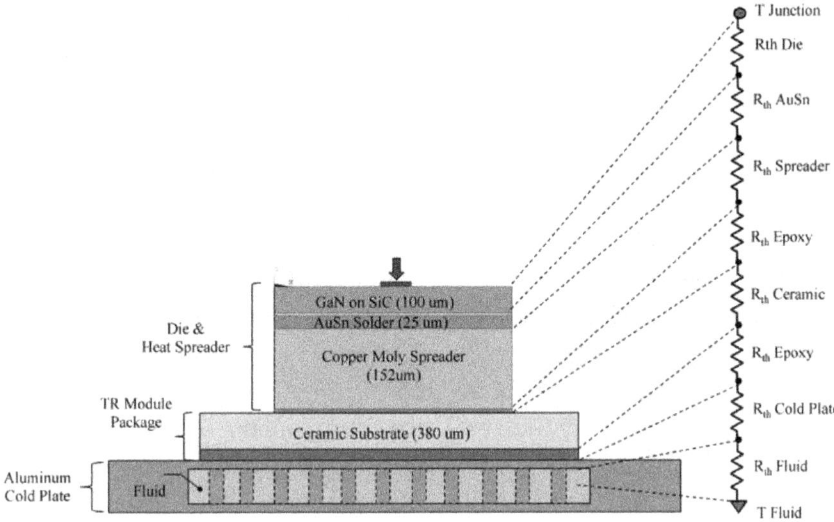

Figure 2. Conventional or "remote" cooling approach showing multiple thermal interfaces separating heat source from heat sink.

2. Motivation and Past Research on Embedded Cooling

To better understand the true nature of the thermal limitations introduced by current cooling techniques, it is instructive to consider a typical liquid cooled system, and its thermal network representation, as shown in Figure 2. This packaging approach is currently the dominant approach for packaging high heat dissipating electronic components, at least for dies with primary heat paths downward (not flip-chipped devices, such as FPGAs, in which the primary heat paths are "upward"). The cooling scheme utilizes the conventional "remote" cooling paradigm, where the die is separated from the coolant by several intervening materials.

The particular packaging materials used in the stack up can vary from one implementation to another, but typically include the following items, from top to bottom: GaN epitaxial layer on a silicon carbide (or silicon, etc.) substrate (i.e., the HPA), a die-attach solder, a metallic (or synthetic diamond) heat spreader, thermally conductive epoxy, a package substrate (high-temperature co-fired ceramic, organic laminate, etc.), another epoxy layer, an aluminum (or copper) cold plate, and, finally, the coolant, which flows in parallel channels within the cold plate.

A thermal finite element analysis was performed on the configuration shown in Figure 2, with the materials shown in the figure, including full 3D spreading within all the materials as well as gate-finger level detail within the die. Temperature-dependent material properties were used for the GaN/SiC HPA, and typical values of thermal conductivity were used for the remainder of the packaging materials.

The temperature rise in selected groups of components, along with the percentage of the total rise from coolant to junction, for a total die dissipation of 18.8 W and an inlet coolant temperature of 30°C, is shown in Figure 3. The peak junction temperature, for the inlet coolant temperature of 30°C, was 208°C. It can be seen from Figure 3 that only 33% of the temperature rise comes from *within* the die (epitaxy and substrate) itself, with the remaining 67% coming from the *packaging materials* beneath the die.

The fact that the thermal bottleneck in conventional cooling lies in the materials between the die and the heat sink highlights the need for investigating alternative cooling approaches, with the goal being to eliminate as many of the thermal interfaces between the die and coolant as possible. Such approaches, sometimes referred to as "Generation 3 (Gen3)" or embedded cooling approaches [9–11], have received considerable attention lately, as die power dissipations have increased to the point where conventional thermal management techniques begin to *severely* limit the performance of otherwise highly capable electronics.

The idea of moving the coolant closer to the heat source to reduce the overall thermal resistance is not new, dating back to at least the early 1980s with the paper by Tuckerman and Pease [12] and, a few years later, the publication of a comprehensive PhD thesis by Tuckerman in 1984 [13]. In their pioneering work, Tuckerman and Pease introduced a "high performance IC heat sink," which consisted of a silicon substrate, with integrated circuits (ICs) on one

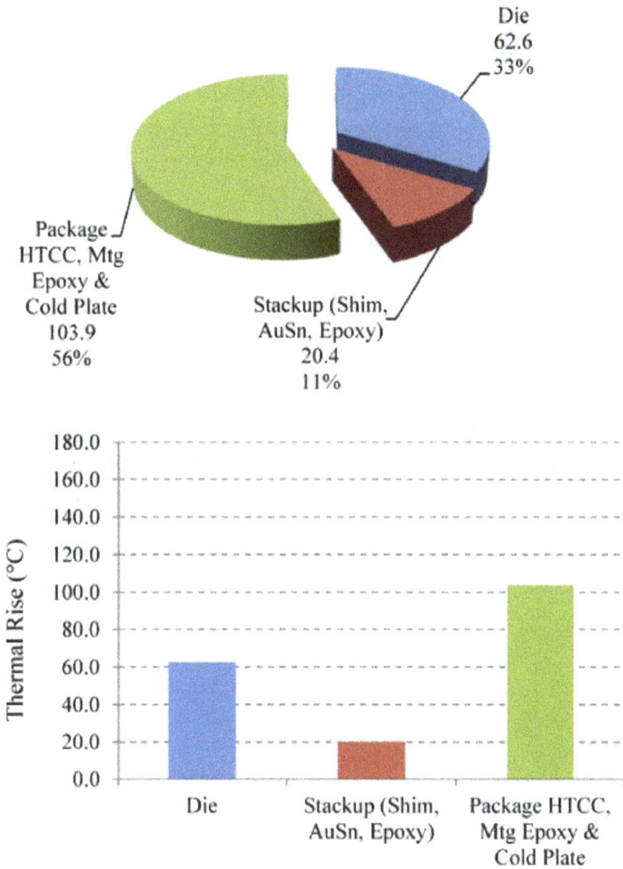

Figure 3. Typical breakdown of thermal rise through conventional remote packaging configurations.

side (which consisted of simple heater circuitry, not functional digital or RF circuits) and chemically etched or mechanically sawn grooves on the other side, to produce a series of parallel fins (with adjacent microchannels) integral to the substrate. They also produced substrates with individual pins rather than fins. The lowermost surface of the silicon substrate (i.e., the tips of the fins or pins) was then closed off with a cover plate bonded to the silicon IC substrate to form a series of flow channels directly beneath the silicon integrated circuitry, through which a suitable coolant could be passed. In addition to an extensive theoretical examination of the fluid flow and heat transfer in such a compact heat exchanger, Tuckerman and Pease also provided experimental results on devices which they fabricated and tested. By applying a known heat dissipation to one surface of the silicon and measuring its temperature rise, they were able to demonstrate a thermal resistance of 0.09°C/W (while dissipating 790 W/cm^2). Tuckerman noted that this was 20 times lower than "state-of-the-art commercial technologies for cooling arrays of ICs" available at the time.

Numerous researchers have subsequently explored the thermal benefits of localized cooling using mini- and microchannels, impingement jets, gap flow, etc., and have made considerable progress toward understanding and summarizing the fundamentals of heat transfer and fluid flow in miniaturized devices and spaces. The review papers by Kandlikar [14] and Thome [15], books by Kandlikar *et al.* [16] and Ohadi *et al.* [17], for example, provide in-depth reviews of the physical processes of single- and two-phase fluid (gas and liquid) transport and heat transfer in micro- and minichannels, carefully pointing out specific physical phenomena which arise in the small-scale world which may differ from those in macrochannels.

The literature bridging the gap between early fundamental fluid dynamics and heat transfer studies and the practical implementations of embedded cooling is too broad to be covered here, so a complete historical review will not be attempted. Instead, a few representative articles from various groups, which have contributed to our current state of the art, will be cited. There is a wealth of information to be gleaned from these references as well as the citations within them, and any omissions of significant contributions are solely due to the limited knowledge of the present author with respect to the worldwide research in this area.

One critical aspect of embedded cooling which was realized early on was the need for large heat transfer coefficients between the fluid and walls. As the cooling fluid moves closer to the heat source, approaching, in the limit, the bottom of die itself, the *surface area* available for heat transfer is drastically reduced. As is well known, the temperature rise in convective heat transfer can be written as

$$\Delta T = \frac{Q}{hA} \tag{2}$$

where Q is the dissipated heat, h is the (average) heat transfer coefficient, and A is the contact area between source and sink. As the area shrinks, h must grow proportionally to maintain the same temperature rise. Therefore, one of the keys to successful embedded cooling is achieving a large heat transfer coefficient to compensate for the decreased area. Increasing h typically involves increasing fluid velocities, often at the expense of increased pressure drop, or, alternatively, utilizing more effective techniques, such as microjet impingement instead of channel flow.

Some relatively early work on "*near* embedded" cooling, with an emphasis on field implementation for cooling an electronics module, was published by Wadsworth and Mudawar [18]. They investigated single-phase heat transfer from an array of simulated chips (heater die), arranged to simulate an electronics module, to a jet of dielectric Fluorinert FC-72 liquid issuing from a rectangular slot jet. Already at that time, the authors noted that thermal management techniques were falling behind in their ability to manage the heat from prevailing electronics modules. Various slot geometries and jet-to-wall standoff distances were tested, and the authors provided correlations for the Nusselt number versus the Reynolds number for many different configurations.

Sung and Mudawar [8] examined a two-phase "hybrid" cooling scheme utilizing microchannel flow fed by jet-impingement orifices with the goal of cooling practical defense electronic assemblies. They investigated both single- and two-phase (boiling) heat transfer using the dielectric HFE-7100 liquid as the coolant. The authors provided analytical predictions for the single-phase heat transfer experiments, which showed good agreement with the experiments. No attempt was made at analytically predicting the two-phase results.

Wang and colleagues [19] at Stanford and Intel, had already started investigating localized cooling of digital VLSI chips for computing applications, utilizing micromachined silicon jets and simulated heater devices. This group designed, fabricated, and tested single- and multi-jet orifices and evaluated their performances with single- and two-phase heat transfer experiments, along with collaborating performance predictions for the single-phase experiments. As with most of the other works utilizing two-phase heat transfer, no numerical/analytical predictions were attempted for the two-phase flow; instead, only experiments were reported. The Stanford group, along with industrial colleagues at Toyota [20] and Ford Motor Company [21], also demonstrated embedded microchannel devices with 3D manifold heat sinks with improved thermal performance and decreased pressure loss compared to conventional microchannel heat sinks due to the 3D nature of the manifold. Their "Embedded Microchannel Manifold Cooler" (EMMC) consisted of a microchannel cold plate etched into silicon, in contact with the heat source, and a fluid-routing silicon-based 3D manifold to deliver and extract the fluid to the cold plate while minimizing pressure drop. The authors considered the challenges of implementing embedded cooling in "large area" applications (20×20 mm^2) with an eye toward real-world applications.

A group of researchers at IBM likewise demonstrated significant advancements in the embedded cooling of digital chips for computing purposes [22–25]. These latter researchers have designed and fabricated effective microcoolers, based on a combination of radial channel flow and jet impingement, and demonstrated the performance improvements in actual state-of-the-art microprocessors. They also developed numerically efficient "compact" modeling techniques to predict the performance of their (two-phase) devices [26] to fill the void in commercial CFD codes, which, for the most part, lack true physics-based models for two-phase flows. Nearly all of the commercial CFD codes which have some two-phase capabilities rely on homogeneous fluid assumptions with correlation-based predictions of void fraction, quality, or other measures of vapor-to-liquid fractions throughout the fluid domain. None of the codes (to the author's knowledge) are able to predict the flow regime (annular, slug, etc.) and are therefore subject to *a priori* assumptions regarding the regime. Consequently, they are, at best, only applicable to specific flow conditions, which may not be realized in practical applications under all operating conditions.

Other researchers have also demonstrated compact microfabricated heat exchangers incorporating elaborate fluid distribution manifolds and multiple distributed jets and returns. The work by the teams at Purdue and IBM [27, 28] is a great example of the utilization of modern silicon microfabrication techniques to enable embedded cooling. This team designed, built, and tested a compact "hierarchical" (3D) manifold microchannel cooler, utilizing an eight-layer integrated multilevel manifold distributor to feed coolant to an array of microchannel heat sinks. They utilized a dielectric fluid (HFE-7100) and demonstrated a volumetric heat removal of 2.87 kW/cm^3 with chip temperatures below 115°C and a modest pressure drop of 138 kPa (20 psi) across the cooler at their highest tested heat flux (660 W/cm^2).

A team of researchers from the University of Maryland and Case Western Reserve University [29, 30] built and tested a "Thin-Film Evaporation and an Enhanced Delivery System" (FEEDS) microcooler in a silicon carbide substrate. They tested their device with both single- and two-phase flows with heat fluxes up to 1,000 W/cm^2. As with other previously cited groups, they provided a good correlation for single-phase predictions but made no attempt at analytical predictions for two-phase operation. This work clearly identifies one of the key potential practical issues with embedded cooling using microscale fluid passages, namely the potential for clogging of the device. They attributed a "notable increase in pressure drop" in their device after several days of testing to partial clogging of the microchannels. The authors state their belief that the "calcium and sodium" found in their device were mostly due to "handling and ambient dust" and that "loose unsintered particles from the manifold or delaminated solder layers could clog the microchannels immediately upon operation." These authors also pointed out some of the practical challenges of trying to push two-phase flow to high vapor qualities in order to maximize the use of the latent heat of the fluid for heat removal. Their experimental results showed drops in heat transfer coefficient for higher heat fluxes due to partial dry-out. The present author can attest to the need for extreme cleanliness procedures during assembly and testing of such microscale devices after also dealing with clogging on several early prototype devices.

Researchers from Rensselaer Polytechnic Institute and Creare [31, 32] dealt with the potential dry-out issue when using high-quality two-phase flows by developing a clever vapor extraction system they termed "Piranha Pin Fins (PPF)." Utilizing silicon microfabrication techniques, they designed hollow, tear-drop-shaped pin fins with an opening on the upstream flow side (the pins resemble a school of piranhas) to serve as conduits for mass (vapor) extraction. The goal of the PPFs was to "proliferate the heat transfer coefficient by increasing flow quality to near unity, while extracting vapor and avoiding critical heat flux."

The group from Georgia Institute of Technology [33–35] designed, fabricated, analyzed, and characterized a hotspot test bed to dissipate high power density through two-phase convective boiling of refrigerant in a microgap with integrated micropin fins and isolation air trenches around resistance heaters. Their main focus was on cooling microprocessors. The test bed dissipated a power density of up to 4.75 kW/cm^2 using R-134a refrigerant as the coolant. Of particular note, this group performed thermodynamic analyses to obtain performance trends across a wide range of candidate coolants, compute pressure drops due to contraction and expansion in inlet/outlet manifolds, and obtain realistic estimates of different contributions to the pressure drops. They did not attempt to model the two-phase heat transfer itself and commented on the lack of tractable first-principles models to predict heat transfer coefficients and other flow and temperature parameters for two-phase flows. Although these authors and others have amply demonstrated the potential improvement in two-phase flows relative to single-phase flows, they also point out that some of the disadvantages are flow instability and maldistribution, in particular when using parallel flow through microchannels. This was one of the driving factors for them to employ microgap flows, which, according to the authors, "minimize or altogether eliminate this disadvantage."

Despite the great amount of progress made over the past three decades, embedded cooling does not appear to have made significant inroads into fielded products but rather has remained mainly in universities and research laboratories. There may be several contributing factors to this lack of (or at least very slow) adoption:

(1) The development of this "high-power" thermal management technique may have preceded the need for such a degree of heat removal, with simpler approaches being adequate for the times.
(2) The manufacturing techniques used for these early microchannel heat sinks (and most compact heat sinks since then), namely etching or individually sawing multiple layers of silicon substrates, followed by a joining process to join the multiple layers, were (and still are) tedious, of low yield, time-consuming, and ultimately very costly.
(3) The approach of manipulating or transforming the die substrate itself to form a compact heat exchanger greatly limits the availability or utility of the concept. Very few open die foundries have the capability or willingness to perform such die processing on either their own products or on customers' custom die designs, so this presents a serious barrier to widespread adoption.
(4) Although the size of the actual "compact heat sinks" (i.e., the etched die substrates themselves) was quite small, the fluidic connections necessary to connect the compact heat sinks to the "outside world" were quite bulky and incompatible with real-world electronics packaging constraints.
(5) Reliability concerns, due primarily to fluid leakage and clogging, are somewhat unique or exacerbated in microscale devices due to the limited space and options for making fluid interconnections and the intolerance of the small channels to debris.

In 2012, the U.S. Defense Advanced Research Projects Agency (DARPA), under the auspices of Prof. Avram Bar-Cohen, decided it was time to revisit embedded cooling technology and funded the Intra/Interchip Enhanced Cooling (ICECool) series of programs [36]. The programs consisted of two separate programs, ICECool-Fundamentals and ICECool-Applications, with the former meant to address fundamental issues related to high-exit-quality, two-phase embedded cooling and the latter meant to address the shorter-term need for improved (embedded) cooling techniques to address current thermal issues hampering further development in electronics and electronic systems used by the Department of Defense (DoD).

In launching these programs, it was realized that significant changes had occurred in several of the previously identified adoption barriers, namely that the need for more effective thermal management techniques had become acute, with developments in thermal management techniques having lagged behind the steady and rapid advancements made in electronic devices, leading to the situation in which many DoD systems had become thermally limited. In addition, new and improved microfabrication techniques have been developed over the past three decades, and this has been seen to have a potentially significant impact on several of the other impediments to the adoption of embedded cooling. The ICECool programs placed nearly equal weight on proving the *reliability* of the embedded cooling approach as they did on demonstrating its performance.

3. State-of-the-Art Embedded Cooling

There were several DoD contractors who participated in the ICECool-Applications program, including Raytheon [37, 38], Northrup Grumman [39, 40], BAE [41], Boeing Defense, and Lockheed Martin [4, 42, 43]. Many of the previously referenced industrial and university research groups were also either prime performers or teamed with several of the DoD contractors. Bar-Cohen *et al.* [11] presented a good review of the thermal-focused DARPA programs, including ICECool.

As one of the performers on the ICECool-Applications program, Lockheed Martin [4, 42, 43] capitalized on advancements made in microfabrication techniques to develop a unique and novel approach for manufacturing the key building block of embedded cooling — the microfluidic manifold (or "compact heat exchanger"). The design abandoned the use of etched and stacked silicon layers and instead

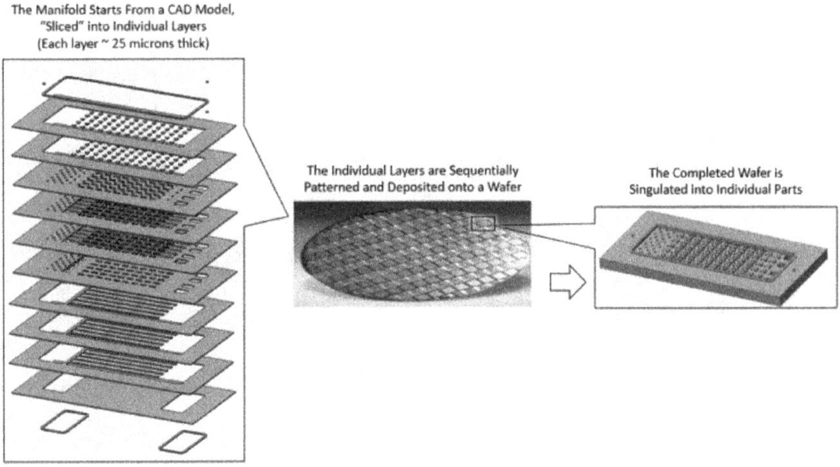

Figure 4. Schematic diagram of the multilayer additive deposition process for fabricating microfluidic manifolds.

utilized a modern additive photolithographic deposition process to manufacture standalone, monolithic microfluidic manifolds.

The manifolds were fabricated using a semiconductor die-like manufacturing process, shown schematically in Figure 4. The design process began with a detailed computer-aided design (CAD) of a desired manifold layout, including placement and routing of internal channels and exterior features, and subsequent optimizations by computational fluid dynamics (CFD)/heat transfer simulations. These simulations were performed to optimize the fluid routing within the manifold and the placement of specific impingement jets to match "hot spots" on the target die. Once the CAD modeling and simulations were complete, the CAD model was sliced into individual layers, and photolithographic master reticles were produced for each layer. From that point onward, the manufacturing process was similar to that used in conventional IC manufacturing, namely exposure of a base substrate with each subsequent layer's reticle, followed by deposition of metal in the pattern defined by the reticle. This process was repeated for each layer, resulting in a finished wafer with multiple copies (100 or more, depending on the size of the individual

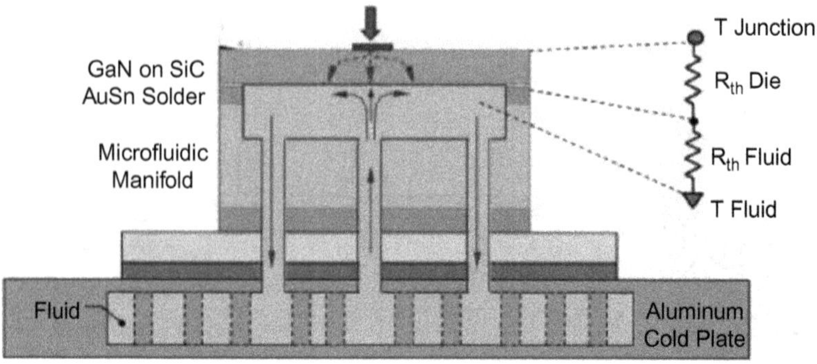

Figure 5. Schematic representation of the embedded cooling concept.

microcoolers) of the manifold on each wafer. Hundreds of such manifolds have been produced in our research efforts, and dozens have been attached to MMICs and characterized to date.

To provide the necessary high heat transfer coefficient, single-phase, submerged jet-impingement heat transfer is utilized, as shown schematically in Figure 5. Multiple microjets are located within the microcooler directly beneath the die, and a multitude of distributed returns are provided to capture, route, and exhaust the heated fluid. Comparing the thermal resistor network shown on the right-hand side of the figure with the thermal resistor network shown in Figure 2, it is seen that most of the thermal interfaces have been successfully removed from the thermal path, leaving only the die and fluid film resistances. The net effect, as validated by numerous measurements on actual devices, is a lowering of the total thermal resistance between junction and coolant by a factor of at least 3×.

An example of one such manifold design and the flow paths within it is shown in Figure 6 (US Patent 10,306,802). The upper image is a cutaway CAD rendering, and the lower image is a side view where the die is on top and the inlet/outlet ports of the manifold are on the bottom. As can be seen, there are multiple parallel supply and return fluid paths on different layers within the manifold and single inlet and outlet openings for interfacing with the next higher assembly. There are a number (≈100×) of openings at the top surface of the manifold

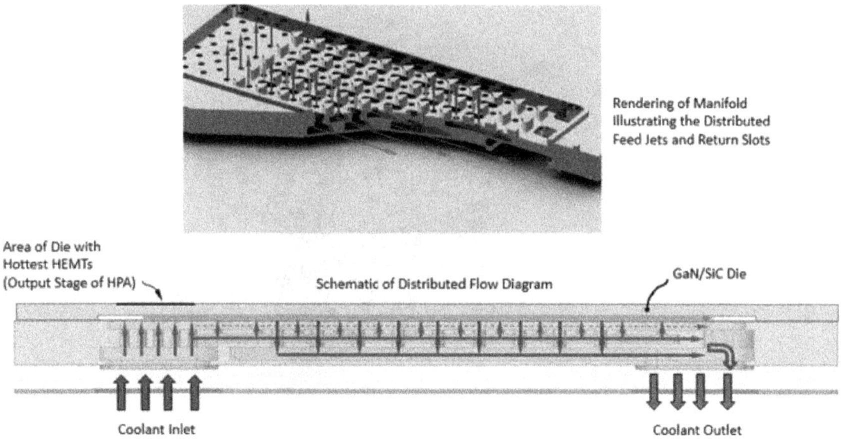

Figure 6. Example microcooler utilizing 3D additive manufacturing capabilities [44].

to create impingement jets of 90 micron diameter, as well as a number of adjacent oval openings to serve as "distributed" return ports to collect and expel the warmed coolant locally.

Removal of the heated fluid adjacent to each jet helps minimize the volume of downflow, preventing bending of the downstream jets and also minimizing the bulk fluid temperature rise along the length of the microcooler. Finally, the manifold has a number of diverter features within the flow gap, intermixed between jets and returns, to direct the gap flow *around* the impingement jets and prevent the channel flow from bending the impingement jets and degrading their performance.

The die is attached to the top surface of the manifold around a "picture frame" border using solder to serve as a robust fluid seal, as illustrated in Figure 7 (solder not shown). The result is a standalone die/heat sink combination which can be handled as a single unit for insertion into its next higher assembly. Each unit can be X-rayed to ensure the integrity of the seal around the perimeter prior to integration into its next higher assembly. Note that there is a raised rim around the central fluid cavity to mechanically block solder ingression into the flow cavity during solder reflow.

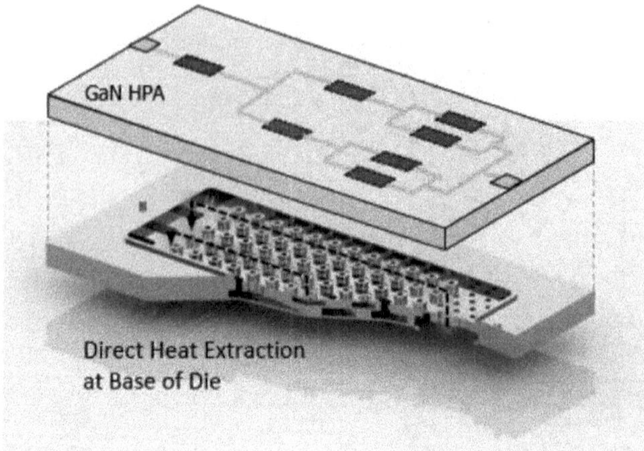

Figure 7. Die attached to manifold.

4. Testing

The RF data presented in Figure 1 clearly demonstrates the negative impact of elevated temperatures on HPA performance and the need for improved thermal management techniques. In this section, we describe two custom test vehicles which were developed in order to *quantitatively* assess the effectiveness of embedded cooling, utilizing manifolds described in the previous section, compared to conventional cooling methods.

4.1. *Thermal demonstration vehicle*

The first test die was a "thermal demonstration vehicle" (TDV) designed to explore the thermal capabilities of the jet-impingement embedded microcoolers. The TDV die was a 4 mil (100 micron) thick GaN/SiC die containing five rectangular tantalum nitride resistors on its upper surface, which were used to provide a relatively uniform 1 kW/cm^2 "background" heat flux (from all five resistors combined), along with four small RF HEMTs to simultaneously provide much higher, i.e., 30 kW/cm^2, local flux hotspots. The overall die footprint was 2.5 mm $\times$ 5 mm, as seen in the top view shown in Figure 8. The resistors and HEMTs were independently supplied with controlled

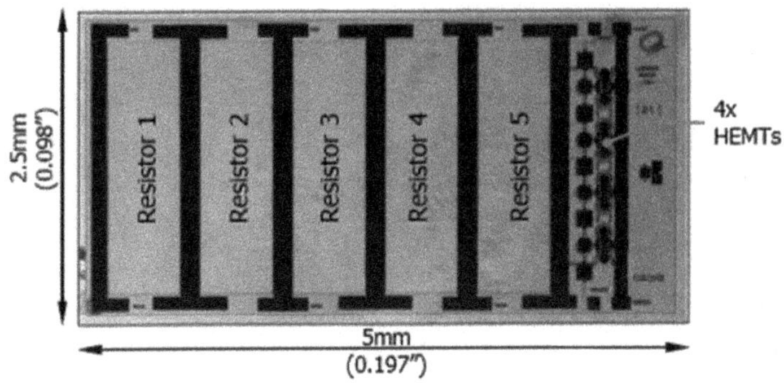

Figure 8. Thermal demonstration vehicle (TDV) die.

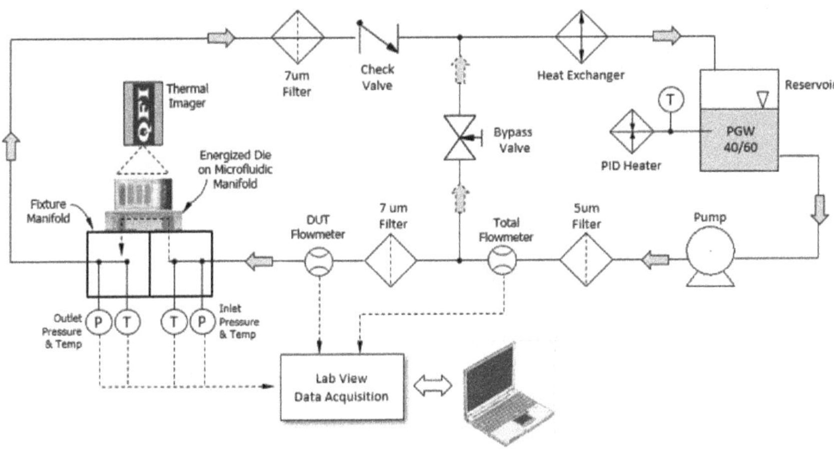

Figure 9. Schematic diagram of test flow station.

DC voltages to obtain a precise heat flux on the upper surface of the die.

The TDVs were attached to their microfluidic manifolds with gold–tin (AuSn) solder in a fashion similar to that shown in Figure 7, incorporated into their test fixtures, which provided external electrical and fluidic ports, and finally integrated into a test flow station, which is shown schematically in Figure 9. The flow station consisted of a fluid reservoir, pump, filters, valves, and other equipment to supply coolant to the TDV manifold. The flow rate and pressure drop across the manifold were measured every second, using a LabView data collection program. The coolant was a 40% propylene glycol and 60% water (by volume) mixture.

The test setup enabled the die to be positioned directly beneath an infrared thermal imager (QFI) to allow for real-time, full-field thermal images of the top surface of the die to be taken. A thermal emissivity calibration map was performed prior to the start of the testing to account for the differing thermal emissivities of the materials on the top surface of the die.

Detailed CFD and conjugate heat transfer (CHT) models of the TDV die/manifold were created in ANSYS Workbench in order to predict (and optimize) the microcooler performance. Figure 10 shows a comparison of the measured (left) and simulated (right) surface temperatures of the TDV while dissipating 1 kW/cm^2 die-level heat flux (131 W total dissipation over the 2 mm × 2.5 mm die) and 30 kW/cm^2 local heat flux in the four HEMTs at a flow rate of

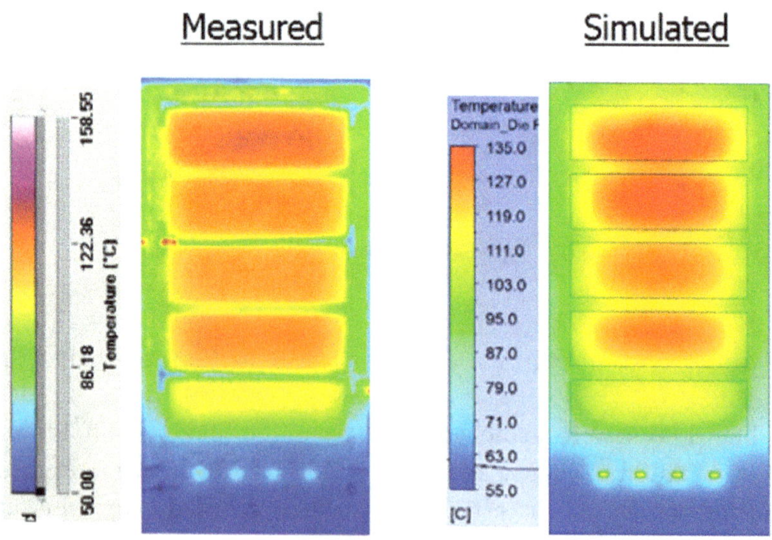

Figure 10. Measured and simulated thermal maps of TDV.

Table 1. Measured vs. simulated temperatures of TDV.

Heat source	Applied heat flux (kW/cm^2)	Measured temp.[a] (°C)	Simulated temp.[a] (°C)	Delta (%)
Resistor 1	2.1	121.5	122.9	1.2
Resistor 2	1.7	116.9	123.2	5.4
Resistor 3	1.9	115.7	118.8	2.7
Resistor 4	2.1	116.4	118.6	1.9
Resistor 5	2.6	98.5	102.5	4
HEMTs	30	119	116	−2.5

Note: [a]Average value over heat source.

100 mL/min. A comparison of the measured and modeled temperature of resistor 4 is given in Table 1. As can be seen from the image and the table, the performance of the microcooler is very predictable using standard CFD tools, with an error on the order of 5%. The discrepancy calculated based on the temperature *rise* relative to the 25°C inlet coolant temperature as opposed to the measured temperature is <6%.

Defining the thermal resistivity as the temperature at the top surface of the die minus the inlet coolant temperature (25°C for these tests) divided by the heat flux (q') applied to the respective surface, i.e.,

$$R_{\text{th}} = \frac{T_{\text{surf}} - T_{\text{inlet}}}{q'} \tag{3}$$

the measured thermal resistance for the resistor surfaces of the TDV was 0.047 cm^2K/W at a flow rate of 100 mL/min, whereas for the HEMTs, it was 0.0031 cm^2K/W. The lower (better) value for the HEMT is due to additional spreading which occurs within the die substrate for smaller heat sources, effectively reducing the heat flux by the time it reaches the lower heat transfer surface. The estimated uncertainty in the computed thermal resistance is ±12%.

The thermal resistivity of the microcooler was measured as a function of flow rate, with the results presented in Figure 11, along with

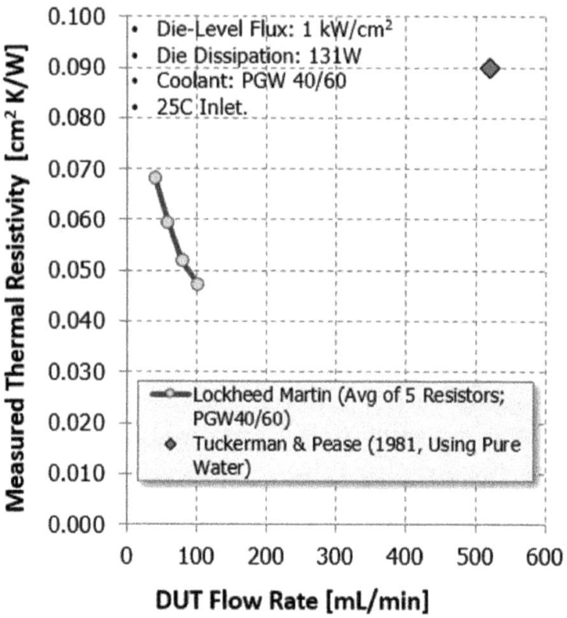

Figure 11. Measured thermal resistivity of microcooler as a function of flow rate. Also shown is a result reported by Tuckerman and Pease for reference.

a single data point representing the results reported by Tuckerman and Pease in their original (now 30+ years old) pioneering work on embedded cooling. It is interesting to note that this microcooler, utilizing single-phase jet impingement, outperforms the original microcooler by a factor of nearly 2× using a flow rate of only 20% of the original. The use of pure water in the current microcooler, as opposed to a diluted propylene gycol — water mixture, would further improve its performance. This is a testament to the advances made in microfabrication, jet impingement, and modeling techniques over the intervening decades.

4.2. *Electrical demonstration vehicle*

Having demonstrated the microcooler's *thermal* capabilities, a second test vehicle was developed to verify and quantify how the improved thermal performance translates to improved RF performance. The test vehicle was an in-house-designed, four-stage broadband (K_u–K_a) GaN/SiC RF MMIC amplifier, utilizing 0.15 μm gate length GaN HEMTs from the Qorvo GaN15 process. The MMIC footprint was roughly 2.5 mm × 5.0 mm, and it was 100 microns thick.

To serve as a baseline for comparison, one such HPA was mounted onto a test fixture and cooled using a *conventional* thermal packaging similar to that shown in Figure 2, including layers down to the package substrate. The RF test fixture was then placed on a temperature-controlled thermal chuck (beneath a thermal imager), and the temperature of the chuck was increased proportionally to the dissipated HPA power in order to simulate the thermal resistance of a conventional cold plate, which was not included in the test setup. The test fixture had input and output RF connectors to allow for the characterization of the RF performance of the MMIC using calibrated RF test equipment.

An identical HPA, from the same wafer as that used with conventional cooling, was mounted onto an embedded microcooler and installed in an RF test fixture similar to that used for the conventionally cooled HPA, but with fluidic ports to supply coolant to the microcooler. The RF output power of the embedded and remotely cooled versions of the electrical demonstration vehicle (EDV) MMIC was then recorded as a function of the input RF drive level. The test station used to test the EDV was the same station which was used to test the TDV (Figure 9), with the addition of RF test equipment (network analyzer, cables, RF input, and RF power meter). The thermal performance of the two cooling approaches was evaluated by taking IR images of the top surface of the MMICs at various drive levels, and the RF performance was measured using RF power measurement equipment.

The measured EDV test results are shown graphically in Figure 12 in the form of (normalized) output RF power versus (normalized) input RF power. The output RF power curves of the embedded-cooled (blue line/symbols) and conventionally cooled (red line/symbol) HPAs are shown on the same graph for comparison. For the embedded-cooled EDV, the microfluidic flow rate was

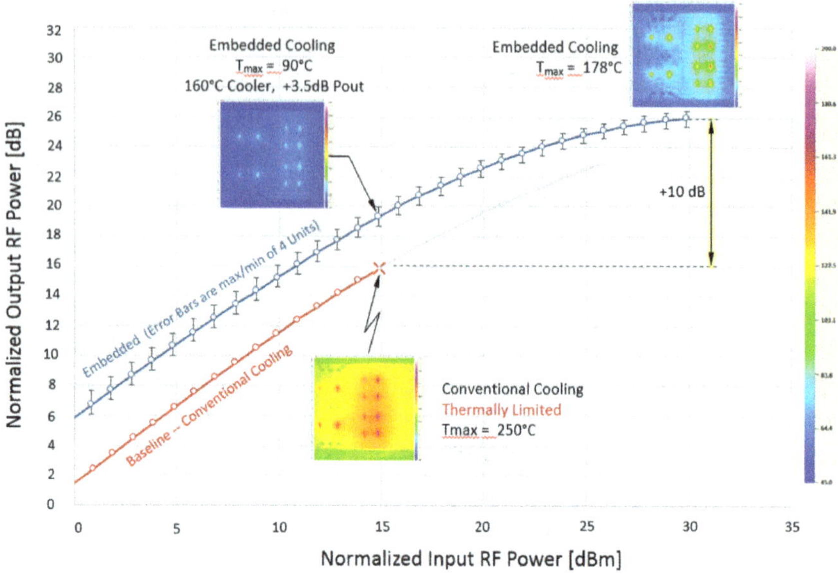

Figure 12. Comparison of remotely cooled and embedded-cooled HPAs.

100 mL/min (of a 40% propylene glycol and 60% water mixture) at a fluid inlet temperature of 45°C.

The data for the embedded-cooled HPA includes the results from four different EDV units, with the solid line representing the *average* output power of the units and the error bars showing the *maximum* and *minimum* measured output powers of the four tested units. Also shown in the figure are several measured thermal IR images recorded at specific operating points during the test, all referencing the same thermal scale shown on the right.

Several observations can be made from the results. The first is that the output power of the HPA when the embedded cooling is in the low drive level (i.e., linear range) is approximately 4 dB higher than the same HPA when conventionally cooled, i.e., the gain has increased by +4 dB. Comparing the thermal maps taken at a (normalized) input drive level of 15 dB shows that the maximum die temperature is only 90°C when embedded cooled but over 250°C when conventionally cooled. This roughly 160°C temperature differential is responsible for the 4 dB difference in gain measured in this (four-stage) HPA.

Another observation to be made from the figure is that the maximum output power of the HPA when cooled conventionally is thermally limited at 16 dB (relative level) if the peak temperature on the die is limited to 250°C. Pushing the HPA any harder would result in degraded reliability and potential short-term burnout. When embedded-cooled, the same HPA can be driven much harder, basically as hard as the test equipment could deliver (a relative input of 30 dB), and therefore produces considerably more output power. The results show that at the maximum available drive level of 30 dB, the output power of the embedded-cooled HPA was over 10 dB higher than the conventionally cooled version of the same MMIC, while its peak temperature is still only 178°C. This means that the HPA is *still not thermally limited*, and given more input power, it could produce more output power, although electrical saturation effects are seen to start at around 30 dB input. Embedded cooling has thus removed the thermal limit of the device and allowed it to achieve its full electrical potential. It is worth noting that at the high end of the drive level, the HEMTs on the embedded-cooled HPA are dissipating an enormous 60 kW/cm^2 or even more heat flux!

Figure 13 shows the computed (normalized) power-added efficiency (PAE) of the conventionally cooled and embedded-cooled HPAs. As can be seen, the PAE of the embedded-cooled HPA, at a relative drive level corresponding to the highest drive level of the conventionally cooled HPA (around 21 dB relative input), is roughly 3 percentage points higher than the PAE of the conventionally cooled HPA at the same drive level. This improvement in efficiency can again be explained thermally, given the 160°C difference in temperature at those drive levels. Since the embedded-cooled HPA is no longer thermally limited, it can be driven harder and operated closer to its optimal PAE point (around 27 dB relative input), which, when compared to the highest possible efficiency of the conventionally cooled HPA, is nearly 5 percentage points higher.

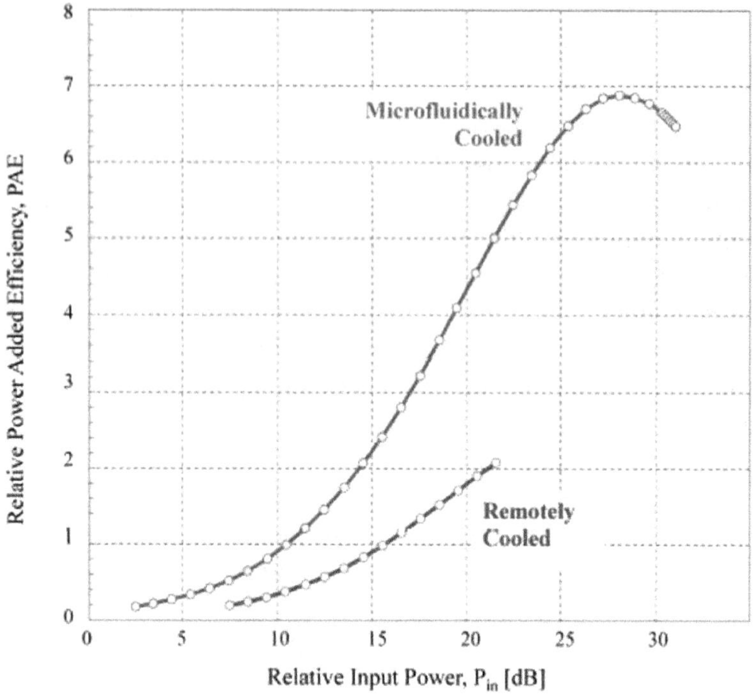

Figure 13. Computed PAE of the conventionally and microfluidically cooled HPAs.

Table 2 provides the full set of measured results for the baseline (remotely cooled) and four embedded-cooled EDV units. It should be noted that EDV unit number 1 was mounted on a slightly different microcooler design compared to units 2–4, with higher internal flow restrictions, which explains the difference in fluid CoP.

The impact of embedded cooling on the thermal and RF performance of HPAs is clear from this table, adding over 4 dB of small signal gain and 10 dB of available output power while, at the same time, improving PAE by nearly 5 percentage points. This is due to the fact that the thermal resistance of the embedded-cooled HPAs is roughly 4.8× lower than that of the remotely cooled HPA, resulting in a greatly reduced junction temperature. In addition to improving RF performance, the reduced junction temperature greatly improves the MTTF as well.

Table 2. Measured data on embedded- and conventionally cooled EDVs.

Metric	Baseline remotely cooled	Unit 1 embedded cooled	Unit 2 embedded cooled	Unit 3 embedded cooled	Unit 4 embedded cooled	Units
Total dissipation	15.0	31.0	34.6	31.0	32.3	[W]
Die heat flux	0.120	0.248	0.277	0.248	0.258	[kW/cm^2]
Volumetric flux[a]	1.8	3.8	4.3	3.8	4.0	[kW/cm^3]
Hotspot flux	26.9	64.4	69.7	65.3	67.8	[kW/cm^2]
Max. junc. temp.	250	178	181	178	191	[K]
HEMT R_{th}[b]	0.0082 ±0.0008	0.0016 ±0.0002	0.0017 ±0.0002	0.0017 ±0.0002	0.0018 ±0.0002	[cm^2 K/W]
Flow rate	N/A	110	105	110	110	[mL/min]
Fluid CoP[c]	N/A	18	39	31	37	[—]
Max RF power	Ref (0 dB)	+10.3	+10.8	+11.4	+11.0	[dB]
Small sig. gain	Ref (0 dB)	+3.5	+4.6	+5.4	+6.0	[dB]
PAE	Ref.	+4.1	+4.5	+5.6	+4.9	[%-points]
Est. MTTF[d]	77 $\times 10^7$	6.9 $\times 10^{11}$	4.5 $\times 10^{11}$	6.9 $\times 10^{11}$	1.2 $\times 10^{11}$	[h]

Notes: [a]Volume includes die, die-attach solder, microfluidic manifold, and an additional volume equal to the manifold to account for an eventual base plate.

[b]Defined as maximum junction temperature minus fluid inlet temperature divided by HEMT (hotspot) flux. Measurement error estimates assume ±10% possible error in IR temperature measurement.

[c]CoP (coefficient of performance) defined as total die heat dissipation divided by pumping power (pressure drop across DUT × flow rate). Assumes a 75% efficient pump/motor combination.

[d]Based on measured Qorvo Arrhenius data for the TQGaN15 process: activation energy = 2.57 eV, MTTF = 3.2 × 10^{10} h at 200°C.

5. Considerations with Embedded Cooling

As mentioned in Section 2, there have been several impediments to the widespread adoption of embedded cooling in fielded systems, despite the enormous demonstrated thermal and RF benefits it offers. One of the biggest impediments is the perceived reliability (or unreliability) of embedded cooling, with three of the main concerns being:

(1) leakage at microfluidic interfaces,
(2) clogging of the microjets (or channels),
(3) erosion caused by the microjets or in microchannels.

All three are legitimate concerns and require special care and attention during the design, fabrication, and servicing stages of the embedded-cooling system. Unfortunately, there are no "silver bullet" methods to guarantee that these issues will not arise. There are, however, some best practices which can help minimize their probability.

With regard to leakage, we have found the use of solders or mechanical (rubber O-rings or gaskets) seals preferable to the use of "hard" epoxies, such as silver-filled, high-thermal-conductivity epoxy. The coverage of solder and its wetting of mating/adjacent surfaces are far superior and more robust than those of epoxies. Solder is also amenable to automated batch processing in conventional equipment used for die attachment and less prone to thermal expansion mismatch failures than hard epoxies. Finally, solders are also less affected by contact with various liquids than are epoxies. If rubber (elastomer) gaskets are used for making seals, it is important to ensure their compatibility with the coolant to avoid interaction. An excellent reference which provides compatibility tables for nearly two dozen common O-ring/gasket materials and hundreds of gases and fluids is the Parker O-ring handbook [48, Chapter 7]. It is also important during the design stage to try to minimize contact between flowing fluid and sealing materials in order to avoid flow erosion of the seals, which could eventually allow fluid leak paths, in addition to breaking off wear particles which can travel downstream and clog microfluidic openings.

In systems which might exhibit catastrophic consequences from leakages, dielectric (electrically non-conductive) fluids could be used as the coolant. Many of the previously mentioned prior works, in fact, utilized such fluids primarily because of their favorable boiling characteristics; however, they also offer the added benefit of not causing electrical shorts if leaks occur. Polyalphaolefin (PAO) is a very common dielectric coolant used on many aircraft platforms, where it also sometimes serves as a lubricant for various mechanical systems. The downside of using dielectric fluids in single-phase heat transfer is that their thermal performance is inferior to that of pure water or even glycol–water mixtures. Incropera [6] provides a detailed discussion of coolant selection, along with several figures of merit with regard to heat transfer capability. It is shown that many of the commercially available "Flourinert" coolants, including FC-43, FC-72, and FC-77, have thermal figures of merit which are only around 0.1–0.25 that of pure water, so the thermal penalty for utilizing dielectric fluids (in single-phase heat transfer) can be significant.

With regard to clogging, our early testing with microfluidics taught us the necessity of implementing clean-room-type processes during design, parts handling, assembly, and testing. We found, as did other groups [30], that the most benign items, such as lint and adhesive-backed tapes, often used to clean fixtures or seal openings while not in use, can generate particles as large as the microchannels or microjets and can easily cause clogging over time if inadvertently introduced into the system. Many of the components within the flow loop, such as pumps, gears, valves, hoses, and gaskets, can also be the *source* of contaminants, especially during wear-in periods or if subjected to high-velocity fluids.

Another potential source of contaminants is the cooling fluid itself. Fresh coolant, right out of the container, invariably contains a multitude of particles of varying materials and sizes. One industrial standard for measuring fluid cleanliness [45] provides a nomenclature to indicate the number of particles greater than a given size per mL of fluid, with three sequential numbers indicating the number of particles greater than 4, 6, and 14 microns. Typical ISO cleanliness, which we have measured for a brand-new coolant (propylene glycol), is ISO 17/16/13, which, while being relatively clean, still allows 1,300 particles of 4 μm and 80 particles of 14 μm per mL of fluid. At even a modest flow rate of 100 mL/min, this means that 130,000 particles sized 4 microns (or larger) and 8,000 particles of 14 microns will flow through the microcooler per minute. A typical fluid cleanliness measurement report, as shown in Figure 14, includes an optical photo of the contaminants along with the particle count histogram.

With proper design and clean handling processes in place, it does appear possible to mitigate clogging within microfluidic devices, at least for time scales on the order of hundreds of hours. As an example, Figure 15 shows test results obtained in our laboratory during a 160 h continuous run with our previously described microcooler. The flow rate and pressure drop across the microcooler were continuously recorded during the entire test, and thermal images of the EDV die were periodically taken to assess its temperature. The flow

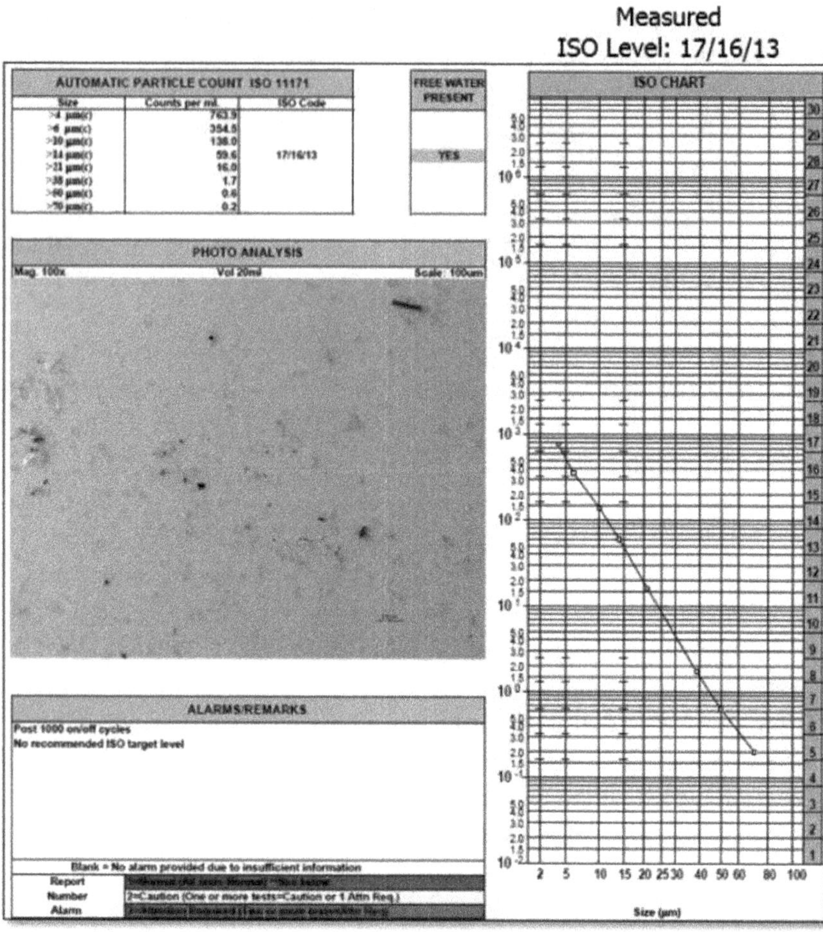

Figure 14. Measured particle counts for fresh coolant prior to first use.

was stopped periodically to perform RF compression tests of the type shown in Figure 12, and the peak output power was recorded. The time instants where RF measurements were made are indicated by the upward-facing arrows in the figure. There were no measurable trends seen in the hydraulic or thermal performance during the entire test, indicating a lack of any noticeable clogging. There were also no significant changes to the RF output power over the duration of the test. While this does not mean that the system could run indefinitely

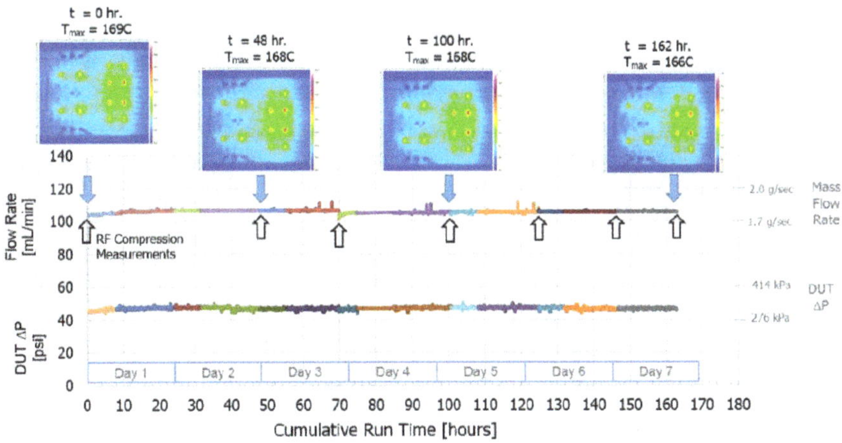

Figure 15. Continuous run for 160 h showing no signs of clogging or performance degradation.

without clogging, it does indicate that a service interval of hundreds of hours is a practical goal for system cleaning and maintenance. Researchers investigating the use of microfluidic devices for biological purposes (cell separation in blood, for example) have devised some clever techniques for preventing clogging using small-amplitude vibrations [46] and electric fields [47].

One of the most effective mitigation paths to prevent clogging, beyond the recommendations already provided, is to increase the size of the fluid passages. It should be noted, however, that having fluid passages which are larger than the smallest particles which pass system filters does not ensure that the devices will not clog. Researchers investigating the clogging mechanism in microchannels have identified several different mechanisms by which small particles can "team up" to clog microchannels larger than any individual particle [49]. There is a trade-off between thermal performance and channel/jet diameter which needs to be made in any design. A general rule of thumb is to use jets and/or channels which are as large as possible and which still support the thermal performance requirements of the application.

Finally, with regard to erosion, the concern with using microjet impingement in particular is that it may cause local pitting/erosion at the jet impact points or at the orifice walls. We have analyzed the potential for erosion both analytically via CFD particle-tracking simulations and experimentally with targeted erosion testing. The analytical results presented by Ditri *et al.* [42] showed that particles below a certain "threshold" size would be trapped by fluid streamlines and never impinge upon the die back side, thus presenting no potential for surface erosion. A similar result was discussed by other researchers investigating potential erosion in microfluidic devices [50]. Our simulations showed that particles of 2 microns or smaller were completely captured by the flow streamlines. Adding two-micron or finer filters to the flow loop would therefore mitigate erosion concerns from fluid particles.

We also performed experiments to confirm that proper controls could prevent erosion. The goal of the erosion test was to experimentally measure the amount of die back-side erosion during (relatively) long-term exposure to flowing fluid and impingement jets in realistic operating conditions. The erosion test consisted of subjecting five die/manifold assemblies to realistic flow conditions for a specified amount of time and periodically sending the units out for inspection of the back-side die profile.

The flow station included five-micron filters to limit the size of potential impinging contaminants within the fluid, as discussed previously (see Figure 9). After an initial baseline surface profile measurement, a total flow time of 1,000 h was accumulated, with periodic stops at 100, 300, 600, and, finally, 1,000 h to send the samples out for surface profilometry to remeasure the back-side profile of the die. The measurement technique utilized a white-light interferometer, with stated depth (pitting) accuracy in the nanometer range and spatial resolution in the micrometer range. Typical of RF HPAs, the die which was used in our testing had an initial four-micron layer of gold on its back side to serve as an RF ground plane.

Figure 16 shows the results of the maximum measured pit depth versus accumulated flow time for each of the five tested dies. The vertical scale spans a total of 4 μm, which is the initial back-side gold thickness. As can be seen, even after 1,000 h of impingement, there is a barely measurable amount of erosion on the back side,

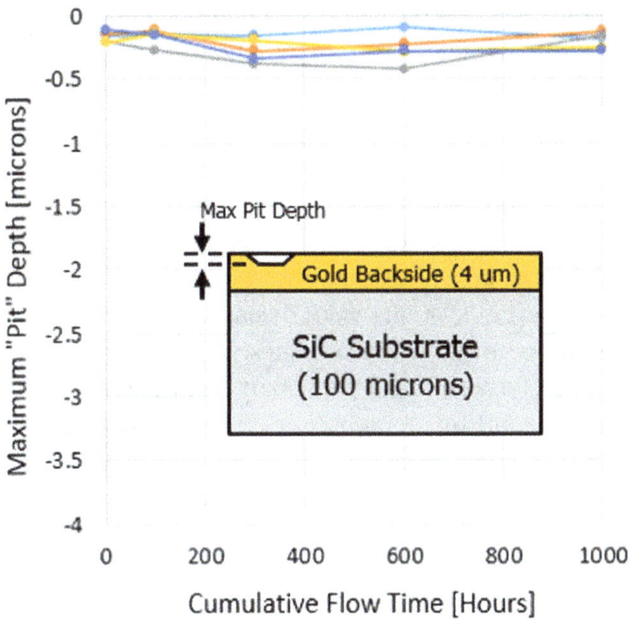

Figure 16. Continuous run for 160 h showing no signs of clogging or performance degradation.

arguably within the measurement repeatability, since some of the curves decrease and then increase again, which is physically unlikely. These results show that prevention of die back-side erosion is possible if suitable filtering and contaminant particle controls are in place. It must be borne in mind, however, that the testing was performed using PGW and the target surface was gold, which, while relatively soft, is also quite immune to corrosion. Other fluid and material combinations may not fare as well, particularly if the fluid is corrosive to the target surface.

6. Conclusion

Embedded cooling has been shown to provide the potential for a greater than 3× reduction in thermal resistance when compared to more traditional remote cooling paradigms when utilizing single-phase submerged jet-impingement heat transfer. CFD and

CHT simulations, performed using standard commercially available software, were shown to compare very favorably with measured data, allowing accurate performance predictions to be made without the need for costly hardware iterations. The impact of reduced thermal resistance on RF performance was experimentally shown to be very significant, with upward of a 4 dB increase in HPA gain, a 10 dB increase in output power, and several percentage points increase in PAE. Finally, some practical considerations for implementing embedded cooling in actual systems were discussed, and recommendations were provided to mitigate such issues. Embedded cooling is thus a very promising thermal management approach for dealing with both the current and future demands of high heat flux applications. Adoption of embedded cooling could lead to very significant reductions in the size, weight, and prime power requirements of those systems.

Acknowledgments

This research was developed with funding from the Defense Advanced Research Projects Agency (DARPA). The views, opinions, and/or findings expressed are those of the author and should not be interpreted as representing the official views or policies of the Department of Defense or the U.S. Government.

The author would like to thank Drs. Ken Plaks, Avi Bar-Cohen, Abirami Sivananthan, and Kaiser Matin from DARPA for their helpful guidance and support of this research.

References

[1] DoD, *Reliability Prediction of Electronic Equipment*, MIL-HDBK-217F, 1991.

[2] D. Via, GaN reliability — Where we are and where we need to go, in: *CS Mantech Conference Proceedings*, 2014, 15–18.

[3] J.A. Mittereder, N.S. Cronk, S.C. Binari, G.D. Via, D. Fanning, H. Tserng, P. Saunier, K. Decker, E. Beam, RF Arrhenius life testing of X-band high voltage GaAs PHEMTs, in: *2009 Reliability of Compound Semiconductors Digest (ROCS)*, Greensboro, NC, 2009, pp. 103–108.

[4] J. Ditri, R. R. Pearson, R. Cadotte, J. W. Hahn, D. Fetterolf, M. McNulty, D. Luppa, GaN unleashed: The benefits of microfluidic cooling, *IEEE Trans. Semicond. Manuf.*, 29 (4), 2016, 376–383.

[5] A. Darwish, B. Huebschman, E. Viveiros, H. Hung, Dependence of GaN HEMT millimeter-wave performance on temperature, *IEEE Trans. Microwave Theory Tech.*, 57 (12), 2009, 3205–3211.

[6] F.P. Incropera, *Liquid Cooling of Electronic Devices by Single-Phase Convection*, John Wiley & Sons, New York, 1999.

[7] I. Mudawar, Assessment of high-heat-flux thermal management schemes, *IEEE Trans. Compon. Packag. Technol.*, 24 (2), 2001, 122–141.

[8] M. Ki Sung, I. Mudawar, Single-phase and two-phase hybrid cooling schemes for high-heat-flux thermal management of defense electronics, in: *2008 11th Intersociety Conference on Thermal and Thermomechanical Phenomena in Electronic Systems*, Orlando, FL, 2008, pp. 121–131, doi: 10.1109/ITHERM.2008.4544262.

[9] A.M. Bar-Cohen, Embedded cooling for wide bandgap power amplifiers: A review, *ASME J. Electron. Packag.*, 141, 2019, 040803.

[10] A. Bar-Cohen, Towards embedded cooling — Gcn-3 thermal packaging technology, in: *Cooling of Microelectronic and Nanoelectronic Equipment: Advances and Emerging Research*, M. Iyengar, K.J.L. Gcislcr, and B. Sammakia (eds.), World Scientific Publishing Company, Singapore, 2014, Chapter 16.

[11] A. Bar-Cohen, J.J. Maurer, D.H. Altman, Embedded cooling for wide bandgap power amplifiers: A review, *ASME. J. Electron. Packag.*, 141 (4), 2019, 040803.

[12] D.B. Tuckerman, R. Pease, High performance heat sinking for VSLI, *IEEE Electron. Device Lett.*, EDL-2, 1981, 126–129.

[13] D.B. Tuckerman, Heat-Transfer Microstructures for Integrated Circuits, Stanford University, 1984.

[14] S.G. Kandlikar, History, advances, and challenges in liquid flow and flow boiling heat transfer in microchannels: A critical review, *ASME J. Heat Transfer*, 134 (3), 2012, 034001.

[15] J.R. Thome, Micro-two-phase electronics cooling: Finally on its way, in: *2019 18th IEEE Intersociety Conference on Thermal and Thermomechanical Phenomena in Electronic Systems (ITherm)*, Las Vegas, NV, 2019, pp. 1354–1363.

[16] S. Kandlikar, S. Garimella, D. Li, S. Colin, M. King, *Heat Transfer and Fluid Flow in Minichannels and Microchannels*, Elsevier B.V., Oxford, 2006.

[17] M. Ohadi, K. Choo, S. Dessiatoun, E. Cetegen, Fundamentals of microchannels, in: *Next Generation Microchannel Heat Exchangers*, Springer Briefs in Applied Sciences and Technology, Springer, New York, 2013.

[18] D.C. Wadsworth, I. Mudawar, Cooling of a multichip electronic module by means of confined two-dimensional jets of dielectric liquid, *ASME J. Heat Transfer*, 112 (4), 1990, 891–898.

[19] E.N. Wang, L. Zhang, L. Jiang, J.M. Koo, J.G. Maveety, E.A. Sanchez, K.E. Goodson, Micromachined jets for liquid impingement cooling of VLSI chips, *J. Microelectromech. Syst.*, 13 (5), 2004, 833–842.

[20] K.W. Jung, F. Zhou, M. Asheghi, E.M. Dede, K.E. Goodson, Experimental study of single-phase cooling with DI water in an embedded microchannels-3D manifold cooler, in: *2019 IEEE 21st Electronics Packaging Technology Conference (EPTC)*, Singapore, 2019, pp. 164–166.

[21] A. Piazza, S. Hazra, K.W. Jung, M. Degner, M.P. Gupta, E. Jih, M. Asheghi, K.E. Goodson, Considerations and challenges for large area embedded micro-channels with 3D manifold in high heat flux power electronics applications, in: *2020 19th IEEE Intersociety Conference on Thermal and Thermomechanical Phenomena in Electronic Systems (ITherm)*, Orlando, FL, 2020, pp. 77–82.

[22] M.D. Schultz *et al.*, Microfluidic two-phase cooling of a high power microprocessor part A: Design and fabrication, in: *2017 16th IEEE Intersociety Conference on Thermal and Thermomechanical Phenomena in Electronic Systems (ITherm)*, Orlando, FL, 2017, pp. 450–457.

[23] M.D. Schultz *et al.*, Microfluidic two-phase cooling of a high power microprocessor part B: Test and characterization, in: *2017 16th IEEE Intersociety Conference on Thermal and Thermomechanical Phenomena in Electronic Systems (ITherm)*, Orlando, FL, 2017, pp. 458–465.

[24] M. Schultz, F. Yang, E. Colgan, R. Polastre, B. Dang, C. Tsang, M. Gaynes, P.R. Parida, J. Knickerbocker, T. Chainer, Embedded two-phase cooling of large 3D compatible chips with radial channels, *J. Electron. Packag.*, 138 (2), 2016, 021005.

[25] T. Brunschwiler *et al.*, Dual-side heat removal by silicon cold plate and interposer with embedded fluid channels, in: *2018 17th IEEE Intersociety Conference on Thermal and Thermomechanical Phenomena in Electronic Systems (ITherm)*, San Diego, CA, 2018, pp. 331–339.

[26] P.R. Parida *et al.*, Thermal model for embedded two-phase liquid cooled microprocessor, in: *2017 16th IEEE Intersociety Conference on Thermal and Thermomechanical Phenomena in Electronic Systems (ITherm)*, Orlando, FL, 2017, pp. 441–449.

[27] D. Back, K.P. Drummond, M.D. Sinanis, J.A. Weibel, S.V. Garimella, D. Peroulis, Design, fabrication, and characterization of a compact hierarchical manifold microchannel heat sink array for two-phase cooling, in: *IEEE Trans. Compon. Packag. Manuf. Technol.*, 9 (7), 2019, 1291–1300.

[28] K.P. Drummond, J.A. Weibel, S.V. Garimella, D. Back, D.B. Janes, M.D. Sinanis, D. Peroulis, Evaporative intrachip hotspot cooling with a hierarchical manifold microchannel heat sink array, in: *2016 15th IEEE Intersociety Conference on Thermal and Thermomechanical Phenomena in Electronic Systems (ITherm)*, Las Vegas, NV, 2016, pp. 307–315.

[29] R. Mandel, S. Dessiatoun, M. Ohadi, Embedded two-phase cooling of high flux electronics using a directly bonded FEEDS manifold, in: *2016 15th IEEE Intersociety Conference on Thermal and Thermomechanical Phenomena in Electronic Systems (ITherm)*, Las Vegas, NV, 2016, pp. 77–84.

[30] D.G. Bae, R.K. Mandel, S.V. Dessiatoun, S. Rajgopal, S.P. Roberts, M. Mehregany, M.M. Ohadi, Embedded two-phase cooling of high heat

flux electronics on silicon carbide (SiC) using thin-film evaporation and an enhanced delivery system (FEEDS) manifold-microchannel cooler, in: *2017 16th IEEE Intersociety Conference on Thermal and Thermomechanical Phenomena in Electronic Systems (ITherm)*, Orlando, FL, 2017, pp. 466–472.

[31] C. Woodcock, F. Houshmand, J. Plawsky, M. Izenson, D. Fogg, R. Hill, S. Phillips, Y. Peles, Piranha pin-fins (PPF): Voracious boiling heat transfer by vapor venting from microchannels — System calibration and single-phase fluid dynamics, in: *Fourteenth Intersociety Conference on Thermal and Thermomechanical Phenomena in Electronic Systems (ITherm)*, Orlando, FL, 2014, pp. 282–289.

[32] A. Kosar, Y. Peles, TCPT-2006-096.R2: Micro scale pin fin heat sinks: Parametric performance evaluation study, *IEEE Trans. Compon. Packag. Tech.*, 30 (4), 2007, 855–865.

[33] X. Zhang, M.H. Nasr, D.C. Woodrum, C.E. Green, P.A. Kottke, T.E. Sarvey, Y.K. Joshi, S.K. Sitaraman, A.G. Fedorov, M.S. Bakir, Design, microfabrication and thermal characterization of a hotspot cooler testbed for convective boiling experiments in extreme-microgap with integrated micropin-fins, in: *2016 15th IEEE Intersociety Conference on Thermal and Thermomechanical Phenomena in Electronic Systems (ITherm)*, Las Vegas, NV, 2016, pp. 85–90.

[34] P.A. Kottke, T.M. Yun, C.E. Green, Y.K. Joshi, A.G. Fedorov, Two phase convective cooling for ultra-high power dissipation in microprocessors, in: *Fourteenth Intersociety Conference on Thermal and Thermomechanical Phenomena in Electronic Systems (ITherm)*, Orlando, FL, 2014, pp. 199–204.

[35] Y.J. Kim, Y.K. Joshi, A.G. Fedorov, Y.J. Lee, S.K. Lim, Thermal characterization of interlayer microfluidic cooling of three-dimensional integrated circuits with nonuniform heat flux, *ASME J. Heat Transfer*, 132, 2010, 041009.

[36] A. Bar-Cohen, J. Maurer, J. Felbinger, DARPA's intra/interchip enhanced cooling (ICECool) program, in: *CS Mantech Conference*, CS Mantech, New Orleans, 2013, pp. 171–174.

[37] D.H. Altman, Development of a diamond microfluidics-based intra-chip cooling technology for GaN, in: *ASME InterPACK*, San Francisco, CA, July 6–9, 2017, pp. V003T04A006.

[38] D.H. Altman, Advanced thermal solutions for GaN RF electronics, in: *IEEE ITherm*, Orlando, FL, May 2017.

[39] V. Gambin, B. Poust, M. Watanabe, D. Ferizovic, A. Oki, M. Wojtowicz, K. Hobart, G. Mandrusiak, Impingement cooled embedded Ddiamond (ICED) GaN HEMTs, in: *ASME InterPACK*, San Francisco, CA, July 2015.

[40] V. Gambin, B. Poust, D. Ferizovic, M. Watanabe, G. Mandrusiak, D. Lin, T. Dusseault, Impingement cooled embedded diamond multiphysics co-design, in: *Intersociety Conference on Thermal and Thermomechanical Phenomena in Electronic Systems (ITherm)*, Las Vegas, NV, May 31–June 3, 2016.

[41] C. Creamer, K. Chu, P.C. Chao, B. Schmanski, T. Yurovehak, S. Sweetland, Microchannel cooled, high power GaN on diamond MMIC, in: *Lester Eastman Conference on High Performance Devices (LEC)*, Ithaca, NY, August 5–7, 2014.

[42] J. Ditri, J. Hahn, R. Cadotte, M. McNulty, D. Luppa, Embedded cooling of high heat flux electronics utilizing distributed microfluidic impingement jets, ASME Paper No. IPACK2015-48689, 2015.

[43] J. Ditri, M. McNulty, S. Igoe, Embedded microfluidic cooling of high heat flux electronic components, in: *Lester Eastman Conference on High Performance Devices (LEC)*, Ithaca, NY, August 5–7, 2014, pp. 1–4.

[44] J. Ditri, J.W. Hahn, M.K. McNulty, Microjet Impingement Heat Sink, US Patent No. 10,306,802, 2015.

[45] ISO International Standard 4406:2017(E) Hydraulic fluid power — Fluids — Method for Coding the Level of Contamination by Solid Particles, Geneva, 2017.

[46] Y. Yoon, S. Kim, J. Lee, *et al.*, Clogging-free microfluidics for continuous size-based separation of microparticles, *Sci. Rep.* 6, 2016, 26531.

[47] D.H. Kang, K. Kim, Y.J. Kim, An anti-clogging method for improving the performance and lifespan of blood plasma separation devices in real-time and continuous microfluidic systems, *Sci. Rep.* 8, 2018, 17015.

[48] *Parker O-Ring Handbook*, 2018 Edition, Catalog ORD 5700/US, Parker Hannifin Corporation.

[49] H. Wyss, D. Blair, J. Morris, H. Stone, D. Weitz, Mechanism for clogging of microchannels, *Phys. Rev. E Stat. Nonlinear Soft Matter Phys.* 74, 2007, 061402.

[50] D. Squiller, I. Movius, M. Ohadi, P. McCluskey, Degradation mechanisms of embedded cooling systems for high heat flux power electronics: Particle erosion of silicon and silicon carbide, in: *2017 16th IEEE Intersociety Conference on Thermal and Thermomechanical Phenomena in Electronic Systems (ITherm)*, Orlando, FL, 2017, pp. 1298–1305.

https://doi.org/10.1142/9789811279379_0006

Chapter 6

Thermal Characteristics and Management Scheme of Self-Emissive and Liquid Crystal Displays

Sungki Kim

Samsung Electronics, Suwon, Korea Intel Corporation, Hillsboro, Oregon, USA
displaycooling@gmail.com

Abstract

Many new technologies have been developed for flat panel displays, and the technical achievements have brought high-performance displays to the market. The demand for higher brightness and image resolutions, however, requires proper thermal management technology, and an inherent increase in heat dissipation from a display requires proper thermal management. Currently, liquid crystal displays (LCDs) dominate the flat panel display market, whereas micro-light-emitting diodes (Micro-LEDs) and organic light-emitting diodes (OLEDs) have emerged as alternative technologies. These displays implement different technologies to reproduce an image, and their thermal phenomena show different characteristics as well. The objective of thermal management in a flat panel display is to ensure picture quality based on thermal reliability. This chapter presents thermal phenomena and characteristics in various types of displays. In addition, the effects of thermal design parameters on a display at the maximum temperature and temperature uniformity are investigated, and thermal management strategies for the various state-of-the-art displays are discussed.

Keywords: Thermal management, OLED, micro LED, flat panel display, LED backlight unit, electronics cooling

Nomenclature

D	Depth of a backlight unit [m]
H	Height of a backlight unit [m]
k_p	Thermal conductivity of material, in-plane direction [W/mK]
k_v	Thermal conductivity of material, through-plane direction [W/mK]
L	Width of a backlight unit [m]
q''	The amount of heat dissipation from an LED [W/m^2]
t	Thickness [m]
t_{total}	Total thickness of a thermal scheme [m]
T	Temperature of LED [°C]
$T_{\max}$	Maximum temperature of LED in a plane [°C]
$T_{\min}$	Minimum temperature of LED in a plane [°C]
x, y, z	Coordinates [m]
X, Y, Z	Dimensionless coordinates, $X = x/L$, $Y = y/H$, $Z = z/D$

Greek Symbols

θ	Dimensionless maximum difference of LED temperature in a plane, $\theta = (T_{\max} - T_{\min})/(T_{\max} - T_{\text{amb}})$.

1. Introduction

Over the past 20 years, demand for higher brightness and image resolutions has driven the development of flat panel display technologies. Many technologies have evolved and diminished in the flat panel display market, and technological achievements, such as high-dynamic-range (HDR) imaging and 8K resolution, have been implemented to improve the performance of the state-of-the-art displays. Higher performance, however, always comes with high heat dissipation from the internal components of the display systems so that further advancements in thermal management schemes have become one of the key factors in state-of-the-art flat panel displays. Over the past decade, liquid crystal display (LCD) technology has emerged as the winner in the flat panel display market over plasma displays. An LCD system is a non-emissive display, in which images can be reproduced by the white light from a backlight unit (BLU) and liquid crystal cells with color filters. Due to the intrinsic characteristics of liquid crystals and the white light from a BLU, an LCD system has technical barriers, such as color gamut, viewing angle, response time, and light leakage. Many studies, however, have focused on optics and materials in LCDs, and the inherent technical barriers have been improved by introducing new technologies, such as quantum-dot light conversion films for color gamut and multilayer laminated lenses for a wide viewing angle. As a result, LCD technologies have matured and dominate the current flat panel display market.

New technologies, namely micro-light-emitting diodes (micro-LEDs) and organic light-emitting diodes (OLEDs), have emerged as promising technologies for the next generation of flat panel displays. These technologies use self-emissive displays (SEDs), which emit light from sub-pixels, and a BLU is not required to reproduce an image. The plasma display, which is the first generation of SEDs to emerge in the past, faced technical issues, such as low color reproduction, electromagnetic radiation, low energy efficiency, and high power consumption, that pushed them out of the market. Later, market demand for a new display led researchers to pursue two potential self-emissive technologies. The two state-of-the-art SEDs can be differentiated based on the method used to emit light: micro-LED displays and quantum-dot or white-light OLED displays. Though the technology used to produce light is different, the thermal characteristics of the two are identical in terms of direct heat generation from the light-emitting device at the front surface of a display.

A substantial portion of the electricity consumption in a display is converted into heat, and the generated heat is transferred to a thermal management device and internal parts. The transferred heat affects the temperature of the internal parts of a display, including the BLU in an LCD and the light-emitting device in an SED, which influences picture quality and lifetime significantly.

Table 1. Internal structure and characteristics of various display technologies in the state-of-the-art flat panel displays.

Light-emitting technology	Illumination type	Light source	Application
Non-self-emissive LCD	Edge light	Single-edge type LED Multiple-edge type LED	TVs, smart phones, industrial signage
	Direct light	Mini-sized LED array Large LED with a diffusing lens	
Self-emissive display	LED package/ Bullet LED	Sub-pixel RGB package/LEDs	Large-format displays
	Micro-LED	Sub-pixel RGB LEDs	TVs, portable devices
	OLED	Sub-pixel RGB organic cells	TVs, smart phones
		QD-blue light organic cells	Small-format displays
		White-light organic cells	

The objective of thermal management in a flat panel display is to ensure picture quality based on thermal reliability. To achieve this objective, hotspots should be mitigated, and the temperature distribution in the display should be made uniform. This chapter presents the thermal phenomena and characteristics of various types of displays and discusses different thermal management strategies for state-of-the-art displays (Table 1).

2. Liquid Crystal Display

Figure 1 depicts a schematic diagram of a typical LCD system. An LCD system consists of an open cell, which has color filters, a liquid crystal layer, thin-film transistors (TFTs), optical parts, and a BLU to produce white light over the display surface. A liquid crystal is not a self-emissive device, and the white planar light always illuminates the backside of the liquid crystal sub-pixels to reproduce an image. The picture quality of an LCD system strongly depends on the white light produced by the light source in a unit, and the performance of

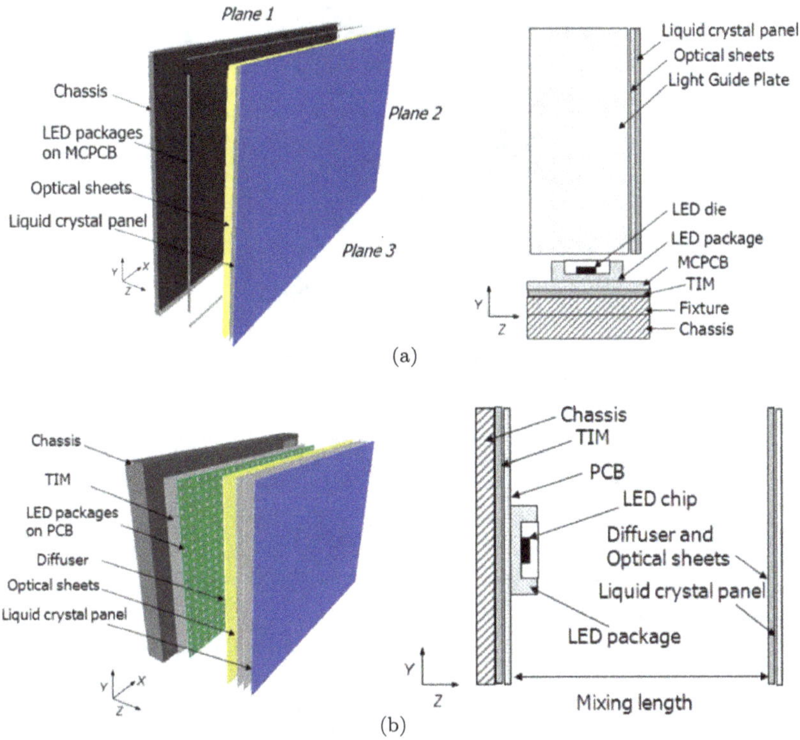

Figure 1. Schematic diagram of LCD displays and backlight structure: (a) edge-type BLU, (b) direct-type BLU.
Source: Copyright (2012) IEEE reprinted, with permission, from *IEEE Transactions on Components, Packaging and Manufacturing Technology*, 2 (11), 2012, 1838–1846.

an LCD system is determined by the proper thermal management of its BLU, which is required to ensure product quality. Currently, most LCD systems use LEDs as a light source, where the temperature is the most important parameter determining the optical characteristics of the white light.

BLUs can be categorized in terms of the alignment method of the light source as an edge-light type or a direct-light type. The thermal characteristics of the edge- and direct-light BLUs are not identical due to the differences in their mechanical structures and the alignment of the light source.

Figure 2 illustrates the distribution of LEDs and temperature measurement points in a BLU for analysis. The temperature of the

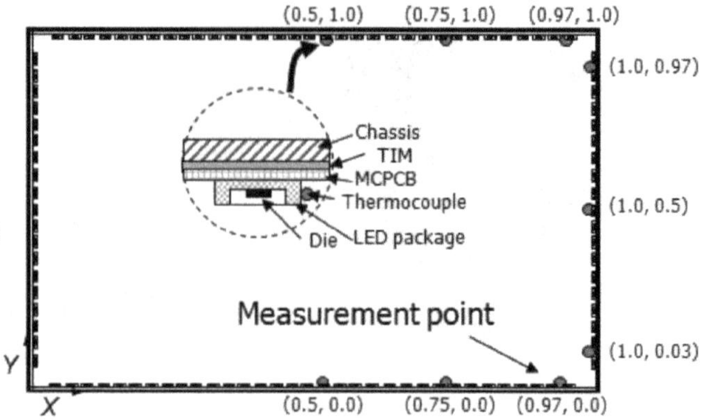

(a)

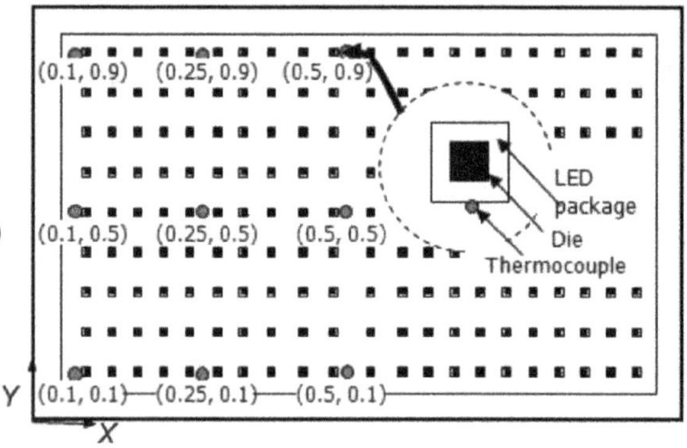

(b)

Figure 2. Dimensionless coordinates and locations of temperature measurement points in the experimental setup: (a) edge-type BLU, (b) direct-type BLU.
Source: Copyright (2012) IEEE reprinted, with permission, from *IEEE Transactions on Components, Packaging and Manufacturing Technology*, 2 (11), 2012, 1838–1846.

light source, i.e., the LEDs, is measured at various locations to evaluate the effect of the amount of heat flux from the sources and the geometrical position. It is impossible to measure the temperature of an LED die with a thermocouple directly because LED light reflects at the metal junction of a thermocouple. The reflection influences the temperatures of an LED package and the neighboring structure, and a thermal resistance model should be used to evaluate the temperature of an LED die from a package case temperature.

The thermal management system for an LCD consists of various materials and should be assumed to be a composite material to analyze its thermal performance. In the following discussion, the effective conductivity of materials is introduced to evaluate the efficacy of thermal design parameters. The assumption is applied to multilayer materials, with the layers combined having negligible contact resistance. We can describe the effective conductivity of a multilayer composite using the following equations:

$$k_p = \frac{\sum_{i=1}^{n} k_1 t_{1,i} + \sum_{i=1}^{m} k_N t_{N,i}}{t_{\text{total}}} \tag{1}$$

$$k_v = \frac{t_{\text{total}}}{\sum_{i=1}^{n} t_{1,i}/k_1 + \sum_{i=1}^{m} t_{N,i}/k_N} \tag{2}$$

where k_p and k_v denote the effective thermal conductivities in the in-plane and through-plane directions, respectively. T_{total} is the total thickness of the multilayer composite, and $t_{1,i}$ and $t_{N,i}$ are the ith thickness of material 1 and material N, respectively. The values of N and M represent the total layer numbers corresponding to n and m, respectively.

A significant temperature difference between LEDs causes a wavelength shift in LED light and a brightness change within the unit. So, the designated performance can be ensured by the uniform temperature of the light source. The uniformity of LED temperatures is evaluated by calculating the temperature difference across a BLU plane. The temperature difference affects the optical performance of an LED, and the relative temperature uniformity should be considered regarding the maximum LED temperature for q. The temperature difference, $(T_{\max} - T_{\text{amb}})$ can be evaluated by the relative temperature difference. In the discussion, θ denotes the dimensionless maximum temperature difference in LEDs in a plane, given by the following equation:

$$\theta = \frac{T_{\max} - T_{\min}}{T_{\max} - T_{\text{amb}}} \tag{3}$$

where $T_{\max}$ and $T_{\min}$ are the maximum and minimum LED temperatures, respectively, in a plane, and T_{amb} is the ambient temperature.

2.1. *Edge-light-type backlight unit*

An edge-light BLU has a light source along one or multiple circumferential planes, and the emitted light from the sources is diffused by optical parts over the display surface. An edge-light-type LCD system has a line-type heat source along the circumferential plane(s), and the heat from the light sources is transferred to internal structures. The line-type heat source is embedded in a very narrow space within a display, with a relatively small number of LEDs embedded in the circumferential distribution. However, the heat flux from each LED is inherently high in an edge-light BLU. The heat flux of an edge-light BLU usually ranges from 8,000 to 24,000 W/m^2 according to the brightness of the display.

The heat transfer model is illustrated in Figure 3. The light source, an LED, is mounted on a metal core printed circuit board (MCPCB) to effectively dissipate heat from the source. The metal core printed circuits are installed on the slender side wall of a BLU perpendicular to the display surface. Heat generated by LEDs is transferred via two paths: to the metal chassis through a metal core printed circuit and to optical parts above the LED die. To enhance the heat transfer from an LED, a high-conductivity material can be applied between an MCPCB and a metal chassis to spread heat effectively.

Table 2 lists the various thermal management schemes and their composite conductivities applied to an edge-type BLU. Assuming

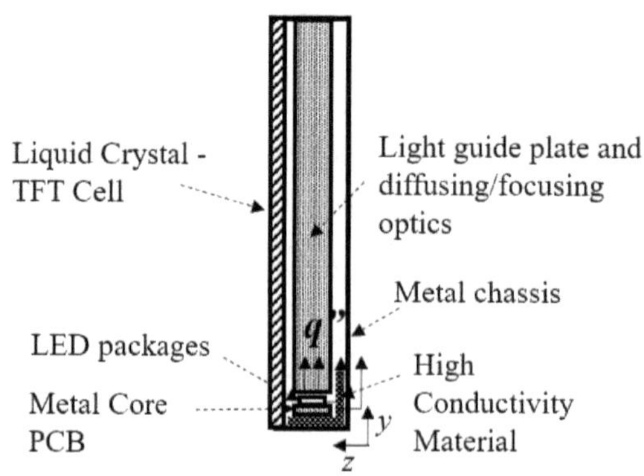

Figure 3. Heat transfer in an edge-type BLU (LED is the heat source).

Table 2. Thermal management schemes and the composite conductivities in an edge-type BLU.

	Metal PCB		TIM		Chassis			Conductivity [W/mK]	
Case	t [m]	k [W/mK]	t [m]	k [W/mK]	t [m]	k [W/mK]	LED pitch [m]	kp	kv
1	1.0×10^{-3}	170	5.0×10^{-4}	1.5	2.0×10^{-3}	110	7.0×10^{-3}	111.64	9.79
2	2.0×10^{-3}	170	5.0×10^{-4}	1.5	1.0×10^{-3}	110	7.0×10^{-3}	128.79	9.88
3	2.0×10^{-3}	170	5.0×10^{-4}	1.5	1.0×10^{-3}	110	1.4×10^{-3}	128.79	9.88
4	2.0×10^{-3}	380	5.0×10^{-4}	1.5	1.0×10^{-3}	110	7.0×10^{-3}	248.79	10.07
5	2.0×10^{-3}	380	5.0×10^{-4}	1.5	1.0×10^{-3}	110	1.4×10^{-3}	248.79	10.07

Source: Copyright (2012) IEEE Reprinted, with permission, from *IEEE Transactions on Components, Packaging and Manufacturing Technology*, 2 (11), 2012, 1838–1846.

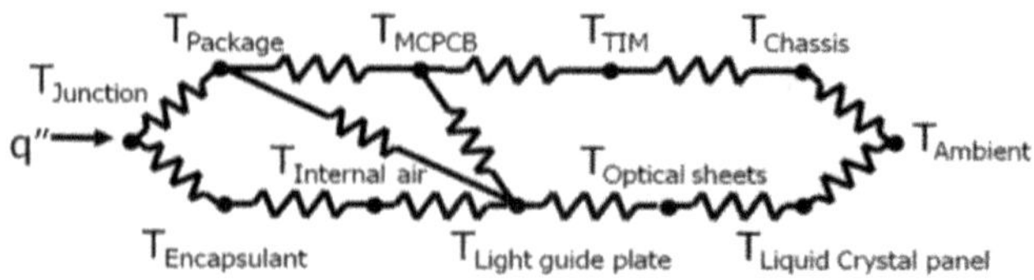

Figure 4. Thermal resistance model of an edge-type BLU.
Source: Copyright (2012) IEEE reprinted, with permission, from *IEEE Transactions on Components, Packaging and Manufacturing Technology*, 2 (11), 2012, 1838–1846.

that the structure of an LCD display is a multi-layer composite material, the performance of a thermal management scheme can be evaluated by analyzing the temperature of the heat source for various anisotropic conductivities.

Figure 4 depicts a thermal resistance model of an edge-type BLU. The model exhibits the complicated paths present in an edge-type BLU due to the complex internal structure of optical components. As described, the heat generated from the heat source has two paths to the ambient. The heat conducted to a chassis increases the temperature of the internal components, with the heat conducted through a light guide plate and optical sheets. In general, the thermal resistance through the optical parts is constant, and the thermal management strategy should focus on the thermal path through a chassis in an edge-type BLU.

Figure 5 exhibits the temperature distribution of an LCD in a four-edge-type BLU. The surface temperature distribution of the display shows isothermal lines in the form of symmetric circles. This shows the typical characteristics of conduction-dominant 2D heat transfer with the heat source on four circumferential planes. The temperature distribution indicates that the heat produced from the light sources at the four edges is transferred to the internal structure. In addition, it is observed that the temperature gradients near Plane 1 (the top horizontal edge) and Plane 3 are identical, suggesting little effect of natural convection from the surface.

Figure 6 portrays the typical isothermal lines on a display's front surface in various edge-type BLUs. The temperature distribution indicates the characteristics of various light source configurations.

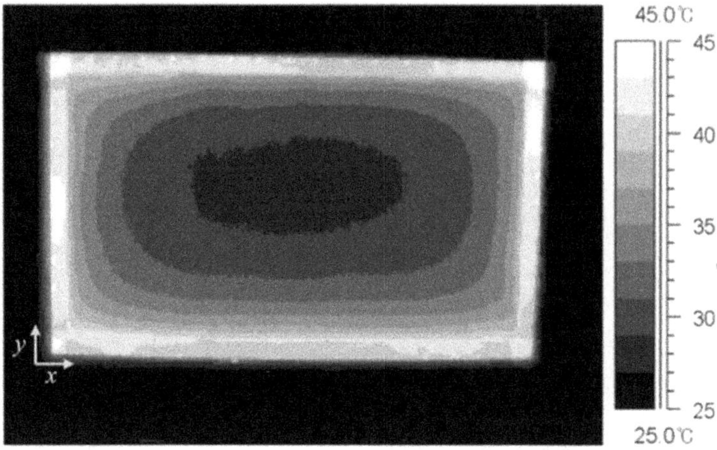

Figure 5. Infrared image of the front surface of an LCD with four-edge-type BLU: The dimensions of the display are 1020 mm in width, 600 mm in height, and 10 mm in depth.

Source: Copyright (2012) IEEE Reprinted, with permission, from *IEEE Transactions on Components, Packaging and Manufacturing Technology*, 2 (11), 2012, 1838–1846.

It is observed that isothermal lines vary with the location and number of heat sources. The heat intrusion from a heat source at the edge plane is conducted into the center through a metal chassis structure and optical parts. The heat transfer forms half-circle-, quarter-circle-, and circle-type isothermal lines on the surface according to the number of the heat-dissipating planes. In the case of an LCD system which adapts an edge-light BLU, the effect of natural convection is meager, and conduction heat transfer dominates the thermal characteristic of the display. In the case of a one-bottom-edge-type BLU, two-dimensional conduction from a bottom-plane heat source dominates heat transfer in the BLU, and the maximum temperature can be found at the center of the bottom-edge plane.

In the case of two-side-edge-type BLU, however, it is distinctive that the two-dimensional effect of conduction becomes diminished at the top plane, and the temperature of the LED follows one-dimensional conduction behavior from the vertical edge plane. This effect is caused by a temperature increase in Plane 3, where heat from two vertical heat sources is transferred by convection.

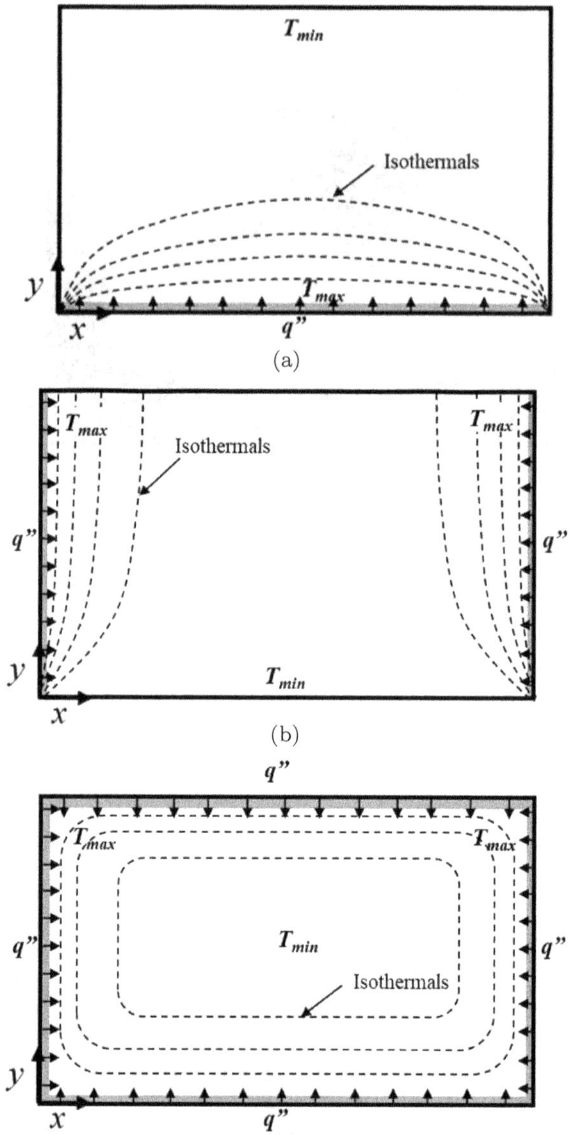

Figure 6. The isothermal lines in LCD with various types of edge-type BLUs: (a) one-bottom-edge-type BLU, (b) two-side-edge-type BLU, (c) four-edge-type BLU (gray color denotes heat source area).

The maximum temperature of the LED is observed at the top-edge plane, and the minimum temperature of the LED is found at the center of the bottom-edge plane, where heat transfer from the heat sources is meager. As shown in Figure 5, the symmetric-circle isothermal lines are observed in the four-edge-type BLU. The maximum temperature points are located at the two top corners, and detailed thermal characteristics are analyzed in the following discussion.

Figure 7 displays the effect of heat flux from LEDs on the temperature of light sources distributed in the four edge planes.

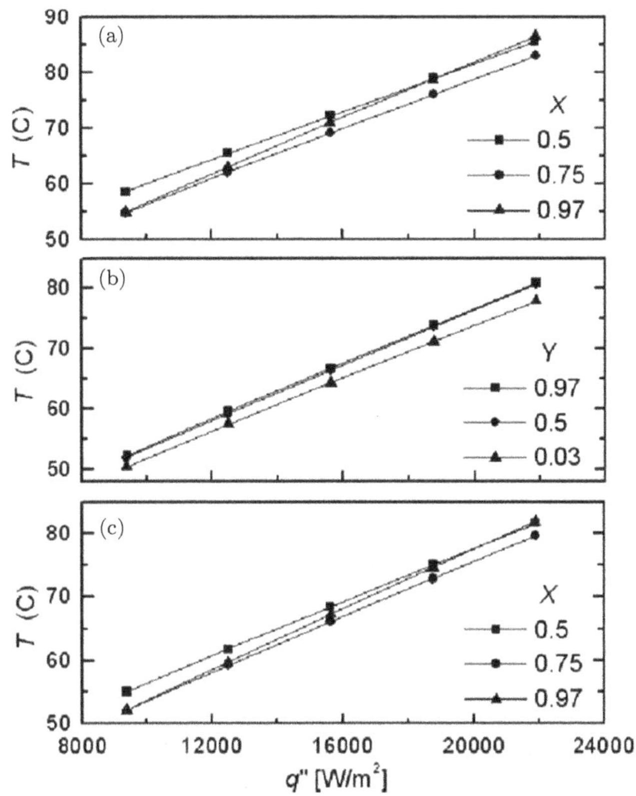

Figure 7. Behavior of an LED temperature, T, at various heat dissipations from an LED in an edge-type BLU in Case 1: (a) Plane 1, (b) Plane 2, (c) Plane 3. *Source*: Copyright (2012) IEEE reprinted, with permission, from *IEEE Transactions on Components, Packaging and Manufacturing Technology*, 2 (11), 2012, 1838–1846.

Figure 5 shows that the LED temperature increases linearly with heat flux in all planes. It is of interest, however, that the effect of heat flux on LED temperature varies according to the plane. The temperature at $X = 0.97$ is increased significantly by the increase of heat flux in the horizontal plane (Figures 7(a) and 7(c)). This indicates corner effect of heat transfer which impacts LED temperatures in an edge-type BLU. It is observed that the increase in the LED temperature with heat flux influences the maximum temperature in the horizontal plane of an edge-type BLU.

The heat flux from an LED is the primary factor affecting the maximum temperature of a light source in a plane. The temperature of the LED increases proportionally to the heat dissipation from LEDs. The maximum temperature is distinctively increased at Plane 1 when the in-plane conductivity of the MCPCB and the heat flux are high. However, the temperature variation behavior is different when the in-plane conductivity is high and the heat flux is modest. This can be attributed to the interaction of heat conduction from LEDs with the adjacent wall so that the counter effect of heat conduction on heat flux and conductivity should be considered in the thermal management of a multi-plane edge-light-type BLU.

The counter effect at the corner significantly increases the LED temperature when the heat flux from the LED is around 18,000 W/m^2. However, in Case 2 in Table 2, in which the in-plane conductivity of the MCPCB is half that of Cases 4 and 5, the maximum temperature is low. (The maximum temperature is found in Cases 3 and 5, in which the heat flux is high.) It is of interest that the effect of the in-plane conductivity of a PCB on the maximum temperature is meager in Cases 2 and 4. This implies that heat flux is the dominant design parameter affecting the LED temperature in an edge-type BLU.

The amount of heat flux from an LED also affects the temperature uniformity in an edge-type BLU. Figure 7 shows that the temperature difference is small in Cases 3 and 5 at Plane 1. This result reveals that the higher in-plane conductivity and heat dissipation from an LED are augmented, and the temperature uniformity is also enhanced in the horizontal plane. However, the performance can be deteriorated by the corner effect, which augments the heat transfer interaction between adjacent walls.

On the vertical walls, the higher heat dissipation from LEDs is favorable for temperature uniformity. But temperature uniformity is not always correlated with the increase in the in-plane conductivity of a PCB or the heat flux from an LED. Figure 8(b) shows that the effect of the two parameters on temperature uniformity is meager in the vertical plane. Case 3 in Table 2, where the in-plane conductivity of a PCB is moderate and the heat flux from an LED

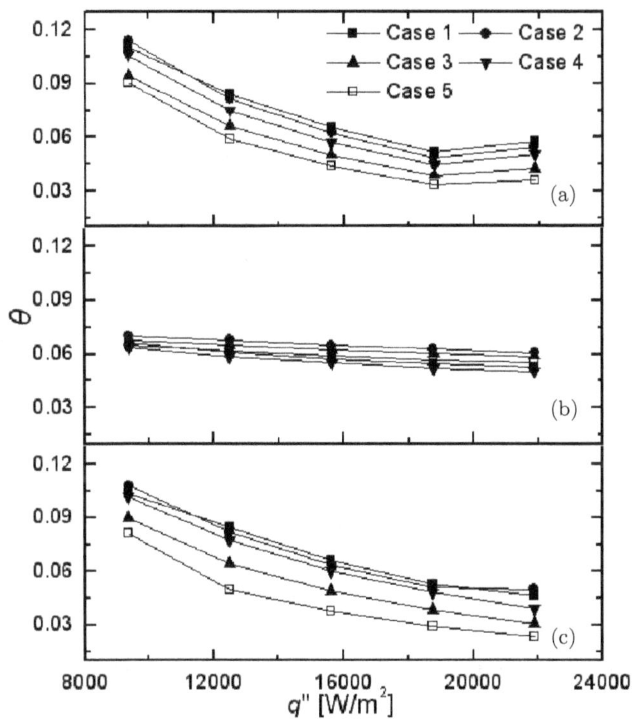

Figure 8. Behavior of the maximum temperature difference in a plane at various heat dissipations from an LED in an edge-type BLU: (a) Plane 1, (b) Plane 2, (c) Plane 3.

Source: Copyright (2012) IEEE reprinted, with permission, from *IEEE Transactions on Components, Packaging and Manufacturing Technology*, 2 (11), 2012, 1838–1846.

is high, exhibits a reduced temperature difference along the plane. So, the higher in-plane conductivity of a PCB and heat flux from a LED enhance the temperature uniformity of LEDs in the horizontal plane; however, this effect is not observed in the vertical plane.

In an edge-type BLU, complicated thermal behavior can be observed for the maximum temperature and temperature uniformity due to thermal flow interaction between adjacent planes. Additionally, the effect of thermal parameters differs from the composite conductivities of heat conducting materials in the unit. Therefore, it is concluded that a careful selection of thermal design parameters should be considered to ensure the display performance and thermal reliability of an LCD system.

2.2. *Direct-light-type backlight unit*

In this section, the effect of thermal design parameters on the maximum temperature and temperature uniformity of light sources in a direct-type BLU is scrutinized. The heat sources are distributed evenly over the front surface, and the amount of heat flux from each light source in a direct-type BLU is smaller than that in an edge-type BLU. Moreover, the material of a PCB, where LEDs are mounted, can have a relatively low conductivity for optimal thermal management. The major design parameters are the amount of heat dissipation from an LED, the conductivities of a PCB, and the thermal interface material, which thermally connects the internal components. Table 3 presents the specifications of typical PCB and thermal interface materials, and Table 4 lists the different thermal management schemes and the composite properties calculated from Eqs. (1) and (2).

As shown in Figure 9, the temperature on the display's front surface shows the characteristics of natural convection heat transfer in a direct-type BLU. Heat generated from an LED is conducted to a PCB through an LED package. The heat interacts with the heat from adjacent LEDs on the plane, and the heat is transported out of the display through a metal chassis structure and an enclosure by conduction. A direct-type BLU has an air cavity in the optical mixing length between a cell and a BLU. The heat from LEDs on a BLU is

Table 3. Conductivities of various PCB and thermal interface materials in a direct-light-type BLU.

		Material	t [m]	Conductivity [W/mK]	
				kp	*kv*
PCB	1	FR4	4.0×10^{-3}	19.5	0.3
	2	FR4 with Cu layers		185.0	0.6
	3	Aluminum Nitride		170	
TIM	1	Si $+AlO_2$	5.0×10^{-3}	1.5	
	2	Graphite		380.0	3.5

Source: Copyright (2012) IEEE Reprinted, with permission, from *IEEE Transactions on Components, Packaging and Manufacturing Technology*, 2 (11), 2012, 1838–1846.

Table 4. Thermal management schemes and the composite conductivities in a direct-type BLU.

Case	Composite materials	Conductivity [W/mK]	
		kp	*kv*
1	PCB 1 + TIM 1	9.5	0.54
2	PCB 1 + TIM 2	179.7	0.61
3	PCB 2 + TIM 1	83.1	0.9.
4	PCB 2 + TIM 2	293.3	1.11
5	PCB 3 + TIM 1	76.4	2.68
6	PCB 3 + TIM 2	286.7	6.2

Source: Copyright (2012) IEEE reprinted, with permission, from *IEEE Transactions on Components, Packaging and Manufacturing Technology*, 2 (11), 2012, 1838–1846.

conducted into the frontal surface of a display system through the air gap in a slender cavity.

Figure 10 depicts the thermal flow model and thermal resistance network of a direct-type BLU. The thermal resistance from an LED to a metal chassis is low, but that from an LED to the display surface is high due to the air cavity in the mixing length. The thermal management strategy should consider the "heat transfer insulation effect" within the air cavity, and optimal design of heat sources, i.e., LEDs, is the first step to ensuring thermal reliability in a direct-type BLU.

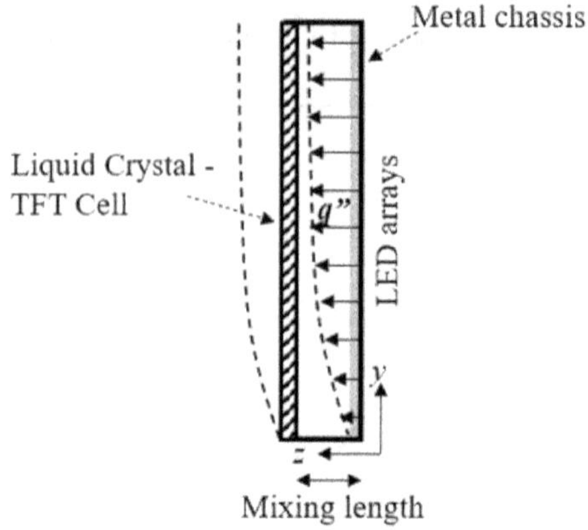

Figure 9. Heat transfer in a direct-type BLU (gray color denotes heat source).

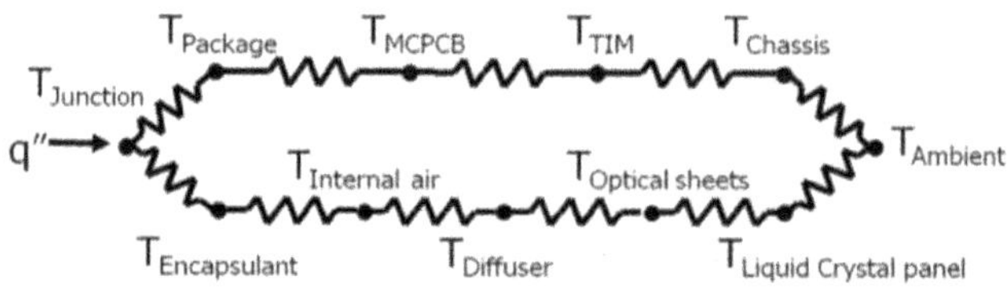

Figure 10. Thermal resistance model of a direct-type BLU.
Source: Copyright (2012) IEEE reprinted, with permission, from *IEEE Transactions on Components, Packaging and Manufacturing Technology*, 2 (11), 2012, 1838–1846.

Figure 11 shows an infrared image and typical isothermal lines on a display's front surface in an LCD with a direct-type BLU. The temperature distribution exhibits non-uniform behavior, as observed in the natural convection of a side-heated vertical wall. The distribution of the heat source, the LED, reflects the thermal characteristics of the display surface, and the relationship varies with the mixing length as well. The cavity in the mixing length functions as a thermal insulation layer, affecting the isothermal lines on the display surface. Therefore, natural convection affects the heat transfer in the cavity, and the display's front surface shows low temperature areas at both bottom sides. In general, the maximum temperature of an LCD with a direct-type BLU can be found at the top center of the display owing to the effect of natural convection.

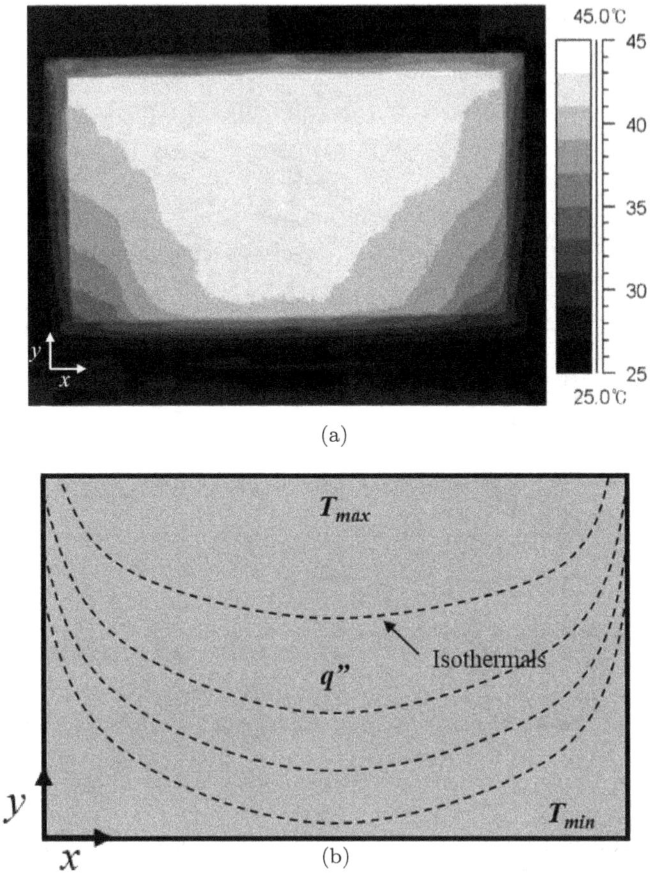

Figure 11. (a) Infrared image of the front of an LCD with direct-light type BLU. (b) Isothermal lines in an LCD with a direct-type BLU (gray color denotes heat source area).

Source: a: Copyright (2012) IEEE Reprinted, with permission, from *IEEE Transactions on Components, Packaging and Manufacturing Technology*, 2 (11), 2012, 1838–1846.

The effect of thermal design parameters is examined in a direct-type BLU in Figure 12. The figure depicts the temperature variation with the heat flux dissipated from a direct BLU. Each case is described in Table 4. It is observed that the temperature increases linearly with the heat flux from an LED, and the maximum value is

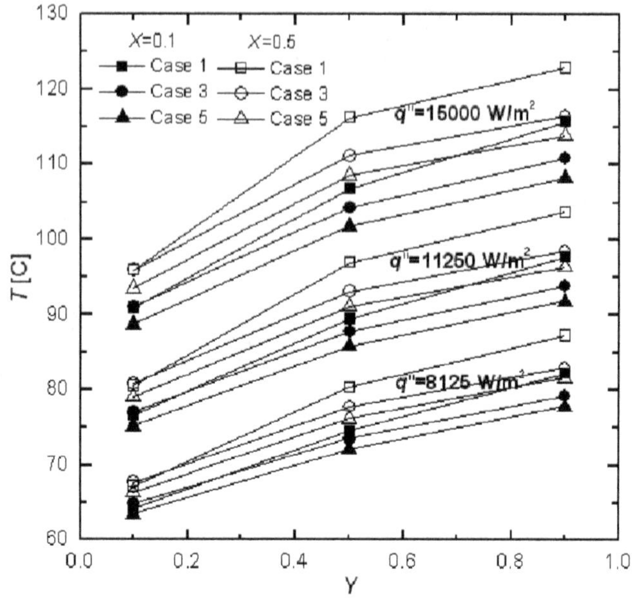

Figure 12. Variation of LED temperature at various heat dissipations from an LED in a direct-type BLU.

Source: Copyright (2012) IEEE Reprinted, with permission, from *IEEE Transactions on Components, Packaging and Manufacturing Technology*, 2 (11), 2012, 1838–1846.

found at the top center of the display. The high through-plane conductivity of a PCB displays a low temperature of an LED, as shown in Case 5. However, the effect of through-plane conductivity on the temperature is enervated in the case where the temperature is low. The in-plane conductivity on temperature is meager in a direct-type BLU, as observed in Case 3, which has an 800% higher in-plane conductivity than that of Case 1. So, it is elucidated that the through-plane conductivity of a PCB and the heat flux from an LED are dominant design parameters affecting the temperature of a direct-type BLU. This is because heat generated by LEDs is uniformly distributed over the display, and the role of in-plane conductivity of a PCB is enervated in the system.

Figure 13 exhibits the behavior of the maximum temperature of an LED, $T_{\max}$, for various composite conductivities. The results elucidate that higher conductivity in both the in-plane and through-plane

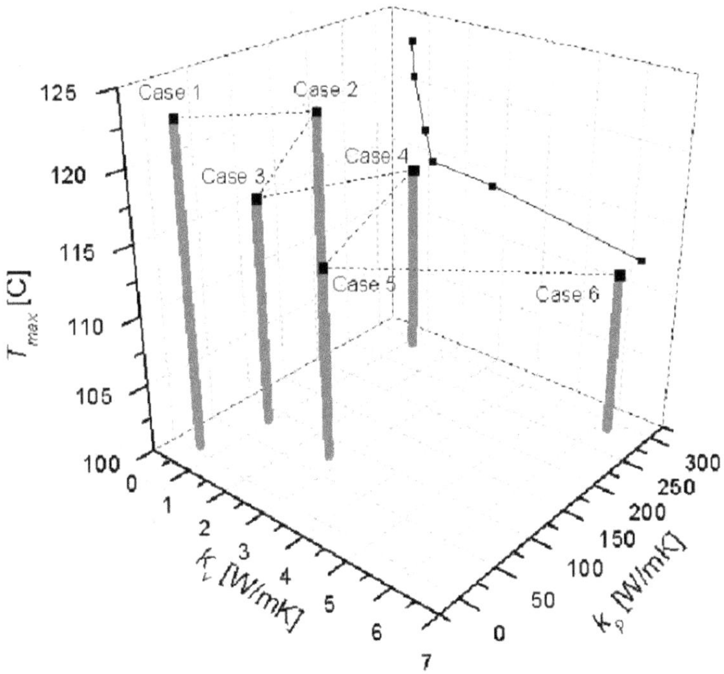

Figure 13. Behavior of the maximum temperature of an LED at various composite conductivities.
Source: Copyright (2012) IEEE Reprinted, with permission, from *IEEE Transactions on Components, Packaging and Manufacturing Technology*, 2 (11), 2012, 1838–1846.

directions results in a low LED temperature. However, it is of interest that a combination of lower in-plane conductivity and moderate through-plane conductivity (Case 5) demonstrates an analogous result on the maximum temperature, compared to the cases of the combination of higher-conductivity materials (Cases 4 and 6). This confirms that a proper combination of materials can enhance thermal management performance, considering composite material conductivities and heat dissipation from the LED in a direct-type BLU.

On the temperature uniformity of a direct-type BLU, the value of the temperature difference reduces with the increase in the in-plane conductivity of a PCB and heat flux from LEDs. The conductivity is the decisive parameter for temperature uniformity, excepting Cases 3 and 5. This demonstrates that temperature uniformity can be achieved with the combined effect of a PCB of low conductivity and a thermal interface material of high conductivity. This means that the performance of a thermal management scheme can be enhanced by the optimal selection of materials for the composite layers of a BLU.

3. Self-Emitting Displays

Self-emitting displays (SEDs) emit light from each pixel, and images can be reproduced without a BLU. So, the structure of an SED can inherently be simple because it does not require the complex optical parts and thermal management scheme of a BLU. Two types of SEDs are available currently: Micro-LED and OLED displays.

In SEDs, heat is generated from the light source directly. Therefore, thermal phenomena in SEDs differ from those in non-emissive displays, which have a remote heat source in a BLU. The most important characteristic of SEDs is that the major heat source, a light-emitting pixel, is exposed to the ambient directly. So, the heat from the source can be actively dissipated into the ambient by natural convection and radiation.

Figure 14 exhibits the basic structure of a micro-LED display. The display uses microscale, three-color LEDs mounted on a substrate. The red-, green-, and blue-color LEDs emit light directly to produce images, and the inorganic device can guarantee higher reliability against heat and ambient conditions. The display area is determined by the resolution of the display, the size of each LED pixel, and the pitch between LED pixels.

An LED has been actively adapted as a light-emitting device for industrial outdoor large-format signage. The LED display uses "bullet type" or "package type" LEDs as light-emitting devices, whose size ranges from 2 to 5 mm and the pitch between the devices ranges from around 1 to 3 mm in general. Although the devices applied to outdoor signage produce much heat for high-brightness performance, the heat flux is relatively moderate, so a complex thermal scheme is not required for the system. Meanwhile, a novel 4K indoor display implements an LED of tens of micrometers in size and pitch on a substrate and has a high-density array of LEDs

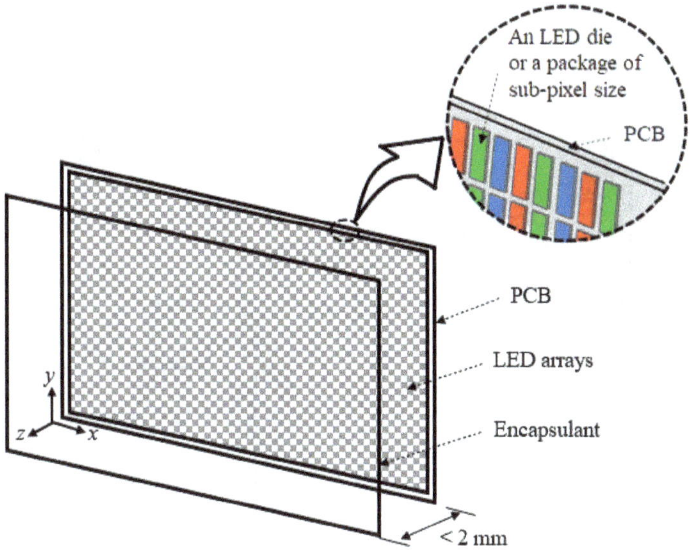

Figure 14. Schematic diagram of the structure of a micro-LED display.

with high heat flux in its light-emitting layer that requires a complex thermal management scheme.

Figure 15 shows a schematic diagram of the structure of an OLED display. The OLED display structure consists of organic material stacks between electrodes, which are deposited on a substrate. Light emits from the emitting layer (EML) in blue or white according to the color reproduction method. The single-wavelength light passes through color filters to produce red, green, and blue lights in each pixel. An encapsulant covers the multi-layer structure, including the EML and a substrate with an operating TFT layer. Heat is a major cause of damage to organic materials in the EML, and the development of thermal management is imperative to ensure the picture quality and reliability of an OLED display.

In the current display market, OLED displays have rapidly expanded their applications, from smartphones to TVs, due to their characteristics of a slim form factor, low energy consumption, and black-color expression. However, OLED displays have critical technical issues, such as the burn-in effect of the organic layer, relatively short lifetimes, and a high luminosity limit. In particular, the high

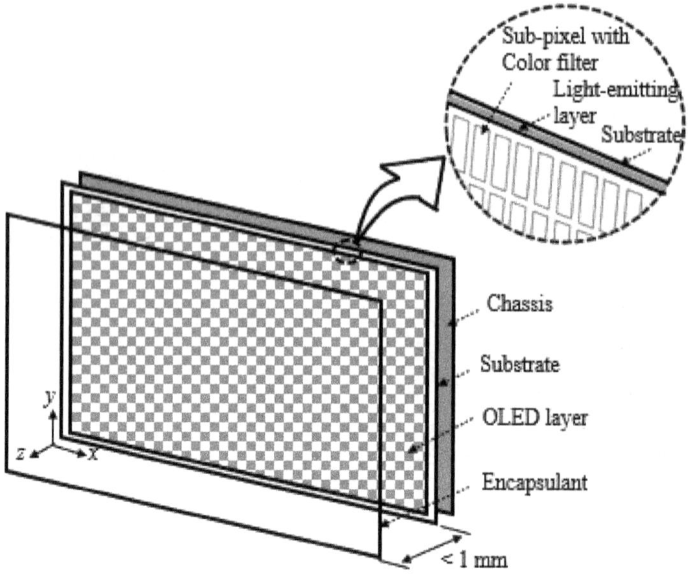

Figure 15. Schematic diagram of the structure of an OLED display.

luminosity limit is crucial because heat dissipation from an emitting layer increases dramatically because of the low quantum efficiency of OLEDs. The heat dissipation impacts the organic layer, inducing failure of color reproduction and light emission performance.

A micro-LED device is more reliable against temperature rise than an OLED device because a light-emitting device is composed of inorganic materials, and the quantum efficiency of an LED is relatively higher than that of an OLED. However, the two SED systems have a planar heat source on the display front, and the characteristics of a micro-LED display system are inherently identical to those of an OLED display system.

3.1. *Heat transfer analysis of self-emissive displays*

Figure 16 depicts a heat transfer model and thermal resistance network of an SED. It is worth noting again that the heat from a BLU in an LCD can be transferred to the ambient through a multi-stage layer structure, which includes a light mixing cavity, optical parts, and a

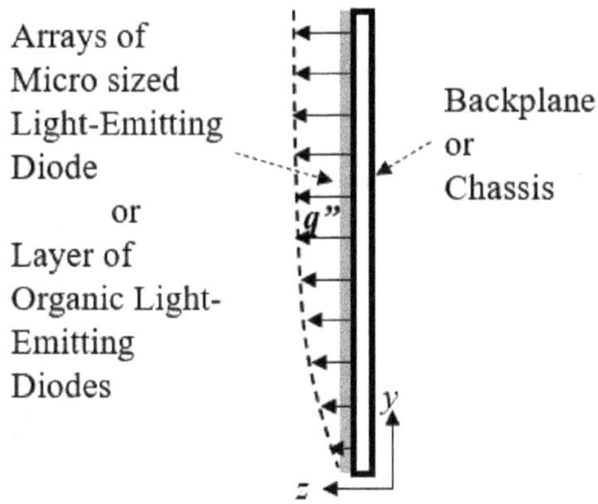

Figure 16. Heat transfer in a self-emissive display (gray color denotes heat source area).

mechanical chassis. In SEDs, meanwhile, the heat transfer path from heat sources to the ambient is relatively simple. The heat generated in a light-emitting layer of an SED has two paths to be transported: to the ambient air in front direction directly and to a backplane or a chassis in backward direction. Due to the high thermal resistance of the front path, most of the heat generated from a light-emitting layer is conducted to a backplane or chassis on the back, which is made with a high-conductive material. The operating circuits, as the second major heat source, also dissipate heat, impacting the light-emitting layer and the neighboring devices. In that case, heat from operating circuits communicates with the light-emitting layer, and it produces hotspots in the light-emitting layer. Then, the hotspot directly affects the wavelength of the emitted light and the lifetime of the light-emitting layer.

The thermal network is totally different from that of an LCD. Figure 17 exhibits the thermal resistance model of a micro-LED and an OLED display. The image can be displayed directly on the display front without a BLU, and the model has simple paths to transfer heat. Figure 18 depicts typical isothermal lines found on the display front of an SED. The maximum temperature is found at the top center, and the minimum temperature is observed at the

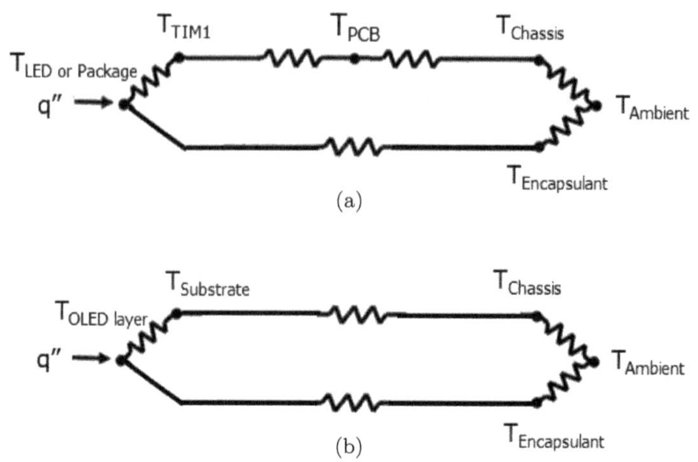

Figure 17. Thermal resistance model of SED: (a) micro-LED (b) OLED.

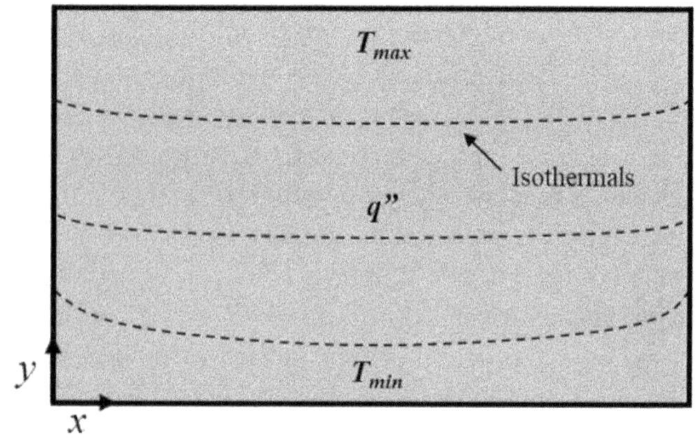

Figure 18. The isothermal lines in SED with a self-emitting device. (Gray color denotes heat source area).

bottom center. The temperature difference within the display front is relatively small, and the temperature distribution shows no abrupt change as found in LCDs with BLUs. The temperature distribution shows convective heat transfer from a vertical side wall with heat dissipation. As shown in Figures 14 and 15, the display panel of an SED is very thin, and the material has a very low heat capacity. These structural characteristics deliver an instant temperature change in the panel, corresponding to the luminosity efficiency. The current market trends for flat panel displays are inclined toward the compactness of the system volume, and one of the advantages of SEDs is that the light-emitting layer can be implemented in a very thin enclosure. However, direct heat dissipation of the light-emitting layer increases the temperature of the layer, and a deficit of convection in the thin enclosure augments the temperature increase through the backside of the display. This yields a dramatic temperature difference within the display system.

The objective of thermal management for a self-emitting device is twofold: mitigating isolated hotspots and maintaining temperature uniformity over the display surface. Therefore, high-performance heat spreaders are required for SEDs to overcome the technical barriers. Although the luminous efficiencies of micro-LEDs and OLEDs are higher than those of LCDs in general, the efficiency becomes worse when the display produces high-brightness images or has a low quantum efficiency. In SEDs, a high luminosity image causes a dramatic drop in light conversion efficiency, and an instant increase in temperature inflicts permanent damage to the emitting device. Therefore, thermal management becomes a key technology for ensuring picture quality and reliability in SEDs.

3.2. *Active thermal management for self-emissive displays*

The isolated hotspot can be observed if a low-efficiency image with a small area is reproduced in SEDs. The light-emitting layer of a micro-LED and an OLED display usually has a low heat capacity because the layer consists of layered materials of less than 1 μm thickness. So, isolated heat dissipation from the light-emitting layer experiences an instant temperature rise when an image of low luminous efficiency is displayed. In another case, heat generated from operating printed circuits also induces an isolated temperature increase at the light-emitting layer, where only limited heat can be dissipated out of the system. A light-emitting device, whether micro-LED or OLED, has a temperature limit against physical and chemical damage, and the device cannot recover even after the temperature drops below the limit. The damage deteriorates the performance and lifetime of SEDs significantly. That is why the performance of the thermal management scheme is an important design parameter in the color-reproducing performance and lifetime of SEDs. The fundamental study scrutinized the effect of flow oscillation on thermal dispersion in a water-filled active heat spreader. In the following, the thermal characteristics of SEDs are discussed based on the experimental results, and an active heat spreader on thermal dispersion is investigated in an emulated 42″ SED.

Figure 19 exhibits the temperature distribution of an emulated SED's surface with a hotspot at the center of the display. It simulates the typical temperature increase in the isolated hotspot in SEDs. As described, most of the heat produced from the light-emitting layer has two heat transfer paths: the display front and the backside heat spreader. The front is a display area where no opaque thermal management scheme can be applied, and most of the heat flows to the heat spreader attached on the backside of the light-emitting layer. If the isolated hotspot persists for a long time in the light-emitting layer, natural convection affects the radial symmetric temperature distribution, and the isolated hotspot shows an upward thermal dispersion phenomenon. This situation can induce irreversible damage to the light-emitting device.

As discussed, the objective of thermal management in an SEDs is to mitigate the isolated hotspot instantly. Here, the efficacy of an active heat spreader is deliberated in SEDs. Figure 20 depicts the effect of an active heat spreader on the hotspot mitigation for an emulated SED. The active heat spreader consists of a thin water-filled rectangular cavity and flow oscillator, and the operating fluid oscillates in the cavity according to the imposing frequency and displacement. The results elucidated that the active heat spreader decreases the maximum temperature of the hotspot and ensures temperature uniformity. This fundamental result implies that thermal dispersion on the light-emitting layer is a prerequisite for a reliable SED.

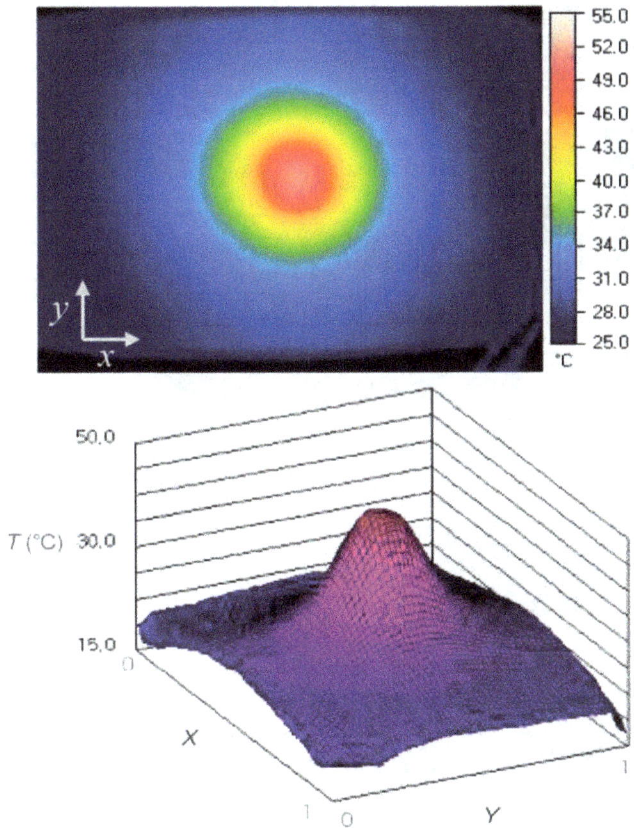

Figure 19. Example of temperature distribution of the front surface in an emulated SED with isolated temperature increase at the center.
Source: Copyright (2009) IEEE Reprinted, with permission, from *IEEE Transactions on Components and Packaging Technologies*, 32 (2), 2009, 235–242.

From the discussion on the active heat spreader, it can be inferred that this enhancement effect by an active heat spreader corresponds to the effect of heat transfer conductivity in a passive heat spreader on hotspot mitigation. In the current market, many SEDs adapt high-conductivity sheet metal or graphite sheet as a passive heat spreader with precise brightness control scheme. It is anticipated that current self-emitting devices will show steady improvement in their luminous efficiency, and the maximum brightness of the system

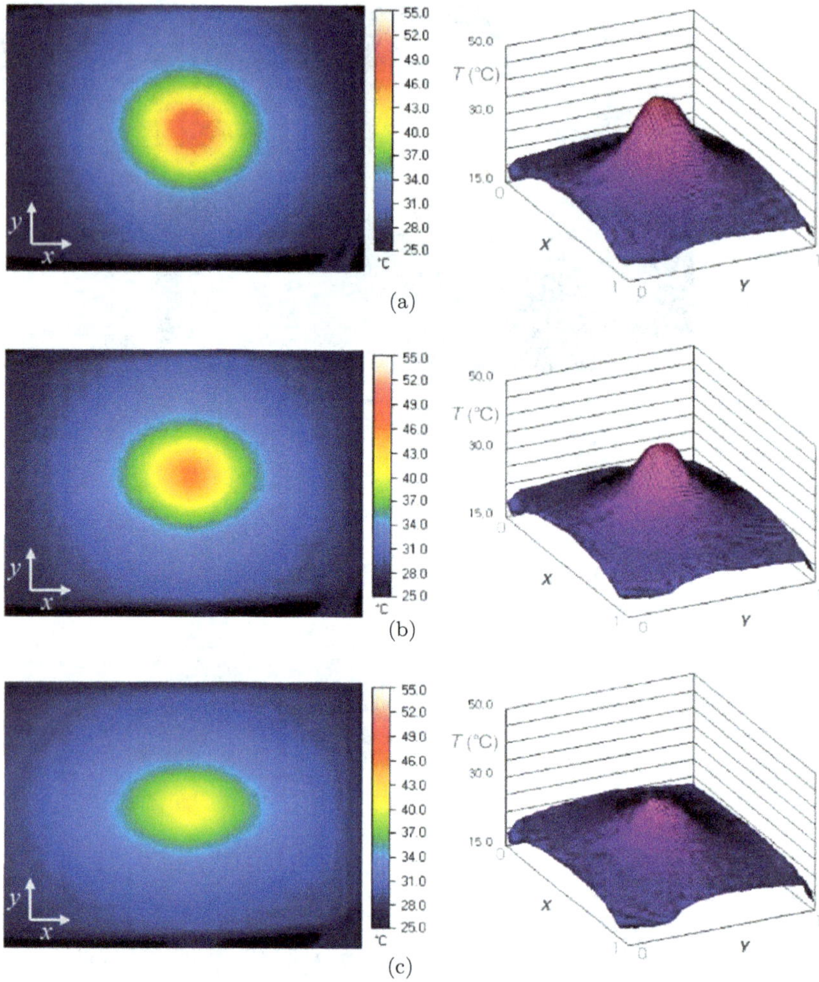

Figure 20. Effect of active heat spreader on temperature distribution in an emulated SED: tidal displacement variation of oscillating flow at frequency of 0.8 Hz: (a) 0.45 cm, (b) 0.98 cm, (c) 2.1 cm.

Source: Copyright (2009) IEEE Reprinted, with permission, from *IEEE Transactions on Components and Packaging Technologies*, 32 (2), 2009, 235–242.

will become a key factor in the competitiveness of SEDs in the near future. Therefore, the choice of thermal management strategy depends on the improvement of heat dispersion at the light-emitting layer in SEDs.

https://doi.org/10.1142/9789811279379_0007

Chapter 7

Advances in Materials Engineering for Conduction-Limited Direct Contact Cooling

Arden L. Moore

Department of Mechanical Engineering,
Louisiana Tech University, Ruston, LA, USA

Abstract

Philosophically, embedded-cooling strategies seek to remove as many barriers as possible between the primary thermal management solution and the component(s) of interest. However, in many cases, such as three-dimensional chip stacks and flexible electronics, conduction processes can still limit primary or secondary heat transfer pathways. Thus, materials development and an enhanced understanding of conduction physics represent opportunities for improving the cooling performance of the overall solution. Here, recent advances in bulk cubic crystals, two-dimensional layered materials, and advances in understanding conduction physics are briefly discussed, with a critical outlook on their potential for improving embedded cooling of electronic devices.

Keywords: Conduction, phonons, interfaces, thermal conductivity, 2D materials

1. Introduction

Since the rise of the computing age, there has been a well-documented drive toward miniaturization and increased density at the transistor level [1, 2], with third-generation FinFET transistors now utilizing 10 nm logic technology and fin widths as small as 7 nm [3]. Almost as well documented have been the increases in power dissipation and power density values across the device, die, and system levels that go along with these scaling trends [4], where CPUs can experience localized heat generation of ~200–400 W/cm^2 at hotspots [5] and RF power amplifiers can reach over 40 kW/cm^2 at their junctions [6]. For decades, as this continual miniaturization progressed, thermal engineers were aided in their quest to accommodate increases in dissipated power by the relatively high thermal conductivity (κ) of the materials associated with integrated circuitry, such as highly crystalline semiconductor substrates and metallic interconnects. These high-κ materials, combined with more modest power densities than those present today, meant that the majority of cooling solution designs focused on packaging, thermal interface materials (TIMs), and the external macroscale heat removal strategy, which predominantly was forced air convection or, more rarely, indirect single-phase liquid cooling. However, over the past decade, there have been a confluence of circumstances in the world of electronics hardware that have pushed existing cooling paradigms to their limits and are now forcing thermal management professionals to focus much more closely on die-level and even junction-level embedded-cooling approaches, including research and development of alternative materials, interfaces, and aggressive means of direct cooling. These challenging circumstances include growing power localization, the transition to 3D chip stacks, the strong reduction of κ from bulk values for thin films and wires due to boundary scattering effects, and the rise in the density of conduction-limiting interfaces within increasingly complex devices. Additional details on each of these phenomena may be found in Refs. [7, 8] for unfamiliar readers.

Even in advanced systems, such as those that incorporate single- or multi-phase cooling via microchannels, impinging jets, vapor

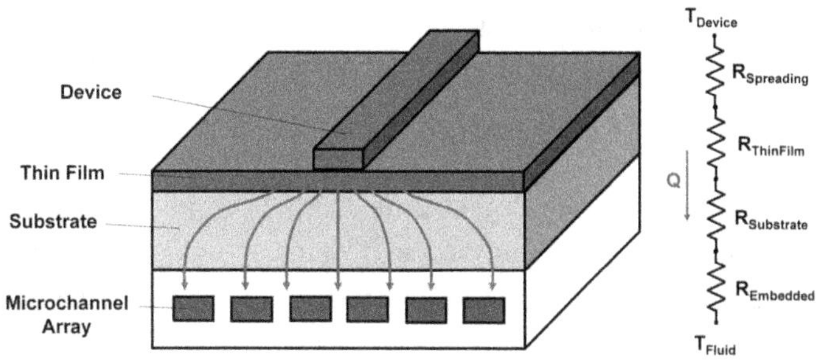

Figure 1. A simplified example of an embedded-cooling scenario and its associated 1D thermal circuit analogue. Interfacial thermal resistances are not depicted but would exist at all material interfaces.

chambers, etc., directly into the packaging, the thermal conduction properties of the constituent materials and interfaces that comprise the die and package are still vital to the overall heat transfer pathways away from junctions. This concept is best conveyed using a simplified example system, such as the one illustrated in Figure 1, in which an embedded-cooling solution interfaces with a heat-generating semiconductor device via its supporting substrate and thin-film materials. Also included in Figure 1 is the example system's 1D thermal circuit analogue. For a given rate of heat transfer Q, the temperature of the device T_{Device} will depend on the sum of the individual thermal resistances between it and the temperature of the working fluid associated with the embedded solution, T_{Fluid}. For this simple system, these include the conduction resistances of the substrate material $R_{\text{Substrate}}$ and any interstitial thin film(s) R_{ThinFilm} which may lie between the substrate material and the device, such as insulating dielectrics or doped channel materials. Interfacial thermal resistances are not depicted here but would exist at all material interfaces and can be significant contributors to the overall thermal resistance. Effective thermal management of T_{Device} then requires the minimization of all constituent resistances in between the device and the embedded-cooling solution. Importantly, the three resistances $R_{\text{Substrate}}$, R_{ThinFilm}, and $R_{\text{Spreading}}$ are all inversely proportional to their respective material's thermal conductivity. In addition, a lateral heat spreading material in contact with the device region would reduce the geometric constraints associated with the lateral spreading resistance, which further reduces the severity and impact of $R_{\text{Spreading}}$.

Although the example depicted in Figure 1 shows a liquid-cooling microchannel array and associates it with a general thermal resistance R_{Embedded}, alternative illustrations replacing it with a backside impinging jet, topside impinging jet (with chip flipped), mounted vapor chamber, etc., would still convey the same important message that even with embedded-cooling systems, conduction processes are important and the constituent materials still matter. Indeed, for many systems with very high effective heat transfer coefficients, such as those often found with phase-change processes, the overall thermal resistance of the system may be dominated by conduction resistances. Thus, improvements in material thermal conductivity and heat spreading capabilities can pay significant dividends to the overall thermal package performance. This then represents an opportunity for fundamental scientific researchers to help solve such pressing issues, where an improved understanding of conduction physics can help material scientists identify pathways for improving thermal conduction in existing materials or even develop entirely new materials with strong potential to outperform the state of the art.

The focus of this chapter is on recent advances in materials development, which may hold significant promise for helping meet the cooling needs of modern electronics across a range of applications in which the embedded-cooling solution's effectiveness is dependent on one or more conduction pathways. The chapter begins by first looking at emerging materials for use as substrates, where the thermal conduction properties of the material upon which the integrated circuits themselves are fabricated can have a significant impact on temperatures at multiple length scales. Next, the growing world of 2D materials is explored, which includes graphene, hexagonal boron nitride (h-BN), and associated heterostructures. While 2D materials have been investigated for use as circuit and sensor elements themselves, the focus here is on their ability to function as transistor-level heat spreaders for traditional and/or flexible electronics. Finally, a discussion on advances in the understanding of conduction physics is presented, as this knowledge helps inform scientists and engineers as to how to better design and develop new materials with advantageous heat transfer characteristics. Finally, an outlook on the future of research in this area is presented, along with examples of possibly innovative solutions currently in their nascent stages but which possess significant long-term potential.

2. Bulk Materials Advances

It should come as no surprise that the substrate material upon which integrated circuit elements are fabricated has a strong influence on device operating temperatures. For decades, silicon (Si) has been the dominant substrate material of choice due to its abundance, processability, amenable bandgap, and other advantageous properties. Along with these factors, high-quality monocrystalline Si possesses a relatively high κ of about 140 W/mK [9, 10]. Other state-of-the-art substrate materials include III–V compound semiconductors, such as aluminum nitride (AlN) and gallium nitride (GaN) with room-temperature κ in the ~200–400 W/mK range [11, 12]. This large degree of variability is due to the sensitivity of κ to impurities and defects, with stringent crystal quality control required to achieve the higher end of κ values [11–15]. Such high- quality crystal growth comes with increased expense, such that lower-cost sintered AlN composites with a κ of ~100 W/mK [16, 17] may be a cost-effective compromise for certain packaging applications, as has been in practice since the mid-1980s [14, 18]. AlN and GaN materials are most commonly associated with light-emitting diodes, optoelectronics, high-power switches, and dielectric layers. Other III–V semiconductors have similarly specialized applications, but possess κ values several times lower than AlN or GaN such that their thermal management potential is limited [19].

For power electronics in particular, the exceptionally high heat flux densities have motivated significant investment into developing synthetic diamond as a potential substrate or thin-film heat spreader of the future [6, 20–37]. The thermal management potential

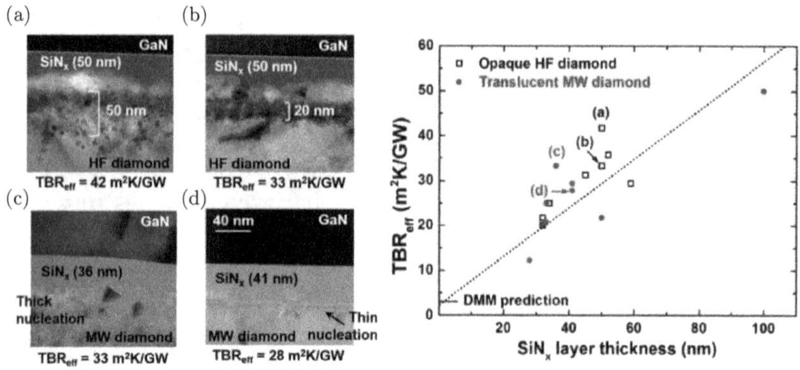

Figure 2. Left: Cross-sectional transmission electron microscope images of representative GaN-on-diamond composite wafers, with ((a) and (b)) utilizing diamond from hot filament (HF) chemical vapor deposition and ((c) and (d)) using diamond from microwave (MW) plasma chemical vapor deposition. The differences in effective thermal boundary resistance (TBReff) are directly tied to the nature of the microstructure near the GaN–diamond interface. Right: Measured dependence of TBReff on SiN_x layer thickness. Relative uncertainty for TBR_{eff} values is approximately 13%.

Source: Reproduced from Ref. [36], with permission of AIP Publishing.

of diamond is unquestionable. Diamond holds the record for κ among known bulk solid materials (2270 and 3450 W/mK at room temperature for natural diamond and isotopically enriched 99.9% ^{12}C diamond, respectively [38]) and possesses an appreciable 5 eV bandgap such that it can serve as a serviceable dielectric when undoped. Unfortunately, natural diamond is scarce and expensive. In addition, the κ of synthetic diamond is even more sensitive to crystal quality than AlN and GaN, so additional research is needed to improve the growth rate, enhance the crystal quality, and reduce the cost of synthetic diamond [39]. Finally, it should be noted that diamond is generally envisioned as a substrate or heat spreading layer only, with additional layers of semiconductor materials attached which function as the actual electronic device, as shown in Figure 2. As a result, there has also been a strong research focus on understanding and improving the thermal transport at heterogeneous GaN-diamond interfaces, as a high interfacial thermal resistance can significantly reduce any benefits offered by high-κ diamond [25, 26, 28, 35–37].

The means by which new high-κ materials are identified are now moving toward a more predictive approach, aided by advances in the understanding of heat conduction physics as well as the emergence of artificial intelligence, machine learning, and materials-by-design ideologies. A prime example of this is the remarkably rapid advances made regarding the material boron arsenide (BAs). In 2013, first-principles calculations performed by Lindsey, Broido, and Reinecke predicted that cubic BAs might exhibit a room-temperature κ of over 2000 W/mK [40], as shown in Figure 3(a). This result, if physically real, would make BAs a comparable thermal conductor to natural diamond near 300 K and, remarkably, a superior thermal conductor to natural diamond at higher temperatures, such as those at which

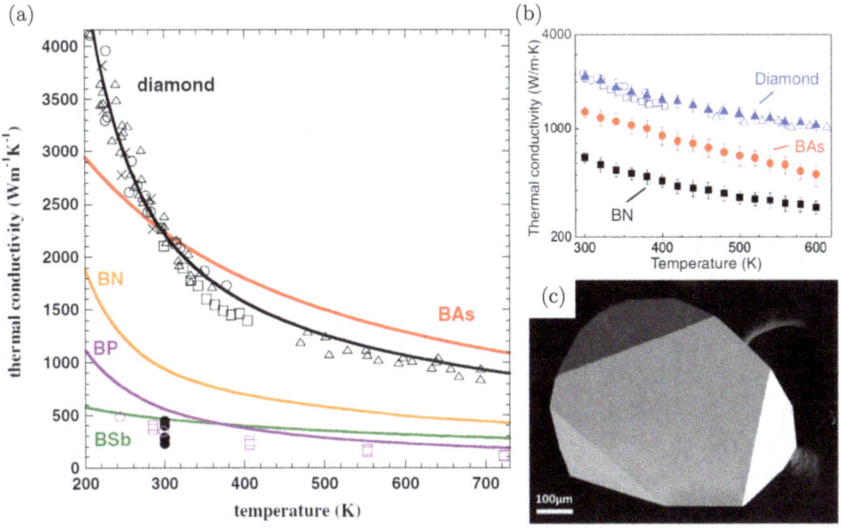

Figure 3. (a) First-principles calculation results for thermal conductivity of several natural boron-based compound semiconductors as compared to that of diamond. Also shown are room-temperature values (solid black circles) for common high-κ materials: gallium nitride, aluminum, aluminum nitride, copper, and silicon carbide. (b) Temperature-dependent κ of diamond, boron arsenide (BAs), and cubic boron nitride (BN). (c) Scanning electron microscope image of the BAs crystal specimen from the work of Li *et al.*

Source: (a) From Ref. [40]. Reprinted with permission from APS. (b) From Ref. [43]. Reprinted with permission from AAAS. (c) From Ref. [44]. Reprinted with permission from AAAS.

power amplifiers typically operate. The root cause of the predicted high κ of BAs is related to the large difference in atomic mass between the two constituent elements B and As. This mass difference creates a large energy gap between acoustic phonon branches and optical phonon branches, which in turn drastically reduces the scattering between the two types of phonons. This is particularly important for the acoustic phonons, which have higher group velocities and tend to carry more of the thermal energy through the atomic lattice compared to the low-group-velocity optical phonons. Another contributing factor to the predicted high κ of BAs is its insensitivity to phonon scattering from isotopic impurities, which typically impacts short-wavelength phonons the most. Since it is the short-wavelength phonons that make a large contribution to the thermal transport in BAs, this isotopic insensitivity as well as the frequency gap mentioned above allow the dominant heat carriers to move efficiently through the atomic lattice.

These predictions sparked a great deal of interest within both the materials science and thermal transport research communities, with many collaborative teams working to first synthesize the elusive material and accurately characterize its thermal transport characteristics to compare with the first-principles results. Synthesizing BAs crystals is not a trivial matter, however, due to volatility issues and the high toxicity of As. In 2015, Lv *et al.* [41] became the first researchers to report the synthesis of BAs crystals a few hundred microns in size and their associated thermal conduction characterization. These early crystals were shown to have significant twinned internal defects, grain boundaries, and As vacancies. Time-domain thermoreflectance (TDTR) results indicated a room-temperature κ of $\sim$200 W/mK, far below the theoretical predictions for pristine monocrystalline BAs. In 2016, Kim *et al.* [42] used a MEMS-based approach to measure the κ of BAs microstructures, with a result of 186 ± 46 W/mK and, likewise, cited the relatively large defect density as a strong limiting factor in thermal conduction. Both of these works pointed toward the need for improved synthesis methods to reduce twinning, isotopes, and other types of defects to which BAs phonon transport is very sensitive. Indeed, rather than discouraging researchers, these preliminary κ values were seen as promising, given that this new material was already outperforming monocrystalline Si despite its high defect densities. These results also seemed to suggest that improvements in crystal quality might pay large dividends in terms of improving thermal conduction. In 2018, three separate research teams reported in the same issue of *Science* the measurements of high-quality BAs single crystals with a κ that approached the levels predicted in 2013. Kang *et al.* [43] utilized TDTR to measure a room-temperature κ of 1300 W/mK, with the temperature-dependent data reproduced in Figure 3(b) along with those of boron nitride and diamond for comparison. Li *et al.* [44], likewise, utilized TDTR to measure their BAs samples (Figure 3(c)) and obtained a room-temperature κ of 1000 ± 90 W/mK. The third of these works, published by Tian *et al.* [45], measured the κ of their BAs samples using multiple methods and obtained an average room temperature value of ~900 W/mK, although a high value of 1160 ± 130 W/mK was observed via TDTR. In addition, this work is notable as it was able to synthesize crystals as large as 4 mm $\times$ 2 mm $\times$ 1 mm, a significant step forward toward realizing sample sizes amenable to semiconductor devices. Together, these works confirm that BAs, when of sufficient crystal quality, is capable of achieving room-temperature $\kappa \sim 7\times$ greater than that of Si. In addition, these works show that modern computational materials science can indeed be a powerful tool for identifying previously untested thermal management materials. In fact, a computational work in 2017 which included higher-order four-phonon scattering processes revised the predicted κ at 300 K for BAs to be closer to ~1400 W/mK [46], which is in strong agreement with the multiple experimental results mentioned above. The challenge now lies in realizing high-quality BAs growth in a manner which is both cost-effective and has high throughput in order to make it more industrially attractive.

It is important to note that the above discussion focused on thermal conduction within bulk materials, i.e., where the characteristic size dimension is several times longer than the range of mean free paths (MFPs), which carry a significant portion of the thermal energy through the material. By nature, materials with high κ tend to have abnormally long MFPs. As a result, when one considers thin films or nanostructures made of these materials where the characteristic dimension is comparable to or smaller than the relevant mean free paths, there can be a significant suppression of κ compared to the bulk values at the same temperature.

The MFPs of hot phonons that contribute significantly to the bulk κ can span from nanometers to several microns, as shown by the calculated [47–51] and measured [52] thermal conductivity accumulation function for bulk Si, which are compiled in Ref. [7]. These are also temperature dependent, as phonon–phonon scattering becomes weaker with temperature, and the population statistics of the phonon population shifts to lower frequencies with decreasing temperature. Consequently, even characteristic feature sizes at the micron level can result in the suppression of κ such that assuming bulk-like values can lead to an overestimate of heat spreading capabilities. Besides Si [53–60], the size-induced reduction of κ from bulk values has been observed experimentally in other electronics materials, including compound semiconductors [61, 62] and copper [63]. For thin films or nanostructures made from diamond, graphite, or Bas, which possess even higher bulk κ and longer phonon MFPs than Si, the onset of suppression effects would be expected to occur at even larger feature sizes. This is not to dissuade the reader from utilizing such materials in thermal management designs, but to call attention to the fact that, with the exception of a high-κ substrate several hundred microns thick, the size of the material does indeed matter.

3. Two-dimensional and Ultrathin Materials as Near-Junction Heat Spreaders

Nearly 10 years ago, the isolation and identification of single-layer graphene (SLG) resulted in an astounding amount of both fundamental and applied research that continues today into its physical properties, synthesis and processing methods, and evaluations of performance in electronic devices [64–67], sensors [68–70], and composites [71–73]. The term graphene refers to a single atomic sheet of sp^2-bonded carbon atoms, similar to the constituent atomic sheets that make up the basal planes of graphite (Figure 4(a)). Today, even few-layer graphene (FLG) may sometimes be referred to as simply "graphene", though many physical properties vary between SLG and FLG, making the distinction important. SLG was the first true so-called 2D material, in which the single atomic layer represents the

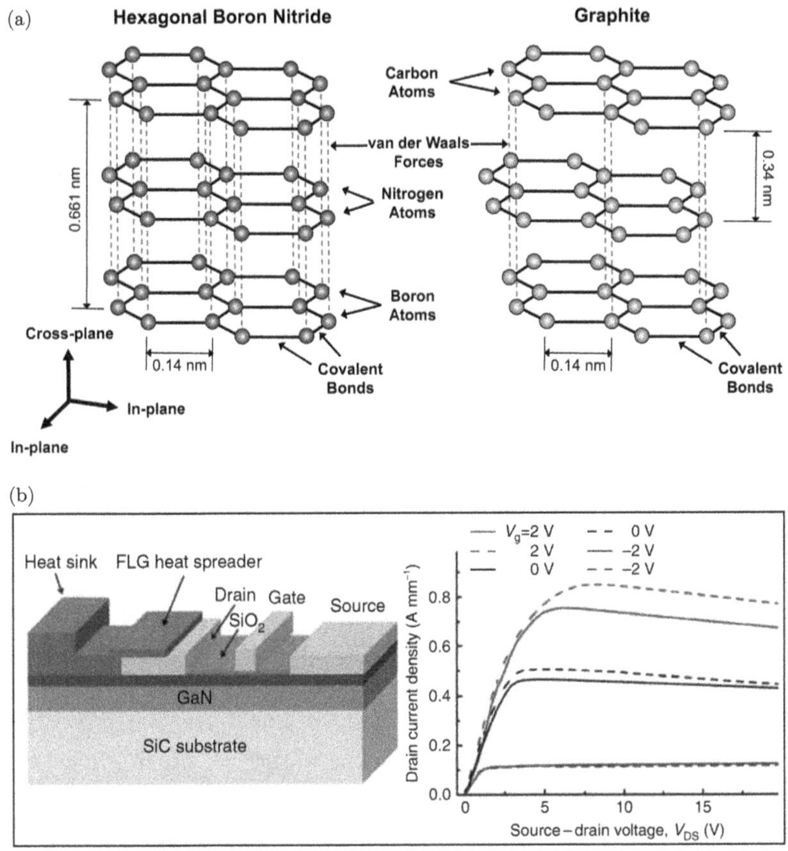

Figure 4. Utilization of 2D/ultrathin materials as near-junction heat spreaders: (a) annotated illustration comparing the atomic arrangements of graphite and hexagonal boron nitride; (b) use of few-layer graphene (FLG) as a heat spreader (left) with a high-power GaN transistor, with the associated improvement in its $I-V$ characteristics (right), where dashed curves are with graphene and solid curves are without graphene.

Source: Reprinted with permission from [28], with permission from Springer.

ultimate confinement possible in the third dimension. Since then, a host of other 2D materials have been discovered, synthesized, and studied via methods developed from the study of graphene, including single layers of the wide-bandgap ceramic h-BN, which in many ways can be viewed as the dielectric analogue of graphene. Bulk h-BN and h-BN composites have been a mainstay of thermal management packaging for decades due to their combination of high κ, dielectric properties, and high temperature stability. In this section, the thermal transport properties of 2D/ultrathin graphene and h-BN are briefly reviewed, with special emphasis on their usage as next-generation thermal management materials as opposed to their electrical or sensing properties. Graphene is considered first, followed by h-BN.

One of the first physical properties of interest to researchers following the isolation of SLG was to determine its κ. Besides being of fundamental scientific interest as to how a single sheet of atoms transmits thermal energy, the high basal plane κ ~1500–2000 W/mK at room temperature reported for pyrolytic graphite [74–77] suggested that SLG may likewise be an exceptionally good thermal conductor. In addition, early calculations predicted an elimination of interlayer phonon scattering in suspended (i.e., freestanding in space as opposed to resting on a surface) graphene due to quantum mechanics, which in turn would result in a higher in-plane κ for suspended SLG than observed in bulk graphite [78]. Further, this viewpoint would predict a trend of increasing κ with decreasing layer thickness for FLG thinner than approximately five atomic layers [79, 80].

Experimental results for suspended SLG have produced κ ~1500–4600 W/mK [79, 81–83] near 300 K. Many of the higher values in this range were obtained based on micro-Raman spectroscopy methods, in which a laser is used as both a non-contact heat source and a means of temperature measurement via calibration of Raman peak position to sample temperature (Figure 3(c) and 4(b)). While these measurements are possible with many commercial Raman systems, this approach can be subject to large uncertainties in the optical absorption and have poor temperature sensitivity compared to other techniques, with especially large uncertainty intervals nearest to room temperature [84]. Non-equilibrium effects can further complicate the accurate interpretation of the Raman-based results, depending on the thermalization length of heat carriers within a given material [85]. Thermal conductivity measurements of suspended 2D materials are also possible using microfabricated sensing platforms, which offer greater accuracy but a much higher degree of difficulty in terms of sample preparation. For a review of suspended graphene thermal conduction measurements performed using microfabricated sensing platforms, see Ref. [83].

The above discussion on thickness-dependent thermal transport is important for two reasons, other than its significance in fundamental physics, especially for engineers who may be considering a graphene-based material for electronics thermal management either in a composite or as a local heat spreader. First, despite the high thermal *conductivity* of some 2D materials, the extremely small cross-section for in-plane transport, as in a heat spreading application, limits the thermal *conductance.* For instances in which spreading heat laterally away from a hotspot is the goal, the thickness/number of layers may need to be increased to obtain sufficient thermal conductance for effective heat spreading. Second, in many practical applications, graphene is not suspended, as done in many experiments focused on fundamental scientific investigation. Instead, it is supported on an insulating substrate or embedded within a polymer, such as when a composite is formed. For example, it has been reported that the κ of SLG supported on the commonly used dielectric layer SiO_2 is suppressed to ~600 W/mK near room temperature [86]. This is still an impressive value, but much lower than the values of over 1000 W/mK associated with suspended SLG. The suppression is attributed to the scattering of phonons, especially out-of-plane ZA phonons which carry a large fraction of the thermal energy, by the SLG–SiO_2 interface. This would then be the atomic analogue of placing a vibrating drum head onto a tabletop while it is in motion. Encasement within another substance would have a similar effect as observed for the SiO_2-supported SLG. For both oxide-encased [87] and oxide-supported graphene [88], it has been shown that κ increases with the number of layers to approach graphite's κ values, which is opposite the trend for suspended graphene. As the number of layers increases, there is a decrease in the influence of interface scattering. Because of long intrinsic phonon MFPs in graphite, it can take 30 layers or more for MLG supported on SiO_2 to reach a κ comparable to that of the basal plane of bulk graphite [88].

We now turn our attention to the 2D form of h-BN, which, as mentioned previously, is a well-established material in the electronics thermal management community in its bulk, sintered, and particulate forms. Like graphite, bulk h-BN has a layered atomic structure (Figure 4(a)), characterized as having strong atomic bonds within the atomic planes and weak van der Waals forces between planes. This strong atomic bonding and light atomic masses help bulk h-BN have a high κ of 350 W/mK along its atomic planes at 300 K [89]. Unlike graphite and graphene, however, h-BN has a wide bandgap of 4–6 eV [90], which allows it to act as an effective dielectric when in contact with or in proximity to electronic devices. Single- and few-layer atomic sheets of h-BN similar to graphene can be obtained via mechanical exfoliation and chemical vapor deposition (CVD) on certain metals, amongst other methods. Also similar to graphene, single- and few-layer samples of h-BN have been isolated to study their thermal transport properties, with results showing that suspended bi-layer h-BN may possess a room-temperature κ of $\sim$484 $+141/-24$ W/m K [91], while few-layer h-BN is more comparable to bulk h-BN with a κ of $\sim$360 W/mK [92].

The use of 2D materials, such as graphene and h-BN, as localized heat spreaders has recently begun, in large part due to their impressive κ as well as the high degree of conformability offered by their extremely thin nature. Besides millimeter-scale hotspots at the die level, individual transistors and other devices can experience localized high-temperature regions that compromise lifetime and reliability as well, especially near the drain contact/channel junction. As a result, heat spreading/thermal management at the junction and device levels are now receiving more attention than in previous decades. Industrial compatibility requires that a potential solution be cost-effective and scalable, which gives passive approaches that do not require electrical power an advantage over active alternatives. The most direct approach is to identify high-κ materials to serve as near-junction heat spreaders, especially those with easy integration and compatibility with existing semiconductor processing practices. This description fits both graphene and h-BN to varying extents, as discussed in the following.

The large in-plane κ of graphene combined with its ability to be grown via CVD at the wafer scale and patterned using existing micro-/nanofabrication processes has led researchers to investigate it as a device-level heat spreading material [93]. Reported reductions in temperature are in the range of $\sim 8 - 45°C$ compared to devices with no graphene present [94–96], although these values span differing power densities, supporting substrates, and measurement approaches such that some context is required to understand how these works compare to one another. Perhaps one of the best examples of FLG being used strictly for thermal management of an individual electronic device is the work by Yan *et al.* shown in Figure 4(b), in which superior $I-V$ curve characteristics associated with an improved thermal state demonstrate how an ultrathin heat spreader can improve not just temperatures but the associated device operation as well. A drawback to utilizing graphene in direct contact with the drain region is that graphene is electrically conductive. Thus, its means of integration must include strategic placement, as illustrated in Figure 4(b), or the use of a performance-degrading insulation layer between the contact and the graphene heat spreader.

Alternatively, as mentioned above, h-BN has a wide enough bandgap to be considered a dielectric such that it can be placed in direct contact with metallic interconnects. This has led to some preliminary studies into utilizing few-layer and ultrathin h-BN films in electronics thermal management applications. While there have been many studies on microscale h-BN particulates being formed into composites for heat spreading, here we are focusing on contiguous planar h-BN films due to their lack of interparticle interfaces and greater degree of novelty. Lin *et al.* [98] placed CVD-grown few-layer h-BN in direct contact with a high electron mobility transistor (HEMT). In their implementation, a 2 nm thick h-BN region on top of the

HEMT was connected to a much thicker h-BN region adjacent to the device to act as a heat sink. It was experimentally demonstrated that the saturation current, cut-off frequency, and maximum oscillation frequency of the HEMT showed improvements with the presence of this thin-to-thick h-BN material. The decrease in temperature of the device associated with the presence of the h-BN was determined indirectly via finite element simulation, with an estimated value of $\sim 32°$. Choi *et al.* [99] applied an 11 nm thick h-BN to an InGaN/GaN quantum well light-emitting diode (LED) and found a faster transient response compared to the bare LED case. In a separate work, Pratik *et al.* [97] used CVD-derived h- BN thin films of uniform thickness similar to what would be applicable to wafer-scale integration and patterning. The h-BN thin films of 77–125 nm thickness were transferred from their growth substrate onto a microfabricated device that simulated a multi-gate transistor geometry and included spatial temperature sensing via a series of platinum resistance thermometers (PRTs). Results showed that the films were capable of lowering microdevice temperatures to an extent that would extend reliability. More generally, the effective thermal resistance seen by the heater line was quantified such that the temperature rise at varying power levels could be ascertained, thus providing a broader applicability of the results to other operating conditions. The best performing h-BN film demonstrated a reduction of the effective thermal resistance seen by the heat source of $15.9\% \pm 4.6\%$ compared to an identical device with no h-BN present.

It is important to point out that identifying an effective near-junction heat spreader is more complicated than simply applying a material with a high in-plane κ. In reality, the performance of a given heat spreader is dependent on the thermal resistance of the entire system as well as the temperature-dependent thermal properties of the constituent materials. As shown by Pratik *et al.* [97], the h-BN thin-film heat spreader was effective near room temperature; however, at temperatures of 400 K and above, the overall thermal resistance was comparable to that of the blank microdevice due to changes in the κ of the substrate and insulating layers such that lateral heat spreading was no longer an efficient means of cooling. In another work focused on heat spreading within biased graphene, Choi *et al.* [100] point out that the effectiveness of a thin heat spreader can be sensitive to many parameters besides simply κ, including dielectric thickness, interface conductance, and gate length, such that a 2D/ultrathin material may work well in one scenario but be ineffective in others. Thus, similar to other areas of electronics thermal management, reduction in a given thermal resistance at the device level is only significant if it attacks a major contributor to the overall thermal resistance of the system. As a result, reports in the literature regarding the effectiveness — or a lack of effectiveness — of 2D/ultrathin heat spreaders should be considered within the context of the means by which they were tested as well as how the interested reader might wish to utilize them within their particular application space.

Another important factor in a 2D/ultrathin material's ability to serve as an effective heat spreader is related to thermal transport in the cross-plane direction. As stated above, part of what makes graphene and h-BN materials attractive for thermal management applications is their large in-plane κ. However, the strong in-plane covalent bonding that helps lead to this condition does not typically occur in the cross-plane direction; as noted in Figure 4(a), in bulk form, the atomic sheets of graphite and h-BN are instead held together via weak van der Waals forces. As a result, the cross-plane κ of such layered materials can be more than an order of magnitude lower than observed for in-plane transport. For example, graphite's cross-plane κ is approximately 6 W/mK [74], over 300 times lower than the in-plane κ found for high-quality samples. A similar degree of anisotropy is found for h-BN, where the cross-plane κ is only ~1.5–2.5 W/mK at 300 K [101]. Thus, for ultrathin heat spreaders composed of multiple atomic layers, the interlayer thermal transport perpendicular to the interface cannot be assumed to be the same as that within the atomic plane and, for sufficiently thick materials, should be accounted for. Often more important than the cross-plane transport within the material itself is the interfacial thermal resistance between the 2D/ultrathin layered material and the device/substrate it is in contact with. For a heat spreader resting atop a heat source, this "vertical" thermal transport across the interface can dominate the cross-plane thermal resistance when the thickness of the 2D/ultrathin material stack is comparable to the cross-plane phonon MFP. For example, the cross-plane phonon MFP for graphite is ~10–20 nm [88, 102], or roughly 30–60 atomic layers. There have been measurements of the thermal interface resistance between SLG and materials commonly used in electronics, such as metals and various dielectrics. Results show that interfacial resistance tends to be in the range of $1.2 \times 10^{-8} - 4 \times 10^{-8}$ m^2 K/W [66, 103–105], which, for context, would be slightly higher than values associated with common metal–dielectric interfaces [106, 107]. However, the interfacial resistance can depend on the 2D/ultrathin material's thickness as well. For example, compared to the highly conformal SLG, FLG becomes stiffer with increasing number of atomic layers/thickness. This makes FLG less conformal in comparison, and thus, there is a reduction in interface adhesion energy [108]. The strong link between interfacial adhesion and interfacial thermal transport would appear to lend some guidance to scientists and engineers as to the potential means of promoting heat transfer, with approaches that would strengthen interfacial adhesion, presumably reducing the thermal interface resistance. However, the reported thermal interface conductance for FLG/silica systems does not exhibit a clear dependence on the layer thickness [103, 104], which may or may not be due to differences in sample preparation and experimental methods. Measurements for 1–10 layers thick graphene sandwiched between Au/Ti and SiO_2 showed a ~4× increase in interfacial thermal resistance when SLG was present, compared to the same stack-up without graphene, but a very weak dependence of interfacial thermal resistance on the number of graphene layers [105]. For h- BN, Li *et al.* [109] measured an interfacial thermal resistance of $\sim 1.6 \times 10^{-8}$ m^2 K/W for monolayer h-BN sandwiched between a metal line and SiO_2, which is comparable to the lower end of the values given above for SLG/SiO_2. The interfacial resistance increased nonlinearly up to $\sim 3.4 \times 10^{-8}$ m^2 K/W for 3.8 nm thick h-BN within the same stack-up [109]. Overall, a clearer understanding of interfacial thermal transport involving 2D/ultrathin materials is required in order to optimize their integration, ensure repeatability, and ultimately control their performance when placed in contact with traditional electronics materials.

More recently, significant research focus has shifted toward heterostructures, in which different 2D materials are stacked one atop another to achieve beneficial electronic transport properties. For example, electron mobility in graphene has been observed to increase significantly when placed on atomically flat h-BN compared to when it is supported on SiO_2 [110]. Besides electrical benefits, the high in-plane κ of h-BN may be attractive for thermal management purposes given that its in-plane κ is over 100 times greater than that of most traditional amorphous dielectrics. This has been explored experimentally by Choi *et al.* [111], who observed via scanning thermal microscopy a measurable decrease in temperatures for biased graphene supported on 35–80 nm thick h-BN as opposed to SiO_2 or flexible willow glass (WG), as shown in Figure 5.

The interfacial thermal resistance between contacting 2D/ ultrathin materials is of obvious importance for the thermal management of such heterostructures, similar to the discussion above regarding the thermal interface resistance between these materials and traditional electronics metals and dielectrics. Using Raman spectroscopy, Lie *et al.* [112] have measured the interfacial thermal transport properties of MoS_2/h-BN and graphene/h-BN interfaces, with reported values of 5.8×10^{-8} and 1.9×10^{-8} m^2 K/W, respectively. They attributed the significantly higher interfacial thermal resistance for the MoS2/h-BN case to weaker transmission of cross-plane phonon modes compared to graphene/h-BN. This not only demonstrates the material-specific aspect of interfacial resistance in 2D heterostructure systems but also represents an opportunity to identify and leverage material combinations with highly compatible phonon characteristics in order to achieve efficient cross-plane thermal transport.

4. Advances in Understanding Conduction Physics

While this chapter is focused on advanced materials for thermal management applications, it is important to emphasize that advances in the study of heat conduction physics have helped elucidate the

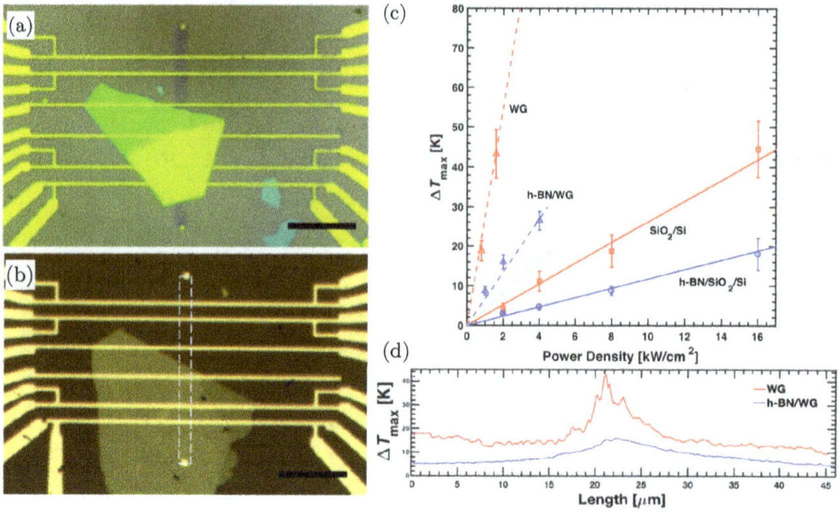

Figure 5. Demonstration of h-BN as a superior dielectric support for self-heating graphene devices compared to SiO_2 or willow glass (WG). Optical images are of the measured device on (a) SiO_2/Si substrate and (b) the WG substrate. The white dashed line shows the location of a patterned graphene strip across the irregularly shaped h-BN flake. All scale bars are 25 μm. (c) Comparison of the maximum temperature rise observed versus power density for differing support materials. (d) Scanning thermal microscopy data of maximum temperature rise versus position across the graphene strip with and without h-BN.

Source: Adapted with permission from Ref. [111]. Copyright (2018) American Chemical Society.

underlying mechanisms behind high κ and low interfacial thermal resistance and, in doing so, provide the insight and means by which materials scientists and engineers can intelligently identify and even synthesize new thermal management materials. In some cases, this research directly applies not just to cooling but to the fundamental operation of next-generation electronics, such as phase-change memory devices and heat-assisted magnetic recording, where controlled localized heating is crucial to their operating principle [113, 114]. The tools used in the study of heat conduction physics may also find uses in the characterization of heat flow, even at the smallest length scales. In particular, scanning thermal microscopy using an instrumented atomic force microscope cantilever can have a spatial resolution of ~10 nm, temperature precision of 50 mK, and detect heat flows as small as 10 pW [113]. Non-contact optical methods have improved as well to the point of offering uses outside of academic laboratories. While infrared imaging at the microscale has become a pillar of electronics thermal characterization, time-domain thermoreflectance (TDTR) and frequency-domain thermoreflectance (FDTR) have now advanced to the point of becoming ubiquitous tools for measuring the thermal conductivity of thin films with micron-scale spatial resolutions. Here, some of the frontiers of conduction physics investigations are briefly highlighted and put into context regarding how they might be of benefit — either near term or further in the future — for applied thermal engineering.

Amorphous materials, such as SiO_2, have been a mainstay of electronics devices almost since their inception, while amorphous Si (a-Si) has become increasingly technologically relevant with the rise of photovoltaics and thin-film transistors in large flat panel manufacturing. More fundamentally, the study of how light and heat move through such materials goes back to the beginnings of civilization such that one might be surprised that scientists are still finding new insights into their physical properties. While measurements of κ across a wide temperature range have been available for a large variety of amorphous materials, emerging studies utilizing cutting-edge simulation techniques in concert with new experimental results are disrupting several long-held conceptions about the physics of how thermal energy is transported in amorphous materials. Since the early part of the 20th century, several models aimed at predicting the κ of amorphous materials assumed that, due to the disordered atomic arrangement, the heat-carrying phonons could be treated as having a wavelength and MFP comparable to the interatomic spacing. These simple models were often able to predict the κ of many amorphous materials with reasonable accuracy but with enough deviation and occasional failings that it was widely accepted that such models were critically limited in their applicability [115]. In the early 1990s, Allen and Feldman [116] reexamined the question of thermal transport in a disordered atomic crystal and proposed an alternative model in which heat was carried via modes which are neither fully localized or propagating but are rather quasi-stationary in space and which overlap with their neighboring modes. In this way, thermal energy is allowed to diffuse through the material, and such non-propagating modes later became known as "diffusons" within the thermal transport community. Later works built upon this view and suggested that vibrational modes in amorphous materials could be classified into three types: propagating modes (propagons), localized modes (locons), and diffusons, as defined previously [117]. Locons are localized vibrations typically centered around an atomic site and thus do not propagate. Propagons, as one might guess, are delocalized and able to propagate similar to phonons in materials with a periodic lattice but are limited to low frequencies [118]. Thus, if ordered in ascending order in terms of their relative "mobility" for carrying thermal energy, these vibrational modes would be given as locons, diffusons, and propagons.

The classification of vibrational modes as propagons, diffusons, and locons is not just an academic exercise but has real meaning to anyone wishing to understand, predict, and engineer thermal transport in amorphous materials. These three modes combine to make up the total κ, but the relative contribution of each influences the temperature and size dependences of κ as well. For example, in a-Si, propagons can have MFPs ranging from tens of nanometers to perhaps even ~1 micron, while diffusons have MFPs closer to ~10 nm or less [119, 120]. This is important, as materials with a non-negligible contribution from propagons to the total κ would experience a size-induced suppression of κ once the characteristic length becomes smaller than the propagon MFP, which runs counter to the typically assumed size independence of κ for amorphous materials. For example, the contribution of propagons to the total κ of a-Si has been determined to be ~35–50% both via calculation [119, 120] and measurement [52], meaning that despite their amorphous nature, a-Si nanostructures could indeed show size-induced suppression of κ due to propagon boundary scattering. In contrast, for amorphous SiO_2, propagons only contribute ~6% to its total κ such that suppression of their MFP due to boundary scattering would not manifest as a significant change in the measured κ, in agreement with most experimental observations for amorphous SiO_2 thin films having a largely thickness-independent κ [119]. These somewhat unexpected results for a-Si have inspired additional experimental studies on amorphous nanostructures that would likely not have been undertaken otherwise, such as that by Kwon *et al.* [121], where a-Si nanotubes and thin films were shown to exhibit strong anisotropic and size-dependent κ (Figure 6(a)), attributes not traditionally associated with amorphous materials or predictions using legacy models. These findings suggest that engineers seeking to predict thermal operating points for devices that contain amorphous materials should not automatically assume bulk-like values of κ if the layer/line dimension is on the micro-/nanoscale, which until recently was an assumption that few questioned. Similarly, it has recently been shown that certain phonons can travel ballistically from one Si region to another through an amorphous SiO_2 layer up to 5 nm in thickness (Figure 6(b)) [122], which goes against the previously held assumptions of phonon MFP consistently being on the order of atomic spacing for such materials. This means that a thin amorphous dielectric sandwiched between two crystalline semiconductors may not be as thermally insulating as generally assumed and that the wave-like nature of heat carriers passing across interfaces requires closer scrutiny from both scientists and engineers.

The above discussion of localized versus non-localized vibrational modes and the ballistic transport of phonons across an interface highlights a more general trend within the conduction physics and materials research communities, which is a greater emphasis on the spectral nature of phonon transport and subcontinuum effects. This shift was recently highlighted in a survey of the field offered by Sood *et al.* [123]. In previous decades, a phonon gas model was commonly employed to interpret experimental results or estimate classical effects, such as boundary scattering in thin films. In such discussions, MFP was often discussed as a singular number that, while acknowledged to be an average, glossed over the fact that many important phonons could have MFPs significantly shorter or longer than this value. Now, experimental methods have been designed which provide spectroscopic data on phonon populations within a given material.

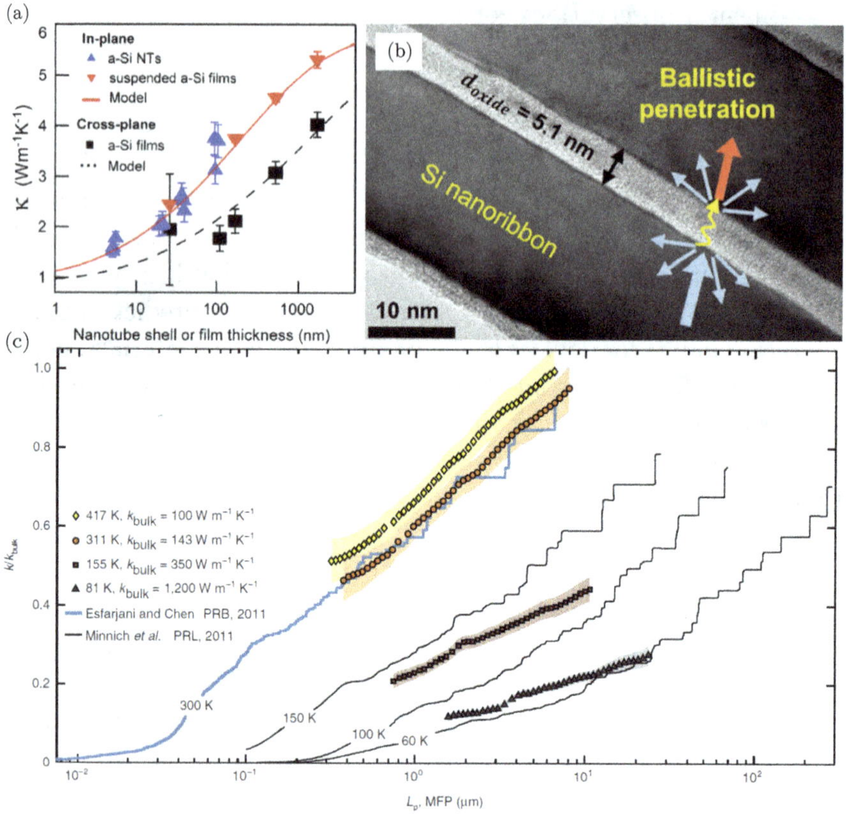

Figure 6. (a) Experimental and modeling results showing size-dependent κ for amorphous Si (a-Si) nanotubes (NTs) and thin films. (b) A conceptual illustration of ballistic phonon penetration through amorphous SiO_2. (c) Normalized thermal conductivity showing accumulation with increasing mean free path (MFP), with experimental data from Regner *et al.* [52] being compared to modeling results from Esfarjani *et al.* [50] and Minnich *et al.* [125].

Source: (a) Reprinted with permission from Ref. [121]. Copyright (2017) American Chemical Society. (b) Reprinted with permission from Ref. [122]. Copyright (2017) American Chemical Society. (c) Reprinted with permission from Ref. [52], with permission from Springer.

An example of this type of data is shown in Figure 6(c), in which the accumulated κ of crystalline Si at 300 K is given as a function of phonon MFP. Previous estimates of the average phonon MFP in Si at 300 K are on the order of ~300–400 nm [56, 124]. However, as shown in the data in Figure 6(c), only about ~50% of the total κ

is attributable to phonons with MFPs in this range or smaller, and a non-negligible fraction of thermal energy is carried by phonons with MFP >1 μm, which has important implications on the thermal transport properties of thin films and nanostructures. As can also be seen in the figure, this MFP spectra representation shows that the accumulation shifts to longer MFPs with decreasing temperature such that size effects on thermal transport would be stronger or observable at larger characteristic lengths compared to at room temperature. As pointed out by Yang and Dames [48], expressing the thermal conductivity accumulation function in terms of MFP spectra has two major benefits. First, plotting such relationships in a similar manner to Figure 6(c) helps researchers and engineers visualize the important MFPs for bulk κ, thus providing a better understanding of what ranges of phonon MFP are important to the transport of thermal energy. Second, building off of such knowledge, the prediction and interpretation of the κ of nanostructures is aided by such knowledge of the bulk MFP spectrum. In addition to computational means of determining the bulk MFP spectrum of a material, experimental methods which have produced MFP spectral results include TDTR [125, 126], broadband FDTR (BB-FDTR) [52], and transient thermal grating [127]. For an in-depth review of this topic, see the recent review by Regner *et al.* [128].

5. Outlook and Opportunities

This brief overview of emerging materials for electronics thermal management is by no means complete but, hopefully, offers the industrial professional a glimpse over the horizon for what new tools may be at their disposal in the next decade or perhaps further. In terms of bulk materials, significant advancements have already been made in the growth, characterization, and integration of synthetic diamond materials. While more work remains in order to overcome the physical and financial barriers that still exist, the idea that it may someday soon be an accessible material for thermal management is exciting. The same goes for BAs, whose ascent onto the scene of high-κ materials has been nothing short of meteoric since the theoretical identification of its potential in 2013. If the momentum behind this material can be maintained to achieve the necessary quality and scalability, many thermal management challenges may be alleviated simply by the utilization of BAs substrates with functional device layers.

The complex world of 2D materials is deserving of its own review, or perhaps several. Here, though, our focus on thermal management potential centered on graphene and h-BN, including heterostructures which combine the two. The impact of suspended versus supported/encased usage on κ was presented, as was the importance of interfacial thermal resistance. Indeed, for many applications, it is the interfacial thermal transport between the 2D material and the materials it contacts which limits or determines the ease with which heat may pass into the high-κ ultrathin layer. While many studies have looked at the intrinsic thermal transport properties of graphene and h-BN, far fewer have examined their potential for electronics thermal management specifically. Thus, there remains much to be determined in regard to optimization, barriers to improvement, and the ease of integration before a final determination can be made as to whether this class of materials has a prominent future for passive near-junction cooling.

Similar to 2D materials, a comprehensive discussion on advances in the understanding of conduction physics would require several niche reviews on a variety of topics. Here, the focus was placed on new insights into thermal transport in materials ubiquitous to electronics, such as amorphous materials and Si. Specifically, the emerging viewpoint of considering the spectral nature of heat carriers represents an important shift — theoretically, numerically, and experimentally — in how we conceptualize the processes which dictate the conduction processes within a given material. This new perspective is not just an academic one, but can and will be used to design new materials, improve interfacial transport, and identify promising new avenues of achieving effective thermal management that may not be intuitively arrived at from the more traditional phonon gas or quasiparticle approach.

These same topics are relevant when identifying near- and long-term research areas of merit. The continual development of synthetic diamond and BAs materials toward industrial practicality is without a doubt worthwhile over the next decade. More broadly, however, the utilization of high physics fidelity predictive materials identification and design, as has occurred for BAs represents an exciting new means through which high-κ materials may be discovered, developed, and engineered. When coupled with fields such as artificial intelligence and machine learning, which can aid in automated identification of promising materials from within a large candidate space, the pace of new materials discovery in the future may greatly accelerate over that of the previous century.

In terms of 2D materials for thermal management applications, much of the existing work has looked at characterizing the intrinsic thermal transport properties of individual materials, either suspended or supported. While such fundamental characterization will no doubt continue as new 2D materials are continually identified and improved, in terms of optimizing the use of 2D materials for heat spreading, it is imperative that more research focus be given to controlling — not just measuring — interfacial thermal resistances. Even for 2D materials with exceptionally high κ, their effectiveness as an on-chip heat spreader may be minimal if interfacial thermal resistances are too high.

Research into conduction physics remains imperative to advancing not just scientific understanding but also to opening up previously unavailable materials and processes by which to control the magnitude, direction, and behavior of thermal energy. The move toward a more spectral viewpoint on phonon populations outlined in this review is a prime example, with the expectation that these experimental and modeling approaches will yield a growing array of spectra for many technologically relevant electronic materials. Looking further out, it is likely that many concepts and engineering techniques utilized in other wave-like branches of physics may be repurposed for controlling thermal conduction within a solid. For example, the idea of a photonic crystal, whose periodic structure dictates by design what types of light are allowed to propagate through it, is currently being adapted to accomplish similar feats for phonons [129–131]. Broadly, this area may be termed phononics and is associated with a precise and intentional control via structure of what vibrational modes may move through a given material. Localized control of heat flow by design may also be used to impart directionality to the flow of heat, such as for thermal metamaterials which allow for "cloaking" of one object from another. Indeed, thermal metamaterials for isolating an electronic component from the heat of a neighbor are an application area already within the purview of thermal engineers, as evidenced by the recent review by Dude *et al.* [132].

Collectively, the advances presented here represent a host of promising new materials and strategies for the thermal engineers of tomorrow. While not all of these paths will lead to success and widespread adoption, lessons will be learned regardless, which will guide future endeavors toward continual improvement of the field. Regardless, it is apparent that the field is evolving to provide future thermal management specialists with a host of different tools to accomplish effective thermal solutions, tools which will span a range of length scales almost as broad as those associated with the electronics themselves.

References

[1] G.E. Moore, Progress in digital integrated electronics, *IEEE Tech. Digest. Int. Electron. Dev. Meeting*, 1975, 11–13.

[2] S. Salahuddin, K. Ni, S. Datta, The era of hyper-scaling in electronics, *Nat. Electron.*, 1, 2018, 442–450.

[3] C. Auth, A. Aliyarukunju, M. Asoro, D. Bergstrom, V. Bhagwat, J. Birdsall, *et al.*, A 10 nm high performance and low-power CMOS technology featuring 3rd generation FinFET transistors, Self-Aligned Quad Patterning, contact over active gate and cobalt local interconnects, in: *2017 IEEE International Electron Devices Meeting (IEDM)*, 2017, 29.1.1–29.1.4.

[4] J.L. Smoyer, P.M. Norris, Brief historical perspective in thermal management and the shift toward management at the nanoscale, *Heat Transfer Eng.*, 40, February 25, 2019, 269–282.

[5] W. Huang, M.R. Stan, S. Gurumurthi, R.J. Ribando, K. Skadron, Interaction of scaling trends in processor architecture and cooling, in: *26th IEEE SEMI-THERM Symposium*, Santa Clara, CA, 2010, 198–204.

[6] A. Bar-Cohen, J.J. Maurer, D.H. Altman, Embedded cooling for wide bandgap power amplifiers: A review, *J. Electron. Packag.*, 141, 2019, 040803.

[7] A.L. Moore, L. Shi, Emerging challenges and materials for thermal management of electronics, *Mater. Today*, 17, 2014, 163–174.

[8] (2019). *IEEE Heterogeneous Integration Roadmap 2019 Edition: Thermal.* Available: https://eps.ieee.org/images/files/HIR_2019/HIR1_ch20_thermal.pdf.

[9] C.J. Glassbrenner, G.A. Slack, Thermal conductivity of silicon and germanium from 3K to the melting point, *Phys. Rev.*, 134, 1964, A1058–A1069.

[10] H.R. Shanks, P.D. Maycock, P.H. Sidles, G.C. Danielson, Thermal conductivity of silicon from 300 to 1400 K, *Phys. Rev.*, 130, 1963, 1743–1748.

[11] G.A. Slack, R.A. Tanzilli, R.O. Pohl, J.W. Vandersande, The intrinsic thermal conductivity of AlN, *J. Phys. Chem. Solids*, 48, 1987, 641–647.

[12] A. Jezowski, B.A. Danilchenko, M. Bockowski, I. Grzegory, S. Krukowski, T. Suski, *et al.*, Thermal conductivity of GaN crystals in 4.2-300K range, *Solid State Commun.*, 128, 2003, 69–73.

[13] G.A. Slack, Nonmetallic crystals with high thermal conductivity, *J. Phys. Chem. Solids*, 34, 1973, 321–335.

[14] W. Werdecker, F. Aldinger, Aluminum nitride: An alternative ceramic substrate for high power applications in microcircuits, *IEEE Trans. Compon. Packag. Manuf. Technol.*, 7, 1984, 399–404.

[15] F. Miyashiro, N. Iwase, A. Tsuge, F. Ueno, M. Nakahashi, T. Takahashi, High thermal conductivity aluminum nitride ceramic substrates and packages, *IEEE Trans. Compon. Packag. Manuf. Technol.*, 13, 1990, 313–319.

[16] K.A. Khor, K.H. Cheng, L.G. Yu, F. Boey, Thermal conductivity and dielectric constant of spark plasma sintered aluminum nitride, *Mater. Sci. Technol. A*, 347, 2003, 300–305.

[17] S.L. Chung, T.I. Tsai, S.C. Huang, High thermal conductivity ceramics from combustion synthesized AlN powder through microwave sintering and reheating, *Int. J. Self-Propog. High Temp. Synth.*, 21, 2012, 45–50.

[18] R.R. Tummala, E.J. Rymaszewski, A.G. Klopfenstein, (eds.), *Microelectronics Packaging Handbook.* New York: Van Nostrand Reinhold, 1989.

[19] L. Lindsay, D.A. Broido, T.L. Reinecke, Ab initio thermal transport in compound semiconductors, *Phys. Rev. B*, 87, 2013, 165201.

[20] A. Bar-Cohen, J.J. Maurer, J.G. Felbinger, DARPA's intra/interchip enhanced cooling (ICECool) program, in: *CS MANTECH Conference*, New Orleans, LA, 2013, 171–174.

[21] J. Anaya, T. Bai, Y. Wang, C. Li, M. Goorsky, T.L. Bougher, *et al.*, Simultaneous determination of the lattice thermal conductivity and grain/grain thermal resistance in polycrystalline diamond, *Acta Materialia*, 139, 2017, 215–225.

[22] J. Anaya, S. Rossi, M. Alomari, E. Kohn, L. Tóth, B. Pécz, *et al.*, Control of the in-plane thermal conductivity of ultra-thin nanocrystalline diamond films through the grain and grain boundary properties, *Acta Materialia*, 103, 2016, 141–152.

[23] E. Bozorg-Grayeli, A. Sood, M. Asheghi, V. Gambin, R. Sandhu, T.I. Feygelson, *et al.*, Thermal conduction inhomogeneity of nanocrystalline diamond films by dual-side thermoreflectance, *Appl. Phys. Lett.*, 102, 2013, 111907.

[24] Z. Cheng, T. Bougher, T. Bai, S.Y. Wang, C. Li, L. Yates, *et al.*, Probing Growth-induced anisotropic thermal transport in high-quality CVD diamond membranes by multifrequency and multiple-spot-size time-domain thermoreflectance, *ACS Appl. Mater. Interfaces*, 10, February 7, 2018, 4808–4815.

[25] J. Cho, D. Francis, D.H. Altman, M. Asheghi, K.E. Goodson, Phonon conduction in GaN-diamond composite substrates, *J. Appl. Phys.*, 121, 2017, 055105.

[26] J. Cho, Z. Li, E. Bozorg-Grayeli, T. Kodama, D. Francis, F. Ejeckam, *et al.*, Improved thermal interfaces of GaN–diamond composite substrates for HEMT applications, *IEEE Trans. Compon. Packag. Manuf. Technol.*, 3, 2013, 79–85.

[27] K.E. Goodson, Thermal conduction in nonhomogeneous CVD diamond layers in electronic microstructures, *J. Heat Transfer*, 118, 1996, 279–286.

[28] V. Goyal, A.V. Sumant, D. Teweldebrhan, A.A. Balandin, Direct low-temperature integration of nanocrystalline diamond with GaN substrates for improved thermal management of high-power electronics, *Adv. Funct. Materials*, 22, 2012, 1525–1530.

[29] J.E. Graebner, S. Jin, G.W. Kammlott, J.A. Herb, C.F. Gardinier, Large anisotropic thermal conductivity in synthetic diamond films, *Nature*, 359, January 10, 1992, 401–403.

[30] J.E. Graebner, S. Jin, G.W. Kammlott, J.A. Herb, C.F. Gardinier, Unusually high thermal conductivity in diamond films, *Appl. Phys. Lett.*, 60, March 30, 1992, 1576–1578.

[31] J.E. Graebner, M.E. Reiss, L. Seibles, T.M. Hartnett, R.P. Miller, C.J. Robinson, Phonon scattering in chemical-vapor-deposited diamond, *Phys. Rev. B Condens. Matter*, 50, August 1, 1994, 3702–3713.

[32] R.B. Simon, J. Anaya, F. Faili, R. Balmer, G.T. Williams, D.J. Twitchen, *et al.*, Effect of grain size of polycrystalline diamond on its heat spreading properties, *Appl. Phys. Exp.*, 9, 2016, 061302.

[33] A. Sood, R. Cheaito, T. Bai, H. Kwon, Y. Wang, C. Li, *et al.*, Direct visualization of thermal conductivity suppression due to enhanced phonon scattering near individual grain boundaries, *Nano. Lett.*, 18, June 13, 2018, 3466–3472.

[34] A. Sood, J. Cho, K.D. Hobart, T.I. Feygelson, B.B. Pate, M. Asheghi, *et al.*, Anisotropic and inhomogeneous thermal conduction in suspended thin-film polycrystalline diamond, *J. Appl. Phys.*, 119, 2016, 175103.

[35] H. Sun, J.W. Pomeroy, R.B. Simon, D. Francis, F. Faili, D.J. Twitchen, *et al.*, Temperature-dependent thermal resistance of GaN-on-diamond HEMT wafers, *IEEE Electron. Device Lett.*, 37, 2016, 621–624.

[36] H. Sun, R.B. Simon, J.W. Pomeroy, D. Francis, F. Faili, D.J. Twitchen, *et al.*, Reducing GaN-on-diamond interfacial thermal resistance for high power transistor applications, *Appl. Phys. Lett.*, 106, 2015, 111906.

[37] Y. Zhou, R. Ramaneti, J. Anaya, S. Korneychuk, J. Derluyn, H. Sun, *et al.*, Thermal characterization of polycrystalline diamond thin film heat spreaders grown on GaN HEMTs, *Appl. Phys. Lett.*, 111, 2017, 041901.

[38] L. Wei, P.K. Kuo, R.L. Thomas, T.R. Anthony, W.F. Banholzer, Thermal conductivity of isotopically modified single crystal diamond, *Phys. Rev. Lett.*, 70, 1993, 3764–3767.
[39] Q. Liang, Y.F. Meng, C.-S. Yan, S. Krasnicki, J. Lai, K. Hemawan, *et al.*, Developments in synthesis, characterization, and application of large, high-quality CVD single crystal diamond, *J. Superhard Mater.*, 35, 2013, 195–213.
[40] L. Lindsay, D.A. Broido, T.L. Reinecke, First-principles determination of ultrahigh thermal conductivity of boron arsenide: A competitor for diamond?, *Phys. Rev. Lett.*, 111, 2013, 025901.
[41] B. Lv, Y. Lan, X. Wang, Q. Zhang, Y. Hu, A.J. Jacobson, *et al.*, Experimental study of the proposed super-thermal-conductor: BAs, *Appl. Phys. Lett.*, 106, February 16, 2015, 074105.
[42] J. Kim, D.A. Evans, D.P. Sellan, O.M. Williams, E. Ou, A.H. Cowley, *et al.*, Thermal and thermoelectric transport measurements of an individual boron arsenide microstructure, *Appl. Phys. Lett.*, 108, May 16, 2016, 201905.
[43] J.S. Kang, M. Li, H. Wu, H. Nguyen, Y. Hu, Experimental observation of high thermal conductivity in boron arsenide, *Science*, 361, 2018, 575.
[44] S. Li, Q. Zheng, Y. Lv, X. Liu, X. Wang, P.Y. Huang, *et al.*, High thermal conductivity in cubic boron arsenide crystals, *Science*, 361, 2018, 579.
[45] F. Tian, B. Song, X. Chen, N.K. Ravichandran, Y. Lv, K. Chen, *et al.*, Unusual high thermal conductivity in boron arsenide bulk crystals, *Science*, 361, 2018, 582.
[46] T. Feng, L. Lindsay, X. Ruan, Four-phonon scattering significantly reduces intrinsic thermal conductivity of solids, *Phys. Rev. B*, 96, October 27, 2017, 161201.
[47] C. Dames, G. Chen, Thermal conductivity of nanostructured thermoelectric materials, in: *Thermoelectrics Handbook: Macro to Nano*, D.M. Rowe (ed.), Boca Raton: CRC Press, 2006.
[48] F. Yang, C. Dames, mean free path spectra as a tool to understand thermal conductivity in bulk and nanostructures, *Phys. Rev. B*, 87, 2013, 035437.
[49] A.S. Henry, G. Chen, Spectral phonon transport properties of silicon based on molecular dynamics simulations and lattice dynamics, *J. Comput. Theor. Nanosci.*, 5, 2008, 1–12.
[50] K. Esfarjani, G. Chen, H.T. Stokes, Heat transport in silicon from first-principles calculations, *Phys. Rev. B*, 84, 2011, 085204.
[51] W. Li, N. Mingo, L. Lindsay, D. Broido, D. Stewart, N.A. Katcho, Thermal conductivity of diamond nanowires from first principles, *Phys. Rev. B*, 85, 2012, 195436.
[52] K.T. Regner, D.P. Sellan, Z. Su, C.H. Amon, A.J.H. McGaughey, J.A. Malen, Broadband phonon mean free path contributions to thermal conductivity measured using frequency domain thermoreflectance, *Nat. Commun.*, 4, 2013, 1640.
[53] A.D. McConnell, K.E. Goodson, Thermal conduction in silicon micro- and nanostructures, *Annu. Rev. Heat Transfer*, 14, 2005, 129–168.

[54] W. Liu, M. Asheghi, Thermal conduction in ultrathin pure and doped single-crystal silicon layers at high temperatures, *J. Appl. Phys.*, 98, 2005, 123523.
[55] W. Liu, M. Asheghi, Thermal conductivity measurements of ultra-thin single crystal silicon layers, *J. Heat Transfer*, 128, 2006, 75–83.
[56] Y.S. Ju, K.E. Goodson, Phonon scattering in silicon films with thickness of order 100 nm, *Appl. Phys. Lett.*, 74, 1999, 3005–3007.
[57] M. Asheghi, M.N. Touzelbaev, K.E. Goodson, Y.K. Leung, S.S. Wong, Temperature-dependent thermal conductivity of single-crystal silicon layers in SOI substrates, *J. Heat Transfer*, 120, 1998, 30–36.
[58] M. Asheghi, K. Kurabayashi, R. Kasnavi, K.E. Goodson, Thermal conduction in doped single-crystal silicon films, *J. Appl. Phys.*, 91, 2002, 5079–5088.
[59] D. Li, Y. Wu, P. Kim, L. Shi, P. Yang, A. Majumdar, Thermal conductivity of individual silicon nanowires, *Appl. Phys. Lett.*, 83, 2003, 2934–2936.
[60] K. Hippalgaonkar, B. Huang, R. Chen, K. Sawyer, P. Ercius, A. Majumdar, Fabrication of microdevices with integrated nanowires for investigating low-dimensional phonon transport, *Nano Lett.*, 10, 2010, 4341–4348.
[61] F. Zhou, A.L. Moore, J. Bolinsson, A. Persson, L. Froberg, M.T. Pettes, *et al.*, Thermal conductivity of inidum arsenide nanowires with wurtzite and zinc blende phases, *Phys. Rev. B*, 83, 2011, 205416.
[62] F. Zhou, A.L. Moore, M.T. Pettes, Y. Lee, J.H. Seol, Q.L. Ye, *et al.*, Effect of growth base pressure on the thermoelectric properties of indium antimonide nanowires, *J. Phys. D: Appl. Phys.*, 43, 2010, 025406.
[63] Y. Yang, W. Liu, M. Asheghi, Thermal and electrical characterization of Cu/CoFe superlattices, *Appl. Phys. Lett.*, 84, 2004, 3121.
[64] E. Pop, V. Varshney, A.K. Roy, Thermal properties of graphene: Fundamentals and applications, *MRS Bulletin*, 37, 2012, 1273–1281.
[65] Y.-M. Lin, A. Valdes-Garcia, S.-J. Han, D.B. Farmer, I. Meric, Y. Sun, *et al.*, Wafer-scale graphene integrated circuit, *Science*, 332, 2011, 1294–1297.
[66] M. Freitag, M. Steiner, Y. Martin, V. Perebeinos, Z. Chen, J.C. Tsang, *et al.*, Energy dissipation in graphene field-effect transistors, *Nano Lett.*, 9, 2009, 1883–1888.
[67] S.-K. Lee, H.Y. Jang, S. Jang, E. Choi, B.H. Hong, J. Lee, *et al.*, All graphene-based thin film transistors on flexible plastic substrates, *Nano Lett.*, 12, 2012, 3472–3476.
[68] F. Schedin, A.K. Geim, S.V. Morozov, E.W. Hill, P. Blake, M.I. Katsnelson, *et al.*, Detection of individual gas molecules adsorbed on graphene, *Nat. Mater.*, 6, 2007, 652–655.
[69] J.D. Fowler, M.J. Allen, V.C. Tung, Y. Yang, R.B. Kaner, B.H. Weiller, Practical chemical sensors from chemically derived graphene, *ACS Nano*, 3, 2009, 301–306.
[70] Y. Shao, J. Wang, H. Wu, J. Liu, I.A. Aksay, Y. Lin, Graphene based electrochemical sensors and biosensors: A review, *Electroanalysis*, 22, 2010, 1027–1036.

[71] F. Yavari, H.R. Fard, K. Pashayi, M.A. Rafiee, A. Zamiri, Z. Yu, *et al.*, Enhanced thermal conductivity in a nanostructured phase change composite due to low concentration graphene additives, *J. Phys. Chem. C*, 115, 2011, 8753–8758.
[72] A. Yu, P. Ramesh, M.E. Itkis, E. Bekyarova, R.C. Haddon, Graphite nanoplatelet-epoxy composite thermal interface materials, *J. Phys. Chem. C*, 111, 2007, 7565–7569.
[73] A. Yu, P. Ramesh, X. Sun, E. Bekyarova, M.E. Itkis, R.C. Haddon, Enhanced thermal conductivity in a hybrid graphite nanoplatelet-carbon nanotube filler for epoxy composites, *Adv. Mater.*, 20, 2008, 4740–4744.
[74] G.A. Slack, Anisotropic thermal conductivity of pyrolytic graphite, *Phys. Rev.*, 127, 1962, 694–701.
[75] Y.S. Touloukian, R.W. Powell, C.Y. Ho, P.G. Klemens, *Thermal Conductivity-Metallic Elements and Alloys* 1. New York: IFI/Plenum, 1970.
[76] C.N. Hooker, A.R. Ubbelohde, D.A. Young, Anisotropy of thermal conductance in near-ideal graphite, *Proc. Royal Soc. A*, 284, 1965, 17–31.
[77] R. Taylor, The thermal conductivity of pyrolytic graphite, *Philosophical Magaz.*, 13, 1966, 157–166.
[78] L. Lindsay, D.A. Broido, N. Mingo, Flexural phonons and thermal transport in graphene, *Phys. Rev. B*, 82, 2010, 115427.
[79] S. Ghosh, W. Bao, D.L. Nika, S. Subrina, E.P. Pokatilov, C.N. Lau, *et al.*, Dimensional crossover of thermal transport in few-layer graphene, *Nat. Mater.*, 9, 2010, 555–558.
[80] L. Lindsay, D.A. Broido, N. Mingo, Flexural phonons and thermal transport in multilayer graphene and graphite, *Phys. Rev. B*, 83, 2011, 235428.
[81] W. Cai, A.L. Moore, Y. Zhu, X. Li, S. Chen, L. Shi, *et al.*, Thermal transport in suspended and supported monolayer graphene grown by chemical vapor deposition, *Nano Lett.*, 10, 2010, 1645–1651.
[82] S. Chen, A.L. Moore, W. Cai, J.W. Suk, J. An, C. Mishra, *et al.*, Raman measurements of thermal transport in suspended monolayer graphene of variable sizes in vacuum and gaseous environments, *ACS Nano*, 5, 2011, 321–328.
[83] I. Jo, M.T. Pettes, L. Lindsay, E. Ou, A. Weathers, A.L. Moore, *et al.*, Reexamination of basal plane thermal conductivity of suspended graphene samples measured by electro-thermal micro-bridge methods, *AIP Adv.*, 5, 2015, 052306.
[84] T. Beechem, L. Yates, S. Graham, Invited review article: Error and uncertainty in Raman thermal conductivity measurements, *Rev. Scientific Instruments*, 86, April 1, 2015, 041101.
[85] S. Sullivan, A. Vallabhaneni, I. Kholmanov, X. Ruan, J. Murthy, L. Shi, Optical generation and detection of local nonequilibrium phonons in suspended graphene, *Nano Lett.*, 17, March 8, 2017, 2049–2056.
[86] J.H. Seol, I. Jo, A.L. Moore, L. Lindsay, Z.H. Aitken, M.T. Pettes, *et al.*, Two-dimensional phonon transport in supported graphene, *Science*, 328, 2010, 213–216.

[87] W. Jang, Z. Chen, W. Bao, C.N. Lau, C. Dames, Thickness-dependent thermal conductivity of encased graphene and ultrathin graphite, *Nano Lett.*, 10, 2010, 3909–3913.

[88] M.M. Sadeghi, I. Jo, L. Shi, Phonon-interface scattering in multi-layered graphene on an amorphous support, *Proc. Nat. Acad. Sci.*, DOI: 10.1073/pnas.1306175110, 2013.

[89] E.K. Sichel, R.E. Miller, M.S. Abrahams, C.J. Buiocchi, Heat capacity and thermal conductivity of hexagonal pyrolytic boron nitride, *Phys. Rev. B*, 13, 1976, 4607–4611.

[90] V.L. Solozhenko, A.G. Lazarenko, J.-P. Petitet, A.V. Kanaev, Bandgap energy of graphite-like hexagonal boron nitride, *J. Phys. Chem. Solids*, 62, 2001, 1331–1334.

[91] C. Wang, J. Guo, L. Dong, A. Aiyiti, X. Xu, B. Li, Superior thermal conductivity in suspended bilayer hexagonal boron nitride, *Scientific Rep.*, 6, April 5, 2016 (online), 25334.

[92] I. Jo, M.T. Pettes, J. Kim, K. Watanabe, T. Taniguchi, Z. Yao, *et al.*, Thermal conductivity and phonon transport in suspended few-layer hexagonal boron nitride, *Nano Lett.*, 13, 2013, 550–554.

[93] Y. Fu, J. Hansson, Y. Liu, S. Chen, A. Zehri, M.K. Samani, *et al.*, Graphene related materials for thermal management, *2D Mater.*, 7, October 22, 2019, 012001.

[94] Z. Gao, Y. Zhang, Y. Fu, M.M.F. Yuen, J. Liu, Thermal chemical vapor deposition grown graphene heat spreader for thermal management of hot spots, *Carbon*, 61, 2013, 342–348.

[95] M.-H. Shih, L.-J. Li, Y.-C. Yang, H.-Y. Chou, C.-T. Lin, C.-Y. Su, Efficient heat dissipation of photonic crystal microcavity by monolayer graphene, *ACS Nano*, 7, 2013, 10818–10824.

[96] Z. Yan, G. Liu, J.M. Khan, A.A. Balandin, Graphene quilts for thermal management of high-power GaN transistors, *Nat. Commun.*, 3, 2012, 827.

[97] K.C. Pratik, A. Rai, T.S. Ashton, A. L. Moore, APCVD hexagonal boron nitride thin films for passive near-junction thermal management of electronics, *Nanotechnology*, 28, 2017, 505705.

[98] Z. Lin, C. Liu, Y. Chai, High thermally conductive and electrically insulating 2D boron nitride nanosheet for efficient heat dissipation of high-power transistors, *2D Materials*, 3, October 18, 2016, 041009.

[99] I. Choi, K. Lee, C.-R. Lee, J.S. Lee, S.M. Kim, K.-U. Jeong, *et al.*, Application of hexagonal boron nitride to a heat-transfer medium of an InGaN/GaN quantum-well green LED, *ACS Appl. Materials Interfaces*, 11, May 22, 2019, 18876–18884.

[100] D. Choi, N. Poudel, S.B. Cronin, L. Shi, Effects of basal-plane thermal conductivity and interface thermal conductance on the hot spot temperature in graphene electronic devices, *Appl. Phys. Lett.*, 110, February 13, 2017, 073104.

[101] A. Simpson, A.D. Stuckes, The thermal conductivity of highly oriented pyrolytic boron nitride, *J. Phys. C Solid State Phys.*, 4, 1971, 1710–1718.

[102] M. Shen, P.K. Schelling, P. Keblinkski, Heat transfer mechanism across few-layer graphene by molecular dynamics, *Phys. Rev. B*, 88, 2013, 045444.

[103] Z. Chen, W. Jang, W. Bao, C.N. Lau, C. Dames, Thermal contact resistance between graphene and silicon dioxide, *Appl. Phys. Lett.*, 95, 2009, 161910.

[104] K.F. Mak, C.H. Lui, T.F. Heinz, Measurement of the thermal conductance of the graphene/SiO_2 interface, *Appl. Phys. Lett.*, 97, 2010, 221904.

[105] Y.K. Koh, M.-H. Bae, D.G. Cahill, E. Pop, Heat conduction across monolayer and few-layer graphenes, *Nano Lett.*, 10, 2010, 4363–4368.

[106] R.J. Stoner, H.J. Maris, Kapitza conductance and heat flow between solids at temperatures from 50 to 300 K, *Phys. Rev. B*, 48, 1993, 16373–16387.

[107] R.J. Stevens, A.N. Smith, P.M. Norris, Measurement of thermal boundary conductance of a series of metal-dielectric interfaces by the transient thermoreflectance technique, *J. Heat Transfer*, 127, 2005, 315–322.

[108] S.P. Koenig, N.G. Boddeti, M.L. Dunn, J.S. Bunch, Ultrastrong adhesion of graphene membranes, *Nat. Nanotechnol.*, 6, 2011, 543–546.

[109] X. Li, Y. Yan, L. Dong, J. Guo, A. Aiyiti, X. Xu, *et al.*, Thermal conduction across a boron nitride and SiO_2 interface, *J. Phys. D Appl. Phys.*, 50, February 9, 2017, 104002.

[110] C.R. Dean, A.F. Young, I. Meric, C. Lee, L. Wang, S. Sorgenfrei, *et al.*, Boron nitride substrates for high-quality graphene electronics, *Nat. Nanotechnol.*, 5, 2010, 722–726.

[111] D. Choi, N. Poudel, S. Park, D. Akinwande, S.B. Cronin, K. Watanabe, *et al.*, Large reduction of hot spot temperature in graphene electronic devices with heat-spreading hexagonal boron nitride, *ACS Appl. Mater Interfaces*, 10, April 4, 2018, 11101–11107.

[112] Y. Liu, Z.-Y. Ong, J. Wu, Y. Zhao, K. Watanabe, T. Taniguchi, *et al.*, Thermal conductance of the 2D MoS2/h-BN and graphene/h-BN interfaces, *Scientific Rep.*, 7, June 3, 2017, 43886.

[113] D.G. Cahill, P.V. Braun, G. Chen, D.R. Clarke, S. Fan, K.E. Goodson, *et al.*, Nanoscale thermal transport. II. 2003–2012, *Appl. Phys. Rev.*, 1, 2014, 011305.

[114] L. Shi, C. Dames, J.R. Lukes, P. Reddy, J. Duda, D.G. Cahill, *et al.*, Evaluating broader impacts of nanoscale thermal transport research, *Nanoscale Microscale Thermophys. Eng.*, 19, April 3, 2015, 127–165.

[115] M.C. Wingert, J. Zheng, S. Kwon, R. Chen, Thermal transport in amorphous materials: A review, *Semicond. Sci. Technol.*, 31, October 6, 2016, 113003.

[116] P.B. Allen, J.L. Feldman, Thermal conductivity of disordered harmonic solids, *Phys. Rev. B*, 48, January 11, 1993, 12581–12588.

[117] P.B. Allen, J.L. Feldman, J. Fabian, F. Wooten, Diffusons, locons and propagons: Character of atomie yibrations in amorphous Si, *Philos. Mag. B*, 79, November 1, 1999 1999, 1715–1731.

[118] H.R. Seyf, A. Henry, A method for distinguishing between propagons, diffusions, and locons, *J. Appl. Phys.*, 120, July 14, 2016, 025101.

[119] J.M. Larkin, A.J.H. McGaughey, Thermal conductivity accumulation in amorphous silica and amorphous silicon, *Phys. Rev. B*, 89, April 14, 2014, 144303.

[120] Y. He, D. Donadio, G. Galli, Heat transport in amorphous silicon: Interplay between morphology and disorder, *Appl. Phys. Lett.*, 98, April 4, 2011, 144101.

[121] S. Kwon, J. Zheng, M.C. Wingert, S. Cui, R. Chen, Unusually high and anisotropic thermal conductivity in amorphous silicon nanostructures, *ACS Nano*, 11, March 28, 2017, 2470–2476.

[122] L. Yang, Q. Zhang, Z. Cui, M. Gerboth, Y. Zhao, T.T. Xu, *et al.*, Ballistic phonon penetration depth in amorphous silicon dioxide, *Nano Lett.*, 17, December 13, 2017, 2017, 7218–7225.

[123] A. Sood, E. Pop, M. Asheghi, K.E. Goodson, The heat conduction renaissance, in: *2018 17th IEEE Intersociety Conference on Thermal and Thermomechanical Phenomena in Electronic Systems (ITherm)*, 2018, 1396–1402.

[124] G. Chen, Thermal conductivity and ballistic-phonon transport in the cross-plane direction of superlattices, *Phys. Rev. B*, 57, June 15, 1998, 14958–14973.

[125] A.J. Minnich, J.A. Johnson, A.J. Schmidt, K. Esfarjani, M.S. Dresselhaus, K.A. Nelson, *et al.*, Thermal conductivity spectroscopy technique to measure phonon mean free paths, *Phys. Rev. Lett.*, 107, August 25, 2011, 095901.

[126] Y. Hu, L. Zeng, A.J. Minnich, M.S. Dresselhaus, G. Chen, Spectral mapping of thermal conductivity through nanoscale ballistic transport, *Nat. Nanotechnol.*, 10, January 6, 2015, 701.

[127] J. Cuffe, J.K. Eliason, A.A. Maznev, K.C. Collins, J.A. Johnson, A. Shchepetov, *et al.*, Reconstructing phonon mean-free-path contributions to thermal conductivity using nanoscale membranes, *Phys. Rev. B*, 91, June 17, 2015, 245423.

[128] K.T. Regner, J.P. Freedman, J.A. Malen, Advances in studying phonon mean free path dependent contributions to thermal conductivity, *Nanoscale Microscale Thermophys. Eng.*, 19, March 7, 2015, 183–205.

[129] M. Maldovan, Sound and heat revolutions in phononics, *Nature*, 503, November 14, 2013, 209–217.

[130] M. Maldovan, Narrow low-frequency spectrum and heat management by thermocrystals, *Phys. Rev. Lett.*, 110, January 11, 2013, 025902.

[131] M. Maldovan, Phonon wave interference and thermal bandgap materials, *Nat. Mater*, 14, July 2015, 667–674.

[132] E.M. Dede, F. Zhou, P. Schmalenberg, T. Nomura, Thermal metamaterials for heat flow control in electronics, *J. Electron. Packaging*, 140, 2018, 010904.

https://doi.org/10.1142/9789811279379_0008

Chapter 8

Monolithic Microfluidic Cooling Using Micropin-Fin Arrays for Local High Heat Flux Remediation: Design Considerations, Experimental Validation, and FPGA Integration

Thomas Sarvey*, Ankit Kaul†, and Muhannad S. Bakir‡

Department of Electrical and Computer Engineering, Georgia Institute of Technology, Atlanta, Georgia, USA

**tsarvey@hotmail.com*

†ankit.kaul@gatech.edu

‡mbakir@ece.gatech.edu

Abstract

With an increase in microelectronic system density, using conventional technologies for cooling continues to become more challenging and is often a limiter of performance and efficiency. One key challenge is to address high background heat fluxes generated across entire chips and packages while mitigating localized hotspots with even higher heat fluxes. In this chapter, cooling of integrated circuits with non-uniform power maps using non-uniform micropin-fin heat sinks is investigated. Two heterogeneous micropin-fin samples were fabricated, and single-phase experiments with deionized water were performed to investigate the effectiveness of local micropin-fin clustering for hotspot cooling. Large background heat fluxes (250 W/cm^2) were cooled in conjunction with even higher hotspot heat fluxes (500 W/cm^2), with the lowest achieved average junction-to-inlet thermal resistance (R_{th}) of 0.092°C/W. Subsequently, to demonstrate the benefit of microfluidic cooling on active silicon, a micropin-fin heat sink was etched into the bulk of an Altera Stratix-V field-programmable gate array (FPGA) built using a 28 nm CMOS process. With a benchmark pulse compression algorithm running on the FPGA, thermal and electrical measurements were made. Deionized water was used as a coolant, with flow rates ranging from 0.15 to 3.0 mL/s and inlet temperatures ranging from 21°C to 50°C. An average junction-to-inlet thermal resistance of 0.07°C/W or a heat transfer coefficient of 1.59×10^4 W/m^2°C was achieved.

Keywords: Bosch process, deep reactive ion etching (DRIE), hotspot, microfluidic cooling, micropin-fin, field-programmable gate arrays (FPGAs), high-performance computing (HPC), microfluidic cooling, thermal management

1. Introduction

Heterogeneous integration of electronic systems using 2.5D and 3D integration strategies combined with continued increases in power dissipation with scaling will collectively impose severe cooling challenges that will not be met by evolutionary advances in cooling. As such, there is significant research in the literature on advanced cooling concepts for emerging microelectronic applications. For a single chip, cooling of the background power dissipation in conjunction with hotspots has historically been addressed using a homogeneous thermal solution. More recently, several other methods of hotspot mitigation have been explored, including dedicated microgap coolers [1], micro-thermoelectric coolers (TECs) [2], liquid jet impingement [3–5], thin-film evaporation [6, 7], and heat spreading through highly thermally conductive materials, such as graphene [5, 8]. While all these approaches have their advantages and disadvantages, they all require relatively complicated fabrication processes. Solutions such as TECs (and, to some extent, heat spreaders) can reduce the temperatures of small hotspots but still require a means of dissipating large background heat fluxes. The microfluidic heat sinks demonstrated in this chapter are fabricated with a single etching step and are also compatible with TSVs and 3D ICs [9].

In the first half of this chapter, heterogeneous micropin-fin arrays are used to cool a large uniform background heat flux (250 W/cm^2) with a hotspot region of higher heat flux (500 W/cm^2). Since flow could be diverted around a high-density pin-fin cluster not spanning the entire width of the channel, two types of clusters were chosen for experimental investigation: local clustering over the hotspot and clustering spanning the entire width of the channel. Observations suggest that a span-wise cluster design performs better than a local cluster design in terms of absolute junction temperature reduction, but with a higher pressure drop penalty. Next, we monolithically integrate the silicon micropin-fin array with a field-programmable gate array (FPGA) to demonstrate electrical benefits. This chapter is based on the results presented in Refs. [10] and [11].

2. Effect of Heat Spread Modeling: Hotspot Thermal Resistance Measurement

Even with a uniform micropin-fin array, heat spreading through the silicon bulk will partially mitigate hotspots. In order to quantify the effect of spreading as a function of base thickness and the ratio of hotspot heat flux to background heat flux, heat transfer simulations were performed using COMSOL 4.3b. A 3D model of a 1 cm $\times$ 1 cm chip was created with a 500 μm $\times$ 500 μm hotspot region at the center. Symmetry boundary conditions were utilized to simulate only a quarter of this geometry. A maximum mesh size of 10 μm was used in the region over the hotspot, and a maximum mesh size of 50 μm was used over the background region, with a maximum growth rate of 1.1 between the two regions.

On the bottom of the chip slice, two heat fluxes were applied, one to the hotspot (q_{hs}) and the other to the background region (q_{bg}). On the top side of the chip, two heat transfer coefficients were applied, one over the hotspot region (h_{hs}), and the other over the remaining background region (h_{bg}). An illustration of the model can be seen in Figure 1. The reference temperature on the convective boundary conditions, h_{hs} and h_{bg}, was set to a constant value of 20°C. Although a microfluidic heat sink would have a gradient at

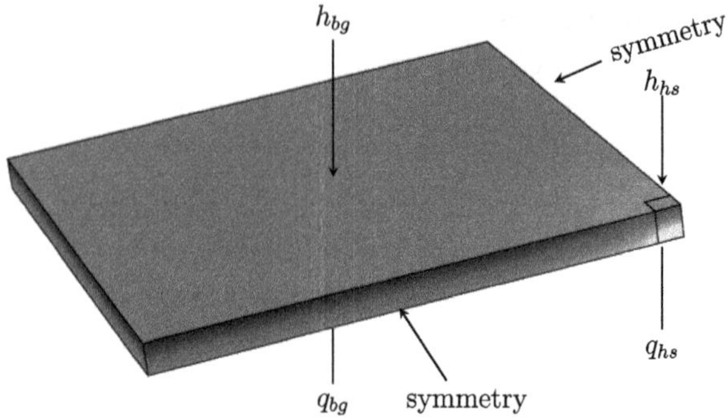

Figure 1. COMSOL heat spreading model.

this boundary temperature due to heating of the fluid, using a linear temperature gradient across the chip does not affect the average temperatures in the simulations. The hotspot heat flux can locally heat the fluid; however, this effect is expected to be small since the hotspot area is 0.25% of the total chip area. Therefore, the hotspot power is a small fraction of the total power, even when the hotspot heat flux density is several times the average heat flux density.

Without heat spreading through the base silicon, the heterogeneous heat sink could be characterized by calculating separate background and hotspot thermal resistances (at specific flow conditions). However, when we define the hotspot thermal resistance as $R_{hs} = (T_{hs} - T_{amb})/q_{hs}$, where T_{hs} is the average hotspot temperature and T_{amb} is the reference temperature in the convective boundary condition, the apparent thermal resistance at the hotspot (and, to a much lesser extent, the background region) depends on the power map applied. When a uniform power map is applied, some heat from the background region is transferred to the high-density cluster, making the apparent thermal resistance greater than it would be with no spreading (with a very thin base). However, as the local heat flux increases, heat begins to leave the dense cluster and find another path by spreading through the base, causing the apparent thermal resistance to decrease. This is illustrated in Figure 2.

An increase in the relative difference between the hotspot and background heat fluxes and the resulting spreading effect lead to an apparent reduction in the hotspot thermal resistance. This can

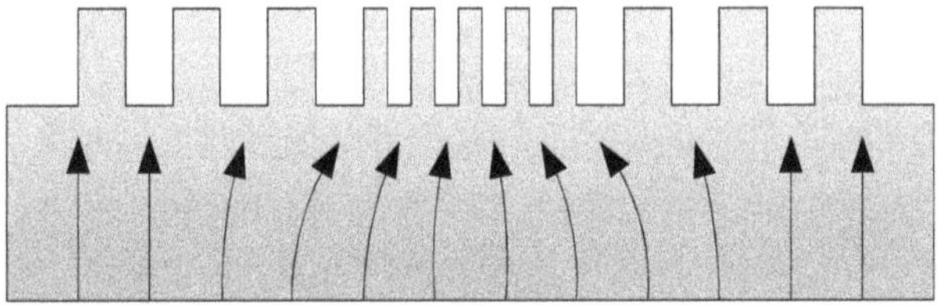

(a)

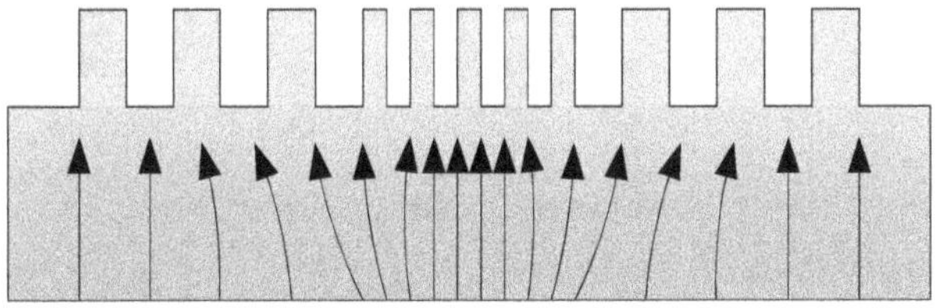

(b)

Figure 2. Direction of heat flow in the substrate for different power maps: (a) uniform heat flux and (b) high hotspot heat flux.

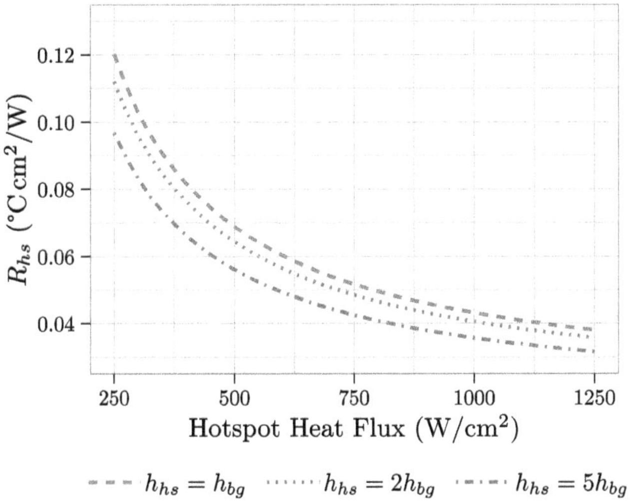

Figure 3. Effective hotspot thermal resistance (R_{hs}) vs. hotspot heat flux (q_{hs}).

be mitigated by locally improving the heat transfer coefficient. To illustrate this, the effective thermal resistance at a centrally located hotspot, h_{Rhs}, is shown in Figure 3, where the silicon base thickness is 300 μm and the background heat flux is 250 W/cm^2. Three lines are shown, where h_{hs} = 100,000 W/(m^2°C), h_{hs} = 200,000

W/(m^2°C), and h_{hs} = 500,000 W/(m^2°C). Due to heat spreading to the surrounding background region, the apparent hotspot thermal resistance changes with the hotspot heat flux. Therefore, in Section 4, hotspot performance will be reported as temperatures relative to background temperatures. It is important to note that temperature measurements were made at the base of the silicon and not at the surface in contact with the fluid; therefore, any measurements that are reported include the effects of spreading.

3. Thermal Test Bed and Heterogeneous Micropin-Fin Samples

In this chapter, the hotspot is a smaller region of the chip with a heat flux considerably higher than that of the background region. By locally clustering micropin fins over a hotspot, it is expected that the local heat transfer coefficient can be increased to deal with the hotspot heat fluxes while minimizing the pressure drop penalty compared to using this high-density clustering over the entire background region.

The heat sink clustering approach discussed above was applied to heterogeneous micropin-fin arrays with local clustering of micropin fins over the hotspots. First, the micropin fins were etched into a 500 μm double-side polished silicon wafer. A pyrex cap was anodically bonded to the micropin fins to seal the channels. Next, a 1.9 μm thick silicon-dioxide layer was deposited on the back side of the wafer, on top of which 0.2 μm thick platinum heaters were deposited. The heaters also acted as resistance temperature detectors (RTDs). Finally, fluid inlet/outlet and pressure measurement ports were etched through the silicon. A diagram of the cross-section of the test chips can be seen in Figure 4.

While enhanced heat transfer is only necessary over the hotspot, local clustering over this region is expected to result in some flow diversion around the higher-density clustering. Therefore, clustering was done in a small region over the hotspot as well as in a region spanning the entire width of the channel to prevent this flow diversion. Two types of micropin-fin clustering were chosen for hotspot cooling:

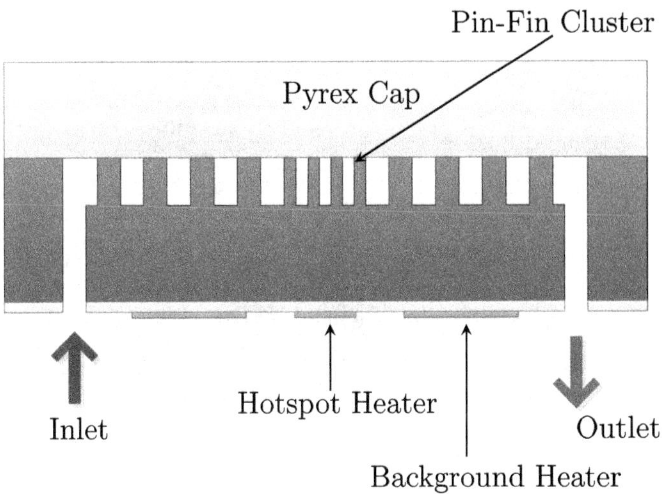

Figure 4. Heterogeneous micropin-fin test chip cross-section.

Figure 5. Cylindrical micropin-fins locally clustered over hotspot.

(1) cylindrical micropin fins with local clustering (Figure 5);
(2) cylindrical micropin fins with span-wise clustering (Figure 6).

Images of the top and bottom of a micropin-fin test device can be seen in Figures 7 and 8, respectively. The background micropin-fin

Figure 6. Cylindrical micropin-fins spanning entire channel.

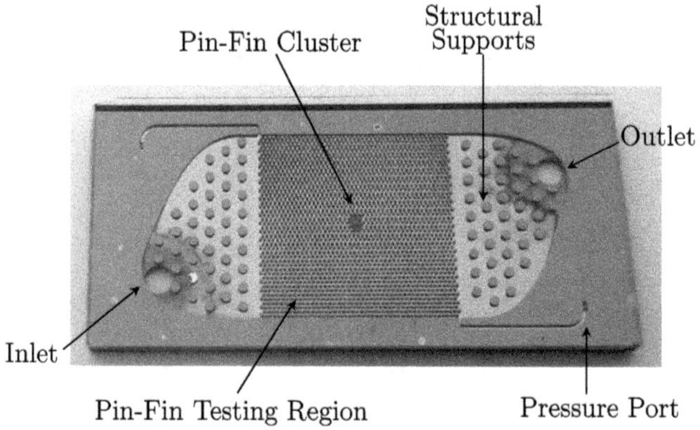

Figure 7. Test chip: top view of etched silicon through pyrex cap.

array covers an area of 1×1 cm^2. The hotspot consists of an area of 500×500 μm^2 at the center of the chip. This region is heated by a dedicated serpentine platinum heater on the bottom of the chip, while the background heat flux is applied through four heaters spanning the chip. Pressure ports at the inlet and outlet sides of the background micropin-fin array provide accurate pressure drop measurements.

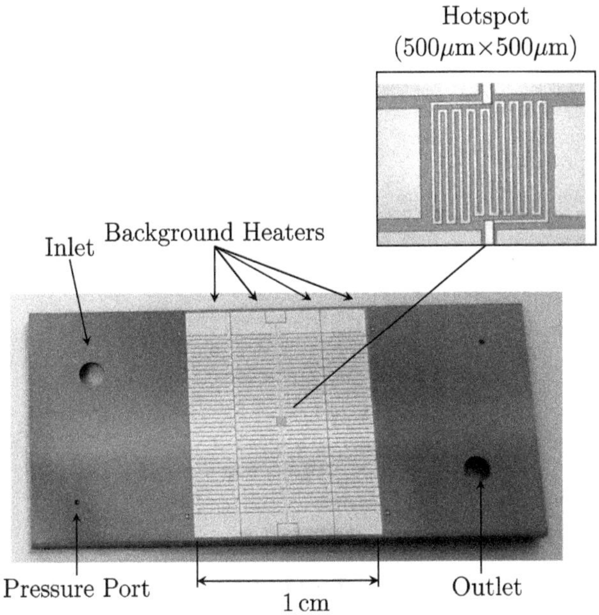

Figure 8. Test chip: bottom view of heater and ports.

Each chip was tested in an open-loop system with deionized water as a coolant, with an inlet temperature of 21 ± 0.5°C. Each experiment was run twice, and the averages of the results from the two runs are reported in Section 4.

4. Experimental Results

Figure 9 shows the rise in junction temperature above inlet temperature ($T_j - T_{in}$) versus axial position in the direction of fluid flow (x) for both types of chips at a flow rate of 208 mL/min and a uniform power of 246 W across the 1 cm × 1 cm chips. With a uniform micropin-fin array, uniform power, constant fluid heat capacity, and no edge effects, one would expect the junction temperature to increase linearly with axial position. A line was fit to the four background junction temperatures using least squares. As can be seen, the temperature does rise linearly, although the first and last temperatures tend to be below the fit line, while those of the central heaters

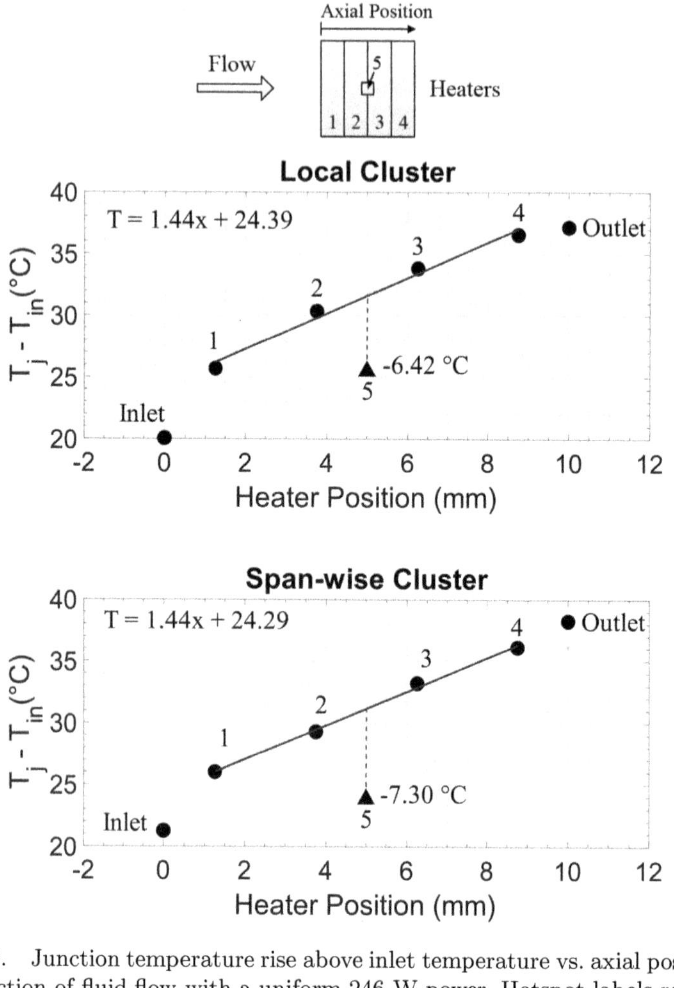

Figure 9. Junction temperature rise above inlet temperature vs. axial position in the direction of fluid flow with a uniform 246 W power. Hotspot labels represent temperature deviation from expected temperature with a uniform micropin-fin density.

tend to be above it. This could be a result of heat spreading to the inlet and outlet plena as well as flow development in these regions.

The slopes of the lines in Figure 9 have units of °C/mm and depend primarily on flow rate and heat flux rather than the micropin-fin dimensions. As expected, both lines have the same slope. By multiplying the slope by the length of the chip, the change in fluid temperature across the 1 cm length of the micropin-fin region can be

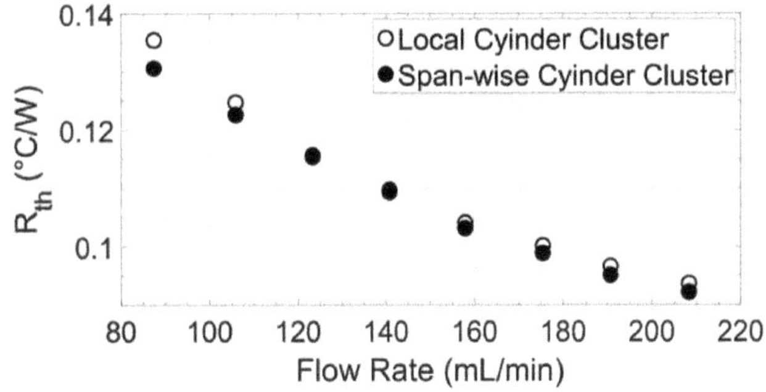

Figure 10. Average background thermal resistance from the silicon to fluid (R_{jf}) vs. flow rate for both dice.

estimated. Since this gradient across the micropin-fin region of the chip only accounts for approximately 85% of the measured $T_{out} - T_{in}$, it is estimated that approximately 15% of the heat flux had spread to the inlet and outlet plena where it was transferred to the water.

The average background junction-to-fluid thermal resistance, $R_{jf} = R_{cond} + R_{conv}$, can be found by subtracting the average fluid temperature from the average junction temperature and dividing by the chip power. Figure 10 shows that the background heat flux is significant for the micropin fins spanning the width of the channel, as they cover a larger surface area and are located directly above sections of the background RTDs. Therefore, it is unsurprising that the average background thermal resistance of the cylindrical sample with span-wise clustering is lower than that of the cylindrical sample with local clustering, despite the same dimensions of the micropin fins over the rest of the sample. The lowest achieved R_{jf} was 0.092 °C/W for the span-wise cylindrical cluster, and the average background heat transfer coefficient was 10,846.4 W/m^2°C. However, as expected, this comes at a higher pressure drop penalty for the span-wise cluster, as shown in Figure 11.

4.1. *Non-uniform heat flux*

Lastly, both devices were tested with a nominal background heat flux of 250 W/cm^2 and a hotspot heat flux of 500 W/cm^2. The results at the highest flow rate of 208 mL/min can be seen in Figure 12. Despite doubling the heat flux relative to the background

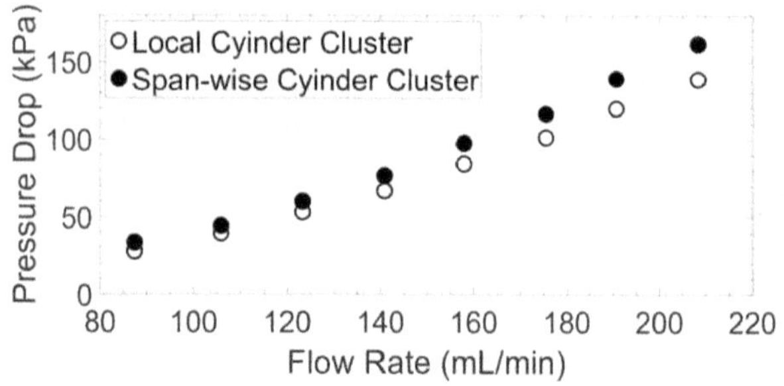

Figure 11. Pressure drop vs. flow rate for both dice.

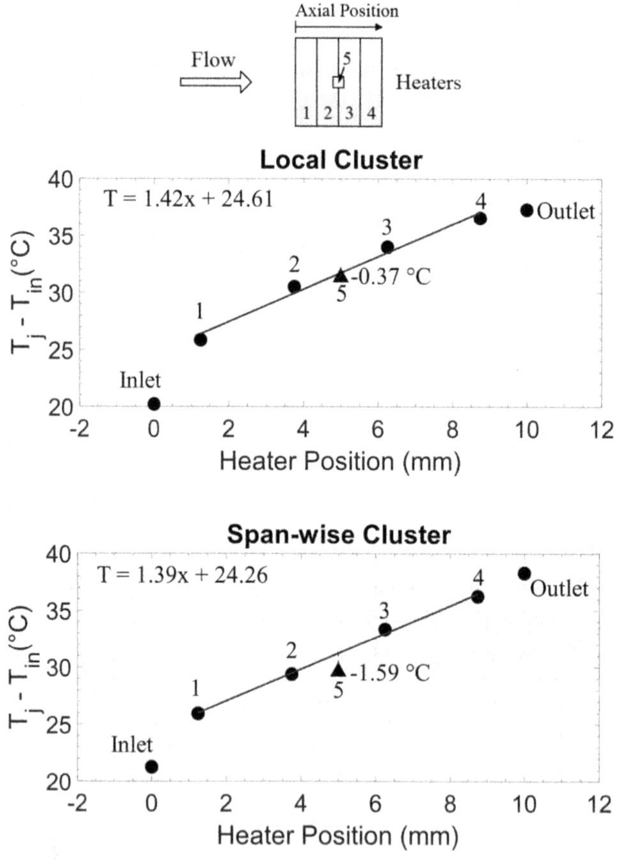

Figure 12. Junction temperature rise above inlet temperature vs. axial position in the direction of fluid flow with 500 W/cm^2 hotspot heat flux and 250 W/cm^2 background heat flux. Hotspot labels represent temperature deviation from expected temperature with a uniform micropin-fin density.

region, hotspot temperatures remain below the background centerline temperatures.

The hotspot temperature is lower than the average background temperature for all uniform power maps (250 W/cm^2 background and hotspot). When the hotspot heat flux is increased to 500 W/cm^2, the hotspot temperature increases but remains below the average background temperature for both device types. The micropin fins with local clustering have a hotspot temperature that very closely matches the background temperature when the hotspot heat flux is twice the background heat flux (maximum difference of 0.25°C, or 0.4%). As seen earlier, the device with high-density micropin fins spanning the entire width of the channel has the lowest hotspot temperatures and would likely be ideal for even higher hotspot heat fluxes relative to background heat fluxes.

The two devices with micropin fins spanning the entire width of the channel had lower hotspot temperatures than those with clustering directly over the hotspots. This could be partially due to flow diversion around the cluster when the cluster does not span the entire width of the channel but also because of heat spreading to the larger high-density cluster when the cluster extends beyond the hotspot region.

5. Microfluidic Integration with FPGA

An Altera Stratix V FPGA, built through a 28 nm process, was post-processed to integrate a micropin-fin heat sink (MFHS) directly into the back of the flip-chip-bonded silicon die, a few hundred micrometers from the active circuitry. This microfluidic-cooled FPGA was then tested with deionized water as a coolant at several flow rates and inlet temperatures. All testing was performed with Altera Stratix V DSP development boards. A comparison was made with a stock board consisting of the default air-cooled heat sink (ACHS). Testing was performed at flow rates ranging from 0.15 to 3.0 mL/s and with inlet temperatures ranging from 21°C to 50°C.

A photograph of the etched die along with the micropin-fin dimensions can be seen in Figure 13. An SEM image of identical micropin

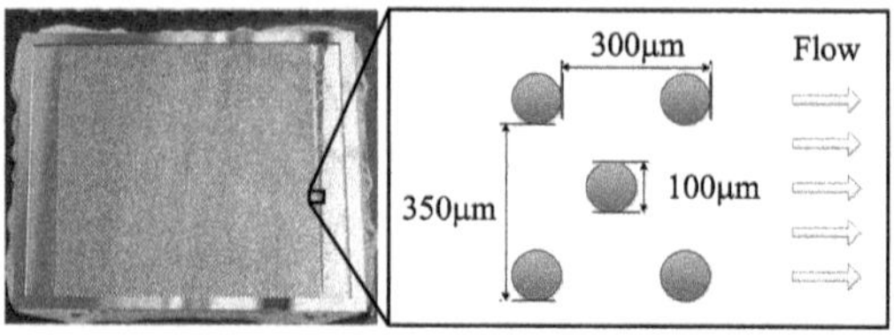

Figure 13. Image of the etched back side of the silicon FPGA die along with the micropin-fin dimensions.

Figure 14. SEM image of micropin fins etched using the same process in a silicon wafer.

fins fabricated through the same process on a silicon wafer can be seen in Figure 14. In general, the aspect ratio and surrounding features are known to affect the profile and depth of etched cavities [12, 13]. Tapering of the micropin-fin sidewalls is visible on the micropin fins closest to the inlet and outlet regions but is minor in the rest of the array. Significant tapering, where the micropin-fin base is narrower than the top, could reduce fin efficiency and thermal performance by limiting the cross-sectional area through which heat can conduct up the micropin fins.

A separate silicon lid was fabricated with an inlet and an outlet port. The lid was first tacked onto the top of the etched FPGA with

Figure 15. Processed FPGA soldered to development board with silicon cap and nanoports for fluid delivery.

high-temperature epoxy in order to provide a smooth surface for resoldering it to the development board. After soldering, the lid was permanently secured with epoxy, and nanoports were attached for the delivery of coolant. A photograph of the resoldered FPGA with nanoports can be seen in Figure 15.

The Stratix V FPGA consists of a silicon die that is flip-chip bonded onto an organic substrate. The back side of the die was used to etch an MFHS in the same bulk silicon as the active circuitry. The Bosch process was used to etch silicon with vertical micropin-fin sidewalls. This batch process could be completed at the wafer level but, in this case, was applied to a single chip at the die level. This fabrication flow was used for this proof of concept due to the relative ease of acquiring packaged parts, but it may be different when optimized for scalability and manufacturing throughput.

6. Testing of FPGA

The FPGA was loaded with a custom pulse compression algorithm designed to mimic common DSP-style use cases of FPGAs and to utilize a large amount of the FPGA resources. The algorithm consisted of nine soft computing cores that could be toggled ON and OFF during runtime. The FPGA was tested in an open-loop system with deionized water as a coolant. The voltage regulator module (VRM) on the board was run at a current higher than its datasheet rating, so air was blown over it in order to prevent overheating.

The FPGA was first tested with a flow rate of 2.4 mL/s, running zero to nine cores in order to vary the FPGA power. The inlet water temperature was 20.5°C, and the ambient air temperature was 19.3°C. Temperature measurements can be seen in Table 1. A stock Stratix V DSP development board was also tested for comparison, using the stock ACHS with which it was bundled.

The pulse compression algorithm uses 80% of the logic, 93% of the memory blocks, and 98% of the DSP blocks on the FPGA. In addition to many subtraction, addition, multiplexing, look-up-table, and memory operations, 346 18-bit multiplications are done every clock cycle. Counting only these multiplications, 934 GOPS are performed when operating all nine cores at 300 MHz.

To capture the thermal gradient produced by the heating of the fluid, measurements were taken with fluid flowing in both directions, as the temperature diode is located at the edge of the chip. Therefore, the temperatures of the microfluidic-cooled FPGA are all reported as a range of two values representing flow in both directions.

Table 1. FPGA thermal and power measurements with MFHS and ACHS.

Cores	MFHS FPGA Power (W)	ACHS FPGA Power (W)	MFHS FPGA Temp. (°C)	ACHS FPGA Temp. (°C)
0	13.2	13.7	21–22	43
1	15.4	16.0	21–23	46
2	17.6	18.3	22–23	49
3	19.8	20.5	22–23	51
4	21.9	22.8	22–23	53
5	24.0	25.1	22–23	56
6	26.2	27.5	22–23	59
7	28.3	29.8	22–24	61[a]
8	30.4	—	22–24	—
9	32.4	—	22–24	—

[a]Temperature warning on board illuminated.

A maximum die temperature of 60°C was set to match the default onboard temperature warning indicator of the Stratix V DSP development kit. Although better thermal results could undoubtedly be achieved with a larger, more powerful ACHS, it should be noted that the stock ACHS ran six computing cores before reaching this maximum temperature, while the microfluidic-cooled FPGA ran all nine cores (a 1.5× improvement in throughput) while maintaining the die temperature below 24°C (with additional power as per Table 1).

The FPGA heat flux is lower than that of many high-power processors, and the low-profile ACHS with which it came is significantly less effective than the best available ACHSs.

6.1. *Variable flow rate testing*

The microfluidic-cooled FPGA was tested with multiple flow rates while running the same pulse compression algorithm with all nine cores. The inlet temperature ranged between 20.3°C and 20.9°C. As can be seen in Figure 16, with an increase in flow rate, the FPGA temperature decreases due to decreasing heating of the fluid and decreasing convective thermal resistance. A minimum average thermal resistance of 0.07°C/W was achieved at a maximum flow rate of

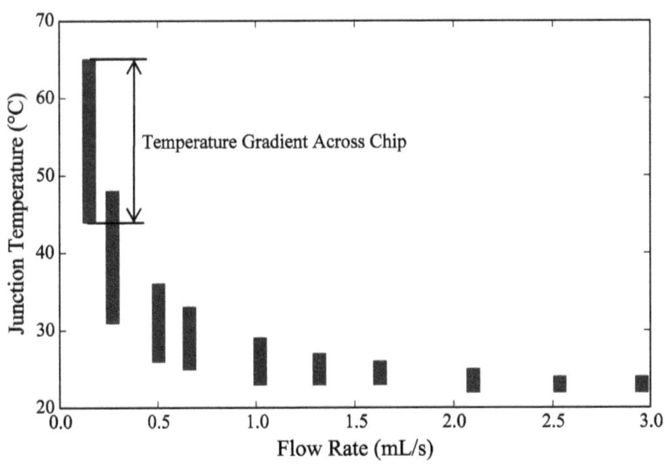

Figure 16. Die temperature vs. flow rate: Die temperature is a range representing the measurements with the temperature diodes near the inlet and outlet.

3.0 mL/s. For a cooled area of ≈9 cm^2, this corresponds to a heat transfer coefficient of 1.59×10^4 W/m^{2}°C.

6.2. *Clock speed*

In addition to increasing performance through increased silicon utilization, improved cooling can also improve performance. To test the dependence of maximum clock frequency on die temperature with the microfluidic-cooled FPGA, the temperature was varied by varying the flow rate with the inlet temperature between 24.3°C and 24.7°C. Due to current limitations of the onboard VRM, seven of the nine cores were run. The outputs from all cores were monitored through the Altera Signaltap tool to detect glitches that occurred in the output waveforms. The maximum clock speeds at which all seven cores operated with no glitches can be seen in Figure 17 as a function of the die temperature measured on the side of the chip closest to the outlet. Decreasing the maximum die temperature from 66°C to 28°C yielded an improvement of 21 MHz, a 6% improvement in clock speed (with an accompanying increase in power).

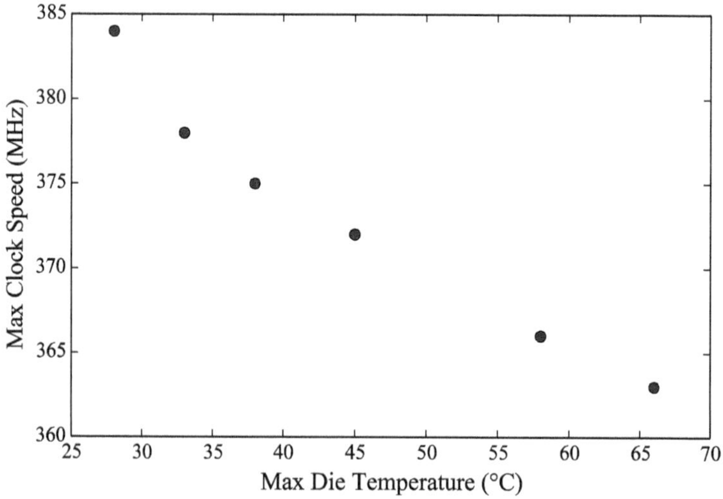

Figure 17. Maximum FPGA clock speed without glitches vs. maximum die temperature.

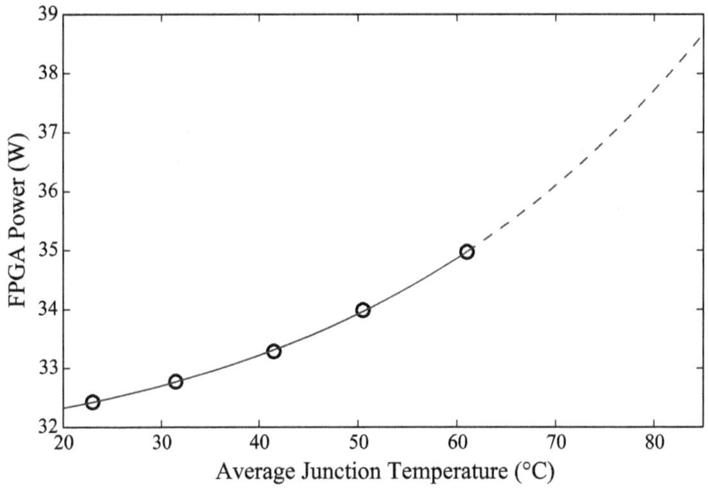

Figure 18. FPGA power vs. average die temperature.

6.3. *Die power*

The total FPGA power versus average die temperature is plotted in Figure 18. The die temperature was varied by varying the inlet temperature at a constant flow rate of 3.1 mL/s since this provided nearly uniform die temperatures. Due to the temperature-dependent leakage power, the measured total FPGA power increases by 2.6%, 4.8%, and 7.8% at 41.5°C, 50.5°C, and 61°C relative to the power dissipation at 23°C. A trend curve using a first-order approximation of the subthreshold leakage current (from Ref. [11]) is also shown in Figure 18. From an efficiency standpoint, the measured increase in FPGA power at elevated temperatures provides another strong motivation for effective cooling.

7. Conclusion

This chapter presented monolithic, heterogeneously designed microfluidic cooling approaches to simultaneously address background and hotspot power densities. Locally clustered and spanwise clustered cylindrical micropin-fin-based microfluidic heat sink designs were explored. In regions above the hotspots, the density of the pin fins was increased to improve locally the effective heat transfer coefficient, which was confirmed experimentally. The device with micropin fins spanning the entire width of the channel had lower hotspot temperatures than the device with local clustering, at the cost of higher pressure drops. This difference in hotspot thermal performance may not only partially result from flow diversion around the local clusters but also likely arise from heat spreading to areas of the high-density clusters that are not directly above the hotspot. This chapter also presented the first monolithic integration of a microfluidic heat sink on an active CMOS die, namely a 28 nm Stratix V FPGA. Both thermal and electrical testing were conducted to validate the efficacy of the monolithic cooling solution.

References

[1] C. Green, P. Kottke, X. Han, C. Woodrum, T. Sarvey, P. Asrar, X. Zhang, Y. Joshi, A. Fedorov, S. Sitaraman, M. Bakir, A review of two-phase forced cooling in three-dimensional stacked electronics: Technology integration, *J. Electron. Packag,*. 137 (4), 2015, 40802.

[2] I. Chowdhury, R. Prasher, K. Lofgreen, G. Chrysler, S. Narasimhan, R. Mahajan, D. Koester, R. Alley, R. Venkatasubramanian, On-chip cooling by superlattice-based thin-film thermoelectrics, *Nat. Nanotechnol.*, 4 (4), 2009, 235–238.

[3] C. Green, A.G. Fedorov, Y.K. Joshi, Fluid-to-Fluid Spot-to-Spreader (F2/S2) hybrid heat sink for integrated chip-level and hot spot-level thermal management, *ASME. J. Electron. Packag.*, 131 (2), 2009, 025002. https://doi.org/10.1115/1.3104029.

[4] D. Nikolic, M. Hutchison, P. Sapin, A. Robinson, Hot spot targeting with a liquid impinging jet array waterblock, *2009 15th International Workshop on Thermal Investigations of ICs and Systems*, 2009, pp. 168–173.

[5] Y. Han, B.L. Lau, X. Zhang, Package-level microjet-based hotspot cooling solution for microelectronic devices, *IEEE Electron Dev. Lett.*, 36 (5), 2015, 502–504.

[6] S. Narayanan, A.G. Fedorov, Y.K. Joshi, On-chip thermal management of hotspots using a perspiration nanopatch, *J. Micromech. Microeng.*, 20 (7), 2010, 075010.

[7] M.H. Nasr, C.E. Green, P.A. Kottke, X. Zhang, T.E. Sarvey, Y.K. Joshi, M.S. Bakir, and A.G. Fedorov, Hotspot thermal management with flow boiling of refrigerant in ultrasmall microgaps, *J. Electron. Packag.*, 139 (1) 2017, 011006.

[8] Z. Yan, G. Liu, J. M. Khan, A.A. Balandin, Graphene quilts for thermal management of high-power GaN transistors, *Nat. Commun.*, 3 (1), 2012, 1–8.

[9] Y. Zhang, A. Dembla, M.S. Bakir, Silicon micropin-fin heat sink with integrated TSVs for 3-D ICs: Tradeoff analysis and experimental testing, *IEEE Trans. Compon. Packag. Manuf. Technol.*, 3 (11), 2013, 1842–1850.

[10] T.E. Sarvey, Y. Hu, C.E. Green, P.A. Kottke, D.C. Woodrum, Y.K. Joshi, A.G. Fedorov, S.K. Sitaraman, M.S. Bakir, Integrated circuit cooling using heterogeneous micropin-fin arrays for nonuniform power maps, *IEEE Trans. Compon. Packag. Manuf. Technol.*, 7 (9), 2017, 1465–1475.

[11] T.E. Sarvey, Y. Zhang, C. Cheung, R. Gutala, A. Rahman, A. Dasu, M.S. Bakir, Monolithic integration of a micropin-fin heat sink in a 28-nm FPGA, *IEEE Trans. Compon. Packag. Manuf. Technol.*, 7 (10), 2017, 1617–1624.

[12] J. Karttunen, J. Kiihamaki, S. Franssila, Loading effects in deep silicon etching, in: *Micromachining and Microfabrication Process Technology VI*, J.M. Karam and J.A. Yasaitis (eds.), Vol. 4174, pp. 90–97, Santa Clara, CA: SPIE, 2000. https://doi.org/10.1117/12.396475.

[13] I.W. Rangelow, Critical tasks in high aspect ratio silicon dry etching for microelectromechanical systems, *J. Vac. Sci. Technol. A Vac. Surf. Films*, 21 (4), 2003, 1550–1562.

https://doi.org/10.1142/9789811279379_0009

Chapter 9

Recent Experimental and Modeling Advances in Two-Phase Embedded Microfluidic Cooling

Todd A. Kingston*, Justin A. Weibel†, and Suresh V. Garimella‡

**Center for Multiphase Flow Research and Education, Department of Mechanical Engineering, Iowa State University, Ames, IA, USA*

†Cooling Technologies Research Center, School of Mechanical Engineering, Purdue University, West Lafayette, IN, USA

‡University of Vermont, Burlington, VT, USA

Abstract

Advances in the packaging approaches and semiconductor materials used for high-power electronic devices will usher in a paradigm shift in the thermal management strategies employed to dissipate extremely high heat fluxes at reasonable operating temperatures. Traditional "remote cooling" systems that have been commonly used to manage thermal loads generated by these devices suffer from parasitic interfacial, conduction, and spreading resistances, which lead to large temperature gradients. The next generation of "embedded cooling" systems will bring coolant very close to the heat source, eliminating these thermal resistances but requiring a solution that can directly manage high local heat fluxes without an intermediate heat spreader. This work highlights recent advances in the experimental characterization and modeling of two-phase embedded-cooling systems. Specifically, it summarizes the development of two-phase hierarchical manifold microchannel heat sinks, a promising high-performance embedded-cooling solution, as well as experimental and numerical studies investigating the underlying microchannel flow boiling phenomena. Static and dynamic flow boiling instabilities among parallel channels, along with the potential for coupling between transient heating conditions and these two-phase flow dynamics, offer unique implementation challenges that may arise in embedded-cooling systems. With continued progress in the understanding of these phenomena and predictive high-fidelity numerical modeling of microchannel flow boiling, embedded cooling systems will alleviate the thermal challenges that currently limit the power and performance of many electronic devices, enabling technological advancements across many industries.

Keywords: Embedded cooling, flow boiling, flow instability, microchannel, microfluidic, two-phase

1. Introduction to Two-Phase Embedded Cooling

The development of high-power electronics, such as solid-state power amplifiers, power-switching devices, and high-performance computing systems, has resulted in the need for innovative thermal management strategies capable of dissipating heat fluxes in excess of 1,000 W/cm^2 to maintain chip temperatures at levels safe for reliable operation [1, 2]. Electronic devices have traditionally been cooled using air or a liquid coolant by attaching a heat spreader and a heat sink to the surface of a packaged die; thermal interface materials are utilized to transfer heat across interfaces within the package and at each mating surface. In this "remote cooling" architecture, the total temperature rise across the thermal management solution is primarily governed by parasitic interfacial, conduction, and spreading resistances between the die and heat sink. With this architecture, large temperature gradients are observed across the stack at high heat fluxes produced by next-generation devices. Significant advances have been made over the past decade to minimize the various thermal resistances along the remote-cooling heat rejection path by developing advanced thermal interface materials [3–7], heat spreaders [8–16], alternative active cooling methods [1, 17], and high-performance air-cooled heat exchangers [1, 18–20]. Nevertheless, despite tremendous advances made in each standalone technology, transformative cooling strategies are still needed to circumvent these thermal resistances inherent to remote cooling, which are prohibitively large for high-power devices.

Microchannel heat sinks have been regarded as a promising technology for dissipating high heat fluxes because the increased surface area and heat transfer coefficients associated with the small channel dimensions provide larger heat rejection areas and more efficient thermal transport, respectively. In a seminal study by Tuckerman and Pease [21], a heat flux of 790 W/cm^2 was dissipated using a microchannel heat sink featuring 50 μm wide and 300 μm deep channels with single-phase water as the working fluid. In recent years, there has been a significant collective effort to further advance the potential of microchannel heat sinks by integrating dielectric fluids and microchannels within the semiconductor substrate directly below the heat-generating devices [22–32]. This "embedded-cooling" architecture removes the interfacial thermal resistances and minimizes conduction and spreading resistances that are characteristic of remote-cooling systems. These embedded, intra-chip heat sinks often consist of many parallel microchannels that are realized using microfabrication techniques. These embedded microchannels hold great promise for high-heat-flux dissipation.

Alongside the non-trivial fabrication, mechanical, and packaging challenges inherent to the approach, embedded microchannel heat sinks also present some unique thermal and hydraulic design challenges. The thin, solid substrate separating the heat-generating semiconductor device and the embedded heat sink significantly reduces heat spreading and exposes the heat sink to very high local heat fluxes. This can result in high local chip temperatures (i.e., hotspots) and large temperature gradients across the device if the cooling solution does not offer sufficiently high heat transfer coefficients. Moreover, the small hydraulic diameters of microchannels can lead to intractably large pressure drops and pumping power requirements for flow over relatively large traversal lengths (e.g., the size of the entire die).

In two-phase cooling strategies, the liquid coolant is allowed to boil inside the embedded channels (i.e., microchannel flow boiling). Because the latent heats of vaporization of common working fluids are orders of magnitude larger than their specific heat capacities, two-phase operation can enable lower chip temperature rises and higher heat dissipation at a given pumping power (or lower flow rates to dissipate the same amount of heat) compared to single-phase operation. Flow boiling also offers enhanced heat transfer coefficients compared to single-phase convection. Two-phase cooling systems can thereby enable reductions in size, weight, and overall power consumption compared to single-phase systems.

To realize these potential performance gains, two-phase cooling systems must evaporate a significant portion of the liquid that enters the heat sink (i.e., the exit vapor quality should be high). However, intermittent dry-out of the liquid film and flow boiling instabilities can lead to reduced performance and premature critical heat flux (CHF) well before a high exit quality can be achieved. Predicting the performance of two-phase cooling systems is also significantly more difficult than that of single-phase systems. Once these challenges are overcome, two-phase embedded-cooling systems offer transformative benefits relative to traditional, single-phase remote-cooling systems and will enable future electronic devices to reach new levels of functionality and performance.

This chapter highlights our recent experimental and modeling efforts that have advanced the state of the art in two-phase embedded cooling. Section 2 presents the development of an intra-chip hierarchical manifold microchannel (MMC) heat sink featuring embedded channels capable of dissipating high heat fluxes exceeding 1,000 W/cm^2 using a dielectric coolant. The design, fabrication, and thermal characterization are presented, with a discussion of the differences between manifold and traditional microchannel heat sinks. Recent experimental and modeling efforts focused on flow boiling instabilities are reported in Section 3. These studies focus on instabilities that are likely to appear in two-phase embedded microchannel cooling systems. Section 4 highlights recent experimental efforts that assess the implications of transient heating conditions on flow boiling characteristics in embedding-cooling systems. Finally, Section 5 reviews recent work that advances the state of the art in high-fidelity modeling of flow boiling in microchannels. A methodology to enable the experimental validation of these modeling efforts is also presented. While each of the sections is self-contained and can be read in any order, when combined together in the order presented, it demonstrates the need for a comprehensive approach combining practical application needs, fundamental understanding, predictive modeling approaches, and experimental validation.

2. High-Heat-Flux Dissipation Using Two-Phase Cooling in Embedded Microchannels

Two-phase cooling in microchannel heat sinks is known to be effective in dissipating high heat fluxes. However, it typically comes with the penalty of impractically large pressure drops because of the relatively long flow lengths of the microchannels (e.g., $L/D > 100$) and accelerational pressure drops associated with liquid-to-vapor phase change. In traditional microchannel heat sinks, fluid is delivered to the microchannels at one end of the heat sink, flows across the entire length of the heat sink, and exits on the opposite end (Figure 1). Thus, the microchannel length is determined by the die size and can result in long flow lengths and high pressure drops. MMC heat sinks, on the other hand, are designed to distribute fluid to multiple inlets and outlets along the heat sink so that the flow length through any single set of microchannels is significantly reduced (Figure 1). This decouples the flow length from the die size and offers the potential for greatly improved performance.

Drummond *et al.* [33] designed, fabricated, and tested a two-phase, heterogeneously integrated hierarchical MMC heat sink featuring high-aspect-ratio microchannels for dissipation of high heat fluxes (Figure 2). The MMC heat sink consisted of a hierarchical acrylic manifold distributor, silicon plenum interface plate, a silicon microchannel plate, and a printed circuit board (Figure 2(a)). The manifold distributor divided a single coolant inlet into nine parallel flow paths that entered a 3×3 array of microchannel heat sinks and then recombined the flow exiting each heat sink into a single coolant outlet. Because of the MMC heat sink design, the effective

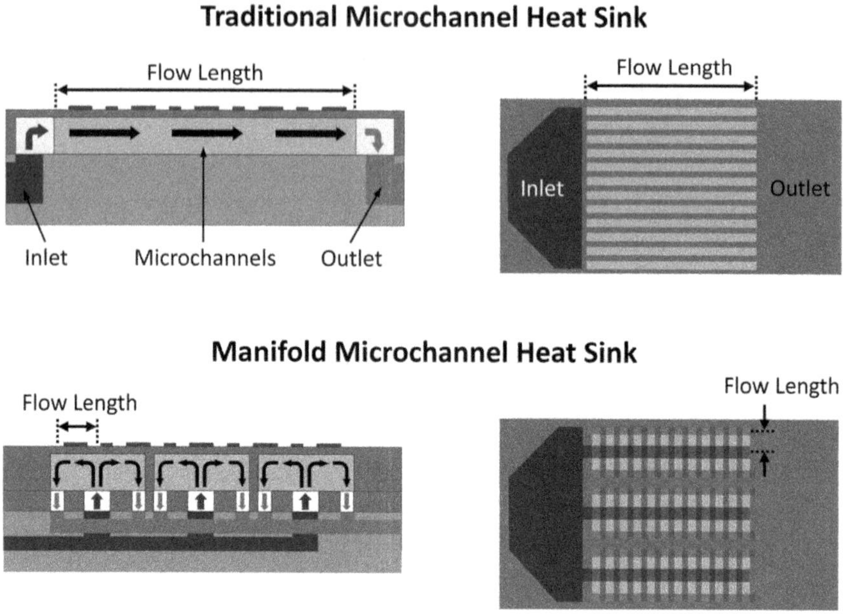

Figure 1. Schematic diagram showing the fluid flow paths of a traditional microchannel heat sink and a manifold microchannel heat sink (MMC). The MMC design results in a shorter flow length in the microchannels relative to the traditional design, which can reduce pressure drop and pumping power.

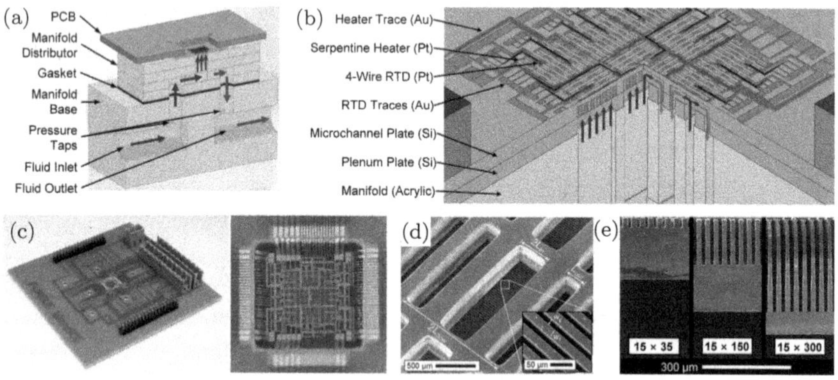

Figure 2. (a) and (b) CAD images of the hierarchical manifold microchannel (MMC) heat sink. (c) Photographs of the silicon test chip with heaters and temperature sensors mounted on the printed circuit board. SEM images of (d) a bonded plenum and microchannel plate and (e) microchannel cross-sections for three different channel depths used in Ref. [33].

Source: Adapted from Refs. [33, 34].

flow length of the 50 parallel microchannels in each of the nine heat sinks was only 750 μm compared to the chip edge length of 5 mm. The high-aspect-ratio microchannels were fabricated using deep reactive-ion etching. Nine serpentine heaters (one above each heat sink) were instrumented on the topside of the silicon microchannel plate (Figure 2(b)). Additionally, the central heater in the 3×3 array contained a centrally located 200×200 μm hotspot heater. Interlaced between the serpentine heater traces were 18 resistance temperature detectors (RTDs). The MMC heat sink was inserted into a flow loop with degassed HFE-7100 as the working fluid. The thermal and hydrodynamic performance of the MMC heat sink was evaluated by incrementing the heat flux while maintaining a fixed inlet sub-cooling of 7°C and an outlet saturation pressure of 123 kPa.

A series of experimental studies was performed by Drummond *et al.* [33, 34]. First, the effect of channel depth (35, 150, and 300 μm, as shown in Figure 2(e)) at a fixed channel width (15 μm) was investigated for channel mass fluxes of 1,300, 2,100, and 2,900 kg/m^{2}s. The boiling curves for each combination of channel geometry and mass flux are shown in Figure 3. Each of the boiling curves typically starts with a linear temperature rise with increasing heat flux, characteristic of the single-phase flow regime, followed by a two-phase flow regime featuring an increased slope. In the single-phase regime in Figure 3, the slope of the boiling curves increases with increasing mass flux (due to higher inlet velocities and longer developing flow

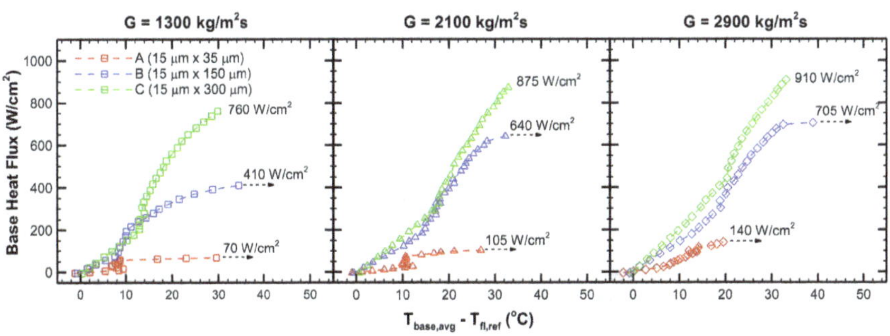

Figure 3. Boiling curves for the hierarchical MMC heat sink at three channel dimensions and three mass fluxes using HFE-7100.

Source: Reprinted from Ref. [33].

lengths) and channel depth (due to more surface area for heat transfer). At the onset of boiling, a gradual transition in the slope of the boiling curves was observed due to the many parallel flow paths (450) that likely each experienced boiling at slightly different heat fluxes. In the two-phase regime, the slopes of the boiling curves are steepest at low heat fluxes when nucleate boiling is dominant. The boiling curves begin to flatten at higher heat fluxes (vapor qualities) as nucleate boiling is increasingly suppressed and vapor blanketing causes local and intermittent dry-out at the wall. The MMC heat sink with the shallowest channels (15×35 μm) had significantly higher chip temperatures at every heat flux and reached CHF at a much lower heat flux than the other two geometries; this was attributed to the reduction in the wetted area. While the temperatures for samples with channel depths of 150 and 300 μm are similar at low heat fluxes, they deviate at higher heat fluxes, and the maximum heat flux dissipated increases with increasing channel depth. From these data, it was concluded that for MMC heat sinks, deeper channels improve the thermal performance but provide a diminishing return due to the decreased fin efficiency for deep channels (as low as 58% for 15 μm $\times$300 μm channels), making the added heat transfer area less effective.

The utility of the embedded-cooling strategy employed by Drummond *et al.* [33] is apparent when considering the effective thermal resistances of the MMC heat sink, which were calculated to be ~5–10 mm^{2}K/W for Samples B and C. This total thermal resistance is similar to the thermal resistance of a single layer of high-performance thermal interface material, which is one of many layered elements in conventional (remote-cooling) electronics package stacks [35]. This illustrates how the thermal resistance associated with embedded-cooling systems is significantly lower than that of remote-cooling systems, regardless of the performance of the attached heat sink.

Next, Drummond *et al.* [34] investigated the effect of channel width and aspect ratio (Figure 4(a)) at mass fluxes of 600, 1,300, and 2,100 kg/m^{2}s using channels with nominal widths of 15 and 30 μm and depths of 150, 300, 400, and 470 μm. For a fixed channel

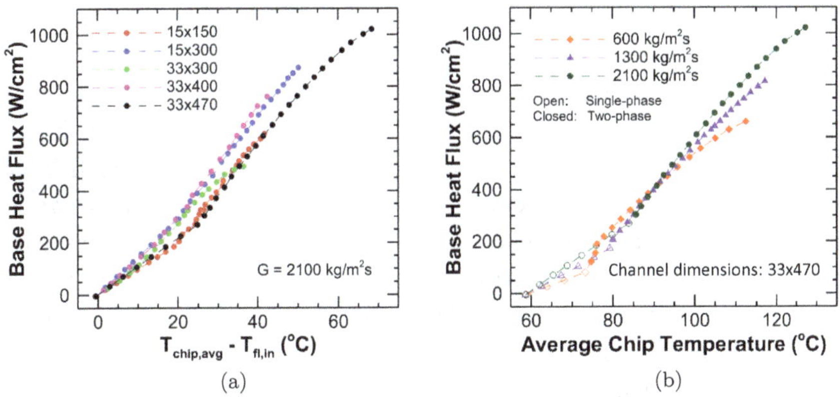

Figure 4. Effect of (a) channel width and aspect ratio and (b) mass flux for channels dimensions of 33 × 470 μm on flow boiling in an MMC heat sink.
Source: Reprinted from Ref. [34].

size of 33 × 470 μm, the flow boiling performance was independent of mass flux for heat fluxes below ~500 W/cm^2 (Figure 4(b)). However, at heat fluxes beyond 500 W/cm^2, lower chip temperatures with increased mass flux were noted. At the highest mass flux of 2,100 kg/m^{2}s, the largest heat flux of 1,020 W/cm^2 was dissipated at a pressure drop of less than 120 kPa and a chip-to-fluid-inlet temperature rise of less than 58°C. This performance is among the highest reported in the literature using dielectric fluids [25].

By comparing heat sinks with different widths at a fixed mass flux (2,100 kg/m^{2}s), Drummond *et al.* [34] concluded that a channel width of 33 μm yielded higher heat transfer coefficients than those with a width of 15 μm, but not necessarily lower thermal resistance. For example, for a fixed channel depth of 300 μm, samples with 15 μm wide channels had lower heat transfer coefficients but had a ~86% larger surface area for heat transfer compared to samples with 33 μm wide channels, which thereby lowered their thermal resistances.

Finally, the effect of hydraulic diameter on thermal performance was investigated by Drummond *et al.* [34] using samples with microchannel cross-sections of 15 × 150 and 33 × 300 μm, which had equal wetted areas and aspect ratios. The sample with the larger hydraulic diameter (33 × 300 μm) provided a higher heat transfer coefficient and lower thermal resistance compared to the sample with the smaller hydraulic diameter due to the increased fluid flow rate needed to maintain a constant mass flux. However, the sample with the smaller hydraulic diameter was able to dissipate a higher maximum heat flux (618 W/cm^2 compared to 494 W/cm^2). In traditional two-phase microchannel heat sinks, heat transfer coefficients are typically unaffected by channel dimensions for a given mass flux. Instead, maximum heat flux dissipation increases with increasing wetted area. The work of Drummond *et al.* [34] shows that, unlike traditional heat sinks, the maximum heat flux dissipated by two-phase MMC heat sinks does not necessarily increase with wetted area because, as will be explained in Section 2.2, as the channel depth becomes large, the incoming liquid cannot reach the base of the channel, which is coated by a vapor blanket.

2.1. *Effects of local hotspot heating*

Drummond *et al.* [34] also conducted experiments for a spatially non-uniform heating profile by generating additional localized heat flux in the centrally located 200 μm $\times$ 200 μm hotspot beyond the uniform background heat flux over the 5$\times$5 mm area. Hotspot heat fluxes were incremented from 0 to 2,700 W/cm^2, with background heat fluxes of 100, 300, 500, 700, and 900 W/cm^2, for mass fluxes of 600, 1,300, and 2,100 kg/m^{2}s. The corresponding heat inputs for these hotspot and background heat fluxes were 0–1.1 and 25–225 W, respectively. The hotspot temperature increased linearly with the hotspot heat flux for all mass fluxes and background heat fluxes. Moreover, regardless of the mass flux, the hotspot temperature was elevated by 16°C at the maximum heat flux of 2,700 W/cm^2 for all background heat fluxes. This is noteworthy for two reasons. First, it shows that the hotspot temperature rise is independent of the background heat transfer coefficient and is instead dictated by heat spreading and conduction resistances in the substrate. Second, it shows that the temperature rise at the hotspot is extremely localized, and the rest of the chip surface temperatures were largely unaffected by the high hotspot heat flux.

2.2. *Effect of channel depth in high-aspect-ratio manifold microchannels*

In traditional microchannel heat sinks, fluid arrives normal to the channel cross-sectional area; increasing the channel depth therefore increases the surface area for heat transfer and typically enables higher heat fluxes to be dissipated. However, for MMC heat sinks, fluid arrives at the top of the channel, and a trade-off between channel length and depth exists, resulting in an optimal channel depth. Moreover, because fluid arrives normal to the heated surface instead of perpendicular to it, the effects of reducing fin efficiency with increasing channel depth/height are exacerbated in the case of MMC heat sinks. To investigate this trade-off, Drummond *et al.* [36] examined the role of two-phase morphology on thermal performance using synchronized optical flow visualizations and infrared temperature measurements of a custom single MMC, which was representative of a repeating unit in their MMC heat sink. Different channel lengths and depths with fixed channel and fin widths (60 μm) were used to construct three different sample geometries (Sample 1: 1,500 μm long and 125 μm deep; Sample 2: 750 μm long and 250 μm deep; and Sample 3: 750 μm long and 1,000 μm deep), as shown in Figure 5(a). The microchannel geometry of Sample 1 closely resembled a traditional microchannel, with a relatively long length and low aspect ratio. In contrast, the geometry of Sample 3 featured a high-aspect-ratio microchannel of extremely short length (channel was deeper than it was long). Sample 2 provided an intermediate case. Subcooled HFE-7100 liquid was delivered to the microchannel at a constant fluid velocity of $\sim$1.1 m/s at the inlet to the channel. Flow visualizations were captured using a high-speed optical camera and synchronized to spatially resolved temperatures on the backside of the channel captured using an infrared camera (Figure 5(b)).

For the channel which closely resembled a traditional microchannel (Sample 1), the observations by Drummond *et al.* [36] were consistent with those reported in the literature for straight microchannels and conventional flow regimes. For the high-aspect-ratio microchannel (Sample 3), the flow morphology exhibited extreme, time-periodic variations (Figure 5(b)). Most notably, incoming liquid could not

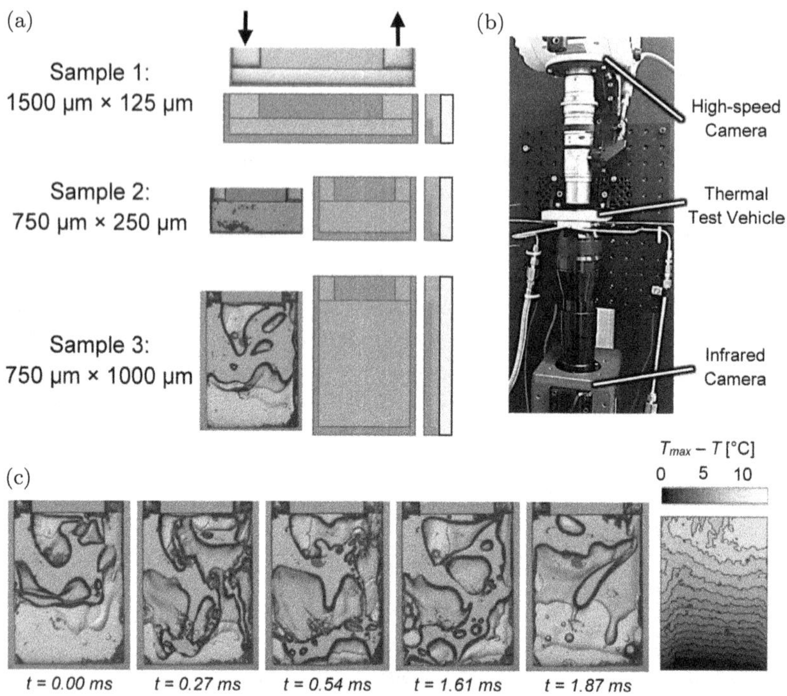

Figure 5. (a) Photographs and schematic diagrams of three manifold microchannel samples used to study the effect of channel length and depth. (b) Experimental setup used to simultaneously capture optical and infrared images of the thermal test vehicle. (c) High-speed images showing the two-phase morphology inside the channel of Sample 3 ($750 \times 1{,}000$ μm) alongside a time-averaged wall temperature map at a heat flux of 115 W/cm^2.

Source: Adapted from Ref. [36].

re-wet the entire base of the channel, and a large region near the bottom of the channel was either intermittently or continuously coated by a vapor blanket, which limited the local heat transfer and increased the wall temperature near the channel base. The microchannel with the intermediate channel depth (Sample 2) offered a balance between flow length and depth that enabled incoming liquid to prevent this formation of a vapor blanket at the base of the channel. Although it featured a significant 4× reduction in wetted area compared to Sample 3, it provided lower wall temperatures at all heat fluxes during two-phase operation. These findings demonstrate the unique flow morphology in two-phase MMC heat sinks and the need to optimize channel geometry for flow boiling conditions.

2.3. *High volumetric heat dissipation*

Recent advances in multi-chip stacking and heterogeneous packaging have resulted in the need to densely integrate fluid flow paths in embedded-cooling architectures to enable not only increased surface heat dissipation (i.e., heat flux) but also increased volumetric heat dissipation (i.e., heat density). Back *et al.* [37] fabricated a compact hierarchical MMC heat sink, with all flow manifold features embedded into the chip, to enable extremely high volumetric heat dissipation (Figure 6). The manifold comprised four silicon wafers, each etched from both sides, for a total of eight feature levels that split a single fluid inlet into gradually finer flow paths. A 9×9 array of microchannel heat sinks occupying a 5×5 mm area was etched into a single silicon die. Each of the 81 heat sinks consisted of 18 high-aspect-ratio microchannels (19 μm wide and 150 μm deep). A 3×3 array of individually addressable heaters and 18 RTDs were patterned on the topside of the silicon microchannel wafer. The silicon layers were hermetically sealed using thermocompression bonding,

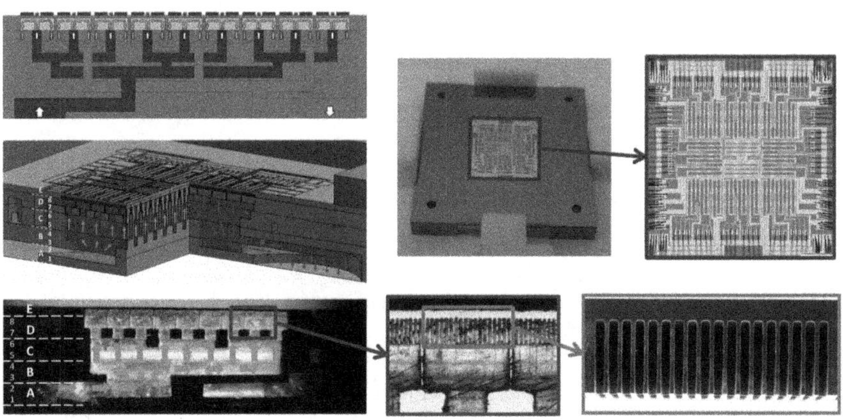

Figure 6. Compact hierarchical manifold microchannel heat sink composed of four thermocompression-bonded silicon wafers with integrated flow features.
Source: Adapted from Ref. [37].

giving the final assembly a compact 5 mm × 5 mm × 2.3 mm working envelope. The test vehicle was inserted into a flow loop with HFE-7100 as the working fluid, and fluid flow rates ranging from 150 to 350 g/min were employed at a fixed inlet temperature of 59°C (sub-cooling of ∼6°C). The heaters were incrementally powered, and steady-state data were recorded at each power level.

Back *et al.* [37] compared the thermal and hydrodynamic performance of the 9 × 9 heat sink array to that of a 3 × 3 heat sink array that featured in their previous work [33, 34] using approximately the same flow rate (350 g/min) and channel geometry (15 μm × 150 μm). Overall, the thermal performance metrics were very similar for the two different test vehicles because of the similar channel size and fluid quality. However, the primary advantage of the 9 × 9 heat sink array was the smaller volumetric envelope of the compact integrated manifold (5 × 5 × 2.3 mm^3), which was significantly reduced compared to the manifold used to feed the 3 × 3 heat sink array (25 × 8 × 10 mm^3). This translated to a maximum volumetric heat dissipation of 2,870 W/cm^3 for the 9 × 9 array, compared to a maximum of only 285 W/cm^3 for the 3 × 3 array. The compact 9 × 9 manifold design required smaller manifold flow features, which, however, increased the total pressure drop relative to the 3 × 3 array by approximately 20–65%. It was concluded that most of this pressure drop occurred in the manifold flow features (and not in the microchannels), and the channel pressure drop accounted for only 20–27% of the total pressure drop, depending on the heat flux.

3. Assessing the Role of Flow Boiling Instabilities on Thermal Performance

Designing heat sinks for two-phase operation and predicting their performance is difficult, in part, because of the use of many parallel flow paths (e.g., over 1,400, as described in the preceding section) and commonly encountered flow boiling instabilities, which can lead to premature CHF [38]. These instabilities have received significant attention through experimental and theoretical analyses [39–42] to better understand their behavior and implications on flow boiling performance, and a number of reviews on the subject are available [43–46].

Two classes of flow instabilities have been widely recognized in the literature: static and dynamic. Flow is said to be subjected to static instability if, when disturbed, its new operating conditions tend asymptotically toward those differing from the initial ones [44]. The range of operating conditions over which static instabilities are observed can be predicted using steady-state considerations [39, 43, 47]. One of the most common static instabilities is the Ledinegg (or flow excursion) instability, which can cause a sudden reduction in flow rate through a heated channel and result in flow maldistribution in parallel channels. Flow is said to be subjected to dynamic instability when there is sufficient interaction between the inertia of the flow and the compressibility of the system, leading to delayed feedback [44]. Dynamic flow boiling instabilities commonly encountered in microscale applications include the rapid-bubble-growth instability, pressure drop instability (sometimes referred to as the upstream compressible volume instability), and parallel-channel instability [48].

3.1. *Modeling Ledinegg-instability-induced flow maldistribution*

Van Oevelen *et al.* [39] developed a methodology to model the two-phase flow distribution in a system of many parallel channels. Their approach built upon the methods previously developed in the literature [47, 49, 50] and enables the prediction of stable flow distributions for a larger number of identical channels, as may be present in a microchannel heat sink. The modeling approach is based on an open-loop flow network, which consists of a pump that supplies sub-cooled liquid to an array of parallel heated channels and then exhausts to a constant-pressure sink. The flow rate distribution and pressure drop in this system of parallel channels are formulated using the conservation of mass for the system, the conversation of momentum for each channel (which includes a pressure drop model), and an equation that describes the characteristic pump curve (e.g., constant mass flow rate or constant pressure). For a typical N-shaped load

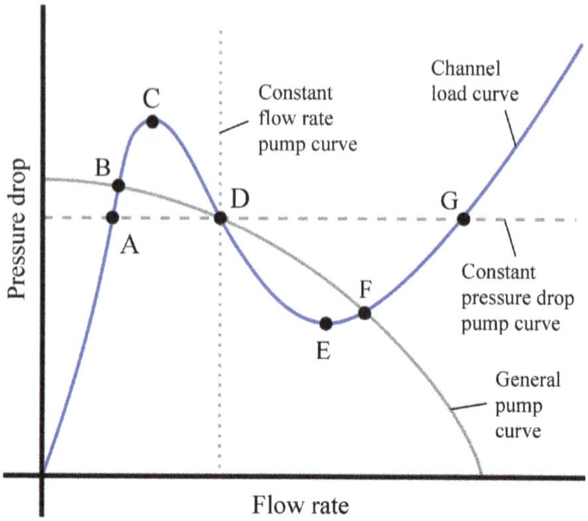

Figure 7. Channel load curve for a channel subjected to flow boiling conditions with constant head input (solid blue line) and various pump curves (gray lines). *Source*: Adapted from Ref. [51].

curve for a heated channel (blue line in Figure 7), any given pressure drop can result in three different flow rates. The cumulative load curve for a system of N parallel channels consists of up to 3^N combinations of individual flow rate distributions at any given pressure drop. Therefore, the computational complexity increases exponentially with the number of channels (N). By assuming identical channel load curves for each channel, Van Oevelen *et al.* [39] reduced the computational complexity needed to solve the system of equations by reducing the number of combinations from 3^N to $(N+2)(N+1)/2$. This enables the flow distribution to be predicted for a large number of parallel channels.

Van Oevelen *et al.* [39] first used their modeling approach to perform a stability analysis on a system of parallel channels subjected to two different pump curves: a constant pressure drop pump curve and a generic pump curve (dashed and solid gray lines in Figure 7, respectively). The stability of the steady-state operating conditions was determined by performing a linear stability analysis. For a constant pressure drop pump curve, the stability of each individual channel is effectively decoupled from that of the other parallel channels, and the flow in each channel is stable if and only if the slope of the channel load curve is positive. For an array of parallel channels subjected to any other general pump curve (including a constant mass flow rate pump curve, as shown by the dotted gray line in Figure 7), the stability is determined by all channels simultaneously. It was shown that in the limit of infinite channels, the system is stable if and only if all channels have a positive channel load curve slope, and unstable otherwise. This behavior is equivalent to that of a system of finite channels with a constant pressure drop pump curve. Thus, the stability behavior for a system of many identical parallel channels operating at a constant flow rate converges asymptotically to the stability of a system subjected to a constant pressure drop with an increasing number of channels. This important finding enables the stability of systems with many identical parallel channels, such as microchannel heat sinks, to be easily evaluated without the need for a full numerical stability analysis.

3.1.1. *Parametric effects on stability and severity of maldistribution*

Van Oevelen *et al.* [39] studied the effect of operating conditions on the range of flow rates over which a negative channel load curve (i.e., pressure drop reduces with increasing mass flow rate) exists for systems with varying numbers of identical parallel channels. The extent of the negative-slope region was characterized as a function of three operating parameters (average channel flow rate, heat input, and inlet sub-cooling) using the dimensionless boiling number (which represents the ratio of heat input to latent heat of evaporation) and the dimensionless sub-cooling number (which represents the ratio of sensible heat capacity of the sub-cooled liquid at the inlet to latent heat of evaporation). The flow rate range for stable flow was found to be fairly insensitive to heat input due to the low accelerational pressure drop contribution predicted by the pressure drop model within this stable region. It was also shown that the boiling number range in which the uniform distribution is unstable increases with increasing sub-cooling number; thus, larger differences between inlet temperature and saturation temperature cause this range to increase. It was shown that maldistribution should not occur when the inlet temperature is sufficiently close to saturation, but this small sub-cooling may be difficult to maintain in practice.

To assess the severity of flow maldistribution, Van Oevelen *et al.* [39] proposed a relative average flow rate starvation metric and then used it to assess the effect of operating conditions in the limiting case where N goes to infinity, which approximates systems with a large number of channels. For low boiling numbers, there is little to no maldistribution. As the boiling number increases, the relative average flow rate starvation first increases, reaches a maximum, and then reduces. The boiling number corresponding to the maximum relative average flow rate starvation increases with the sub-cooling number.

3.1.2. *Effect of lateral thermal coupling*

Van Oevelen *et al.* [51] expanded their prior two-phase flow distribution modeling methodology to include the effect of thermal conduction in the walls between neighboring parallel channels to study its effect on the Ledinegg instability. A model for the thermal-hydraulic behavior inside the system of parallel channels was integrated into a system flow network. The thermal model accounted for heat transfer in the system by considering internal convection in each channel, axial and lateral conduction in the solid walls, and heat loss to the ambient. The effective lateral (between channels) and external (to ambient) thermal conductances, respectively, determined the strength of thermal coupling between neighboring channels and the ambient heat loss; values for these conductances were inputs to the model. While this modeling approach is straightforward in principle, the number of operating points that need to be evaluated scales exponentially with the number of channels, resulting in high computational costs. Thus, two channels were used to explore the effects of thermal coupling on flow distribution.

In the first of the two cases considered, thermal coupling between the two channels was incorporated for a lateral thermal conductance of 148 W/mK, while in the other case, the channels were assumed to be thermally isolated from each other (lateral thermal conductance was zero). It was shown that flow maldistribution occurs in a flow rate range corresponding to two-phase flow, and while this maldistribution cannot be avoided, it can be reduced by incorporating thermal coupling. In the thermally coupled case, the range of flow rates for which maldistribution was observed was smaller than in the thermally isolated case (9.5–24.2 mg/s compared to 5.3–38.9 mg/s), and the flow rate ratios of the two channels were closer to unity (i.e., the maldistribution was less severe); this is shown graphically in Figure 8. For example, for a fixed total flow rate of 20 mg/s (as indicated by the black dots in Figure 8), only one stable solution exists for both cases, and both have a non-uniform flow distribution. At this flow rate, the flow distribution in the thermally isolated case was 98.7% in channel 1 and 1.3% in channel 2, whereas in the thermally coupled case, it was 89% in channel 1 and 11% (*an eightfold increase*) in channel 2. This significant finding first revealed that channel-to-channel thermal coupling allows for heat to be redistributed, thus mitigating the deleterious effects of the Ledinegg instability and

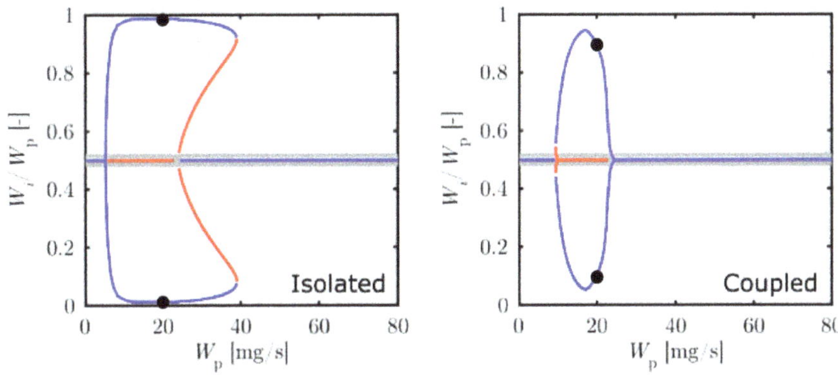

Figure 8. Relative flow rate distribution (W_i/W_p) as a function of total flow rate (W_p) for two identical heated parallel channels that are either (a) thermally isolated or (b) thermally coupled. The colors of the lines indicate the stability of each operating point (blue is stable, red is unstable). Uniform distributions are highlighted with gray shading. The black dots correspond to the operating condition at a 20 mg/s total flow rate.

Source: Adapted from Ref. [51].

moderating large differences in wall temperature, vapor quality, and flow maldistribution. The importance of thermal coupling on maldistribution of two-phase flow in parallel-channel heat sinks was also demonstrated by Patankar and Salamon [52], who showed that there exists a thermal-coupling threshold above which no maldistribution is observed and that this threshold increases quadratically with the number of channels.

The effect of the total uniform heat load on the channels on the flow distribution behavior in the thermally coupled case was also investigated by Van Oevelen *et al.* [51]. This study revealed that the occurrence and severity of flow maldistribution are dependent on the magnitude of the heat input to the microchannel array (Figure 9). For low values of heat input, no maldistribution occurs, whereas at higher values, maldistribution can occur for a specific case-dependent flow rate range. There exists a specific heat load threshold that must be exceeded for maldistribution to be possible, and this threshold depends on the strength of the lateral thermal coupling. In the lower limit, corresponding to isolated channels, the threshold goes to zero; i.e., maldistribution is always possible, regardless of the heat load. In the limit of extremely high lateral thermal coupling, the threshold reaches an upper limit, meaning that maldistribution cannot be suppressed indefinitely by increasing the strength of lateral thermal coupling. This threshold heat flux increases as the lateral wall conductance increases, converging to a constant value in the limit of very high lateral conductance.

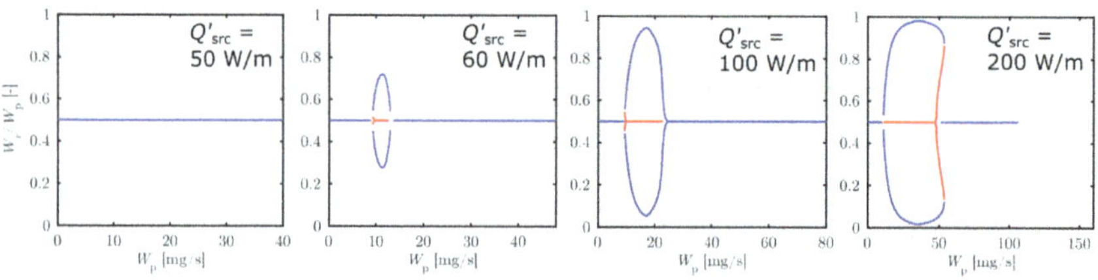

Figure 9. Relative flow rate distribution (W_i/W_p) as a function of total flow rate (W_p) in two identical heated and thermally coupled parallel channels for a range of heat inputs (Q'_{src}).

Source: Reproduced from Ref. [51].

The validity of the modeling approach developed by Van Oevelen *et al.* [51] was assessed by comparing it with experimental data from Flynn *et al.* [53], who measured wall temperature profiles in a system of two parallel channels fabricated in a silicon substrate. The wall temperatures were obtained for both a thermally coupled case, with silicon between the channels, and an approximately thermally isolated case, where the silicon substrate between the channels was etched away. For modeling purposes, lateral thermal conductances for the thermally coupled and isolated cases were estimated by Van Oevelen *et al.* [51] through experiments. The experimentally measured and predicted wall temperatures in the axial direction for both the thermally isolated and thermally coupled cases were in good agreement, thus confirming the accuracy of the modeling approach.

3.2. *Ledinegg-instability-induced temperature excursion*

An experimental investigation of the Ledinegg instability in two identical, thermally isolated parallel microchannels (Figure 10) was performed by Kingston *et al.* [54]. The borosilicate glass microchannels had an inside diameter of 500 μm and a wall thickness of 100 μm. The outside surface of the channels was coated with a 100 nm thick

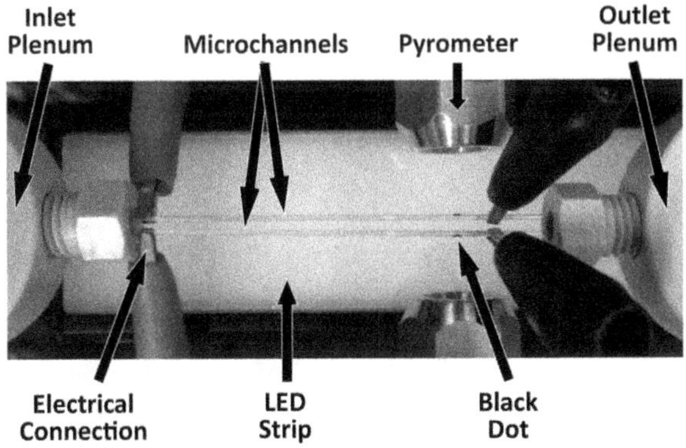

Figure 10. Experimental test section outfitted with two parallel 500 μm diameter borosilicate glass microchannels used to study flow boiling instabilities. Heating is applied via Joule heating using the optically transparent ITO layer on the outside of the microchannels. The flow direction is from left to right.

layer of indium tin oxide (ITO), providing an optically transparent and electrically conductive layer to enable visualization of the two-phase flow while subjected to Joule heating. Degassed, dielectric liquid (HFE-7100) was delivered to the parallel channels using a constant-pressure source in an open-loop system. Both channels were uniformly subjected to the same power, which was increased in steps. Flow visualization was synchronized with measurements of pressure drop, average mass flux through the two channels (the individual mass fluxes through each channel could not be measured), and individual wall temperatures to characterize the thermal-fluidic effects of the Ledinegg instability.

At low power levels, single-phase flow was observed in both channels, and equal wall temperatures and heat transfer coefficients were reported (Figure 11). From the thermal signatures, it was concluded that there was an even distribution of mass flux to both channels from the inlet plenum during single-phase flow (i.e., no flow maldistribution).

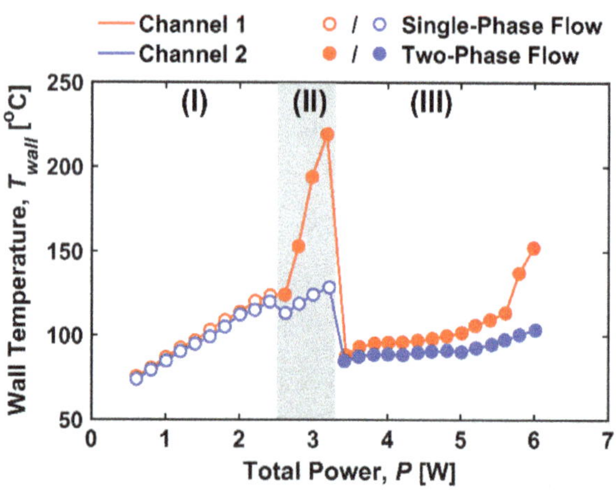

Figure 11. Wall temperature of two parallel microchannels as a function of total power: open symbols (○) indicate single-phase flow, and closed symbols (•) indicate two-phase flow in each of the channels. The labeled regions indicate operating conditions, where: (I) flow through both channels is single-phase, (II) boiling is occurring in channel 1 only, and (III) boiling is occurring in both channels. The Ledinegg-instability-induced temperature excursion is shown in Region (II).
Source: Adapted from Ref. [54].

At a slightly higher power level (2.6 W), boiling incipience occurred in one channel, resulting in boiling in that channel and single-phase flow through the other. Boiling incipience in one of the channels triggered the Ledinegg instability, which caused flow maldistribution between the two channels and a hydrodynamic excursion with increasing power, which induced a temperature excursion. Specifically, the wall temperature in the channel undergoing boiling increased sharply from 124°C at a power level of 2.6 W to 219°C at 3.2 W (Figure 11). An initial reduction in the wall temperature of the channel undergoing single-phase flow was observed at incipience, from 120°C at a power level of 2.4 W to 113°C at 2.6 W, and was followed by a linear increase with power to a wall temperature of 129°C at 3.2 W. The thermal isolation between the two channels prevents the equalization of the wall temperatures and mass flux in the two channels and exacerbates the effects of the Ledinegg instability.

The growing temperature difference between the two channels observed by Kingston *et al.* [54] continued with increasing power until boiling incipience occurred in the second channel at a power level of 3.4 W. This caused the flow resistances of the two channels to become approximately equal again, which reduced the flow maldistribution between the two channels and suppressed the temperature excursion induced by the Ledinegg instability. The wall temperatures of both channels reduced significantly as the flow became more evenly distributed, which was reflected in a large increase in the heat transfer coefficient. This illustrates the thermal implications of the Ledinegg instability in thermally isolated parallel channels and provides a new understanding of the mechanisms leading to the observed temperature excursion.

In a recent work, Miglani *et al.* [55] quantified the steady and transient flow maldistribution between two thermally isolated parallel microchannels resulting from the Ledinegg instability. A unique experimental test facility provided first-of-its-kind measurements of the flow rate through each individual microchannel. It was shown that when fluid in one channel is boiling and the other has single-phase flow, up to 96.5% of the flow is directed into the channel with single-phase flow, and the boiling channel received a mere 3.5%.

Additionally, Miglani *et al.* [56] experimentally showed that increasing the thermal coupling between the two channels, thereby enabling heat exchange between them, suppressed the wall temperature difference, reduced the operating window over which flow maldistribution was observed, and dampened the severity of the flow maldistribution. These experimental data for thermally isolated and coupled cases were quantitatively compared to model predictions using the approach of Van Oevelen *et al.* [51], and excellent agreement was obtained.

3.3. *Experimental characterization of dynamic flow boiling instabilities*

Dynamic flow boiling instabilities are governed by complex interactions between the inertia of the flow and the compressibility of the system, resulting in delayed feedback and making it difficult to predict their behavior. Significant efforts have therefore been directed toward the experimental characterization of these dynamic instabilities to enable improved understanding. One dynamic flow boiling instability that is unique to the microscale and not observed in macroscale systems is the rapid-bubble-growth instability [57]. This instability is characterized by rapid spherical growth of a vapor bubble, until it reaches a size comparable to the channel hydraulic diameter, followed by explosive axial growth in the downstream and, often, upstream directions, which commonly leads to local flow reversal (Figure 12) [57]. It is the small hydraulic diameters in microscale systems that confine vapor bubbles during boiling, causing them to significantly alter the operating characteristics of the system. Two underlying mechanisms contribute to the rapid-bubble-growth instability: (i) a large liquid superheat, and (ii) the large pressure spike that is generated and propagates through the surrounding liquid due to the rapid growth of the confined bubble [57]. Another commonly observed flow boiling instability at the micro- and macroscale is the pressure drop instability. This instability is characterized by time-periodic axial oscillations of the boiling boundary between the single-phase liquid and the two-phase mixture, which occur in unison for

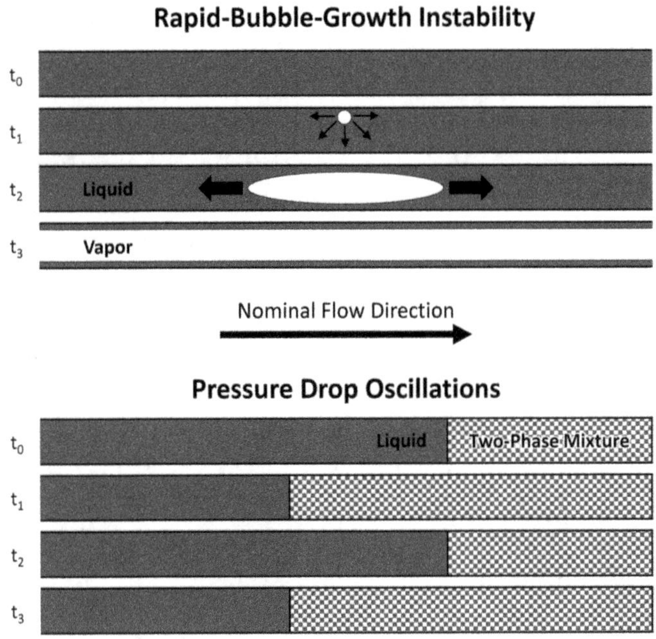

Figure 12. Schematic diagram illustrating the evolution of the two-phase morphology inside a microchannel during the rapid-bubble-growth instability and pressure drop oscillations. Single-phase liquid enters the left side of the channel.

each channel in parallel-channel systems (Figure 12) [58]. This instability is attributed to the interaction between the vapor within the heated channel and the compressibility of the flow path upstream of the channel [58]. The pressure drop instability often results in time-periodic oscillations in the pressure drop, mass flux, and wall temperature, which can cause premature CHF.

Different types of flow boiling instabilities are often intertwined, making it difficult to design or incorporate mitigation strategies. In an attempt to simplify the problem and study a reduced-order system, many researchers have turned to studying flow boiling instabilities in a single-microchannel configuration. While this has been a useful approach, it can be difficult to extrapolate the knowledge gained from single-channel flow instability studies to parallel microchannel heat sinks because of the potential differences in the hydrodynamic boundary conditions. For example, the constant mass flux used in many single-channel studies is fundamentally different from the hydrodynamic boundary conditions of an individual channel in a parallel-channel system. In microchannel heat sinks, which often feature hundreds of parallel channels, the hydrodynamic behavior of any one individual channel does not significantly influence the full set of channels. Thus, the resulting hydrodynamic boundary conditions for that single channel are similar to those under a constant pressure drop. Under this arrangement, the flow rate through an individual channel in a parallel-channel system can fluctuate significantly.

Kingston *et al.* [59] experimentally investigated the rapid-bubble-growth instability and the pressure drop instability in a single microchannel subjected to a temporally constant and spatially uniform heat flux. Degassed HFE-7100 liquid was driven through the single microchannel using an open-loop test facility, which featured a constant pressure difference between a pressurized upstream reservoir and an ambient-pressure reservoir. This hydrodynamic boundary condition enabled the mass flux through the channel to vary in time, much like that of an individual channel in a parallel-channel heat sink, while eliminating the confounding influence of the parallel-channel instability. The circular cross-section borosilicate glass microchannel used by Kingston *et al.* [59] had an inside diameter of 500 μm and a wall thickness of 100 μm. A 100 nm thick layer of ITO was applied to the outside surface of the microchannel, making it optically transparent but still electrically conductive. This enabled the simultaneous visualization of the two-phase flow and uniform Joule heating. High-frequency (2,500 Hz) measurements of the inlet and outlet pressures of the microchannel, mass flux, inlet and outlet fluid temperature, and wall temperature were synchronized to high-speed optical images to enable time-resolved characterization of the dynamic flow boiling instabilities.

Five different combinations of flow inertia and inlet liquid sub-cooling were investigated by Kingston *et al.* [59]. The effect of flow inertia was studied by varying the nominal single-phase mass flux (200, 400, and 800 kg/m^{2}s) while holding constant the inlet liquid sub-cooling (5°C). Similarly, the effect of inlet liquid sub-cooling was studied at different levels (5°C, 15°C, and 35°C) at a single flow inertia level (400 kg/m^{2}s). At each combination, the heat flux was incremented.

3.3.1. *Rapid-bubble-growth instability at the onset of boiling*

Kingston *et al.* [59] first used this experimental test facility to study the rapid-bubble-growth instability at the onset of boiling and quantify its underlying mechanisms. At the onset of boiling, a small vapor bubble formed, became circumferentially confined by the microchannel wall, and quickly grew axially both upstream and downstream from the nucleation site in an explosive-like manner, as shown in Figure 13. This event also generated a large local pressure spike within the microchannel that propagated upstream and downstream through the liquid. This caused a local flow reversal that starved the heated channel of incoming liquid and caused the wall temperature to increase.

The effects of flow inertia and inlet liquid sub-cooling on the severity of the rapid-bubble-growth instability were also quantified by Kingston *et al.* [59]. It was found that low flow inertia exacerbated the rapid-bubble-growth instability and resulted in severe temperature spikes because the heated channel was starved of incoming liquid replenishment for longer durations. These severe temperature

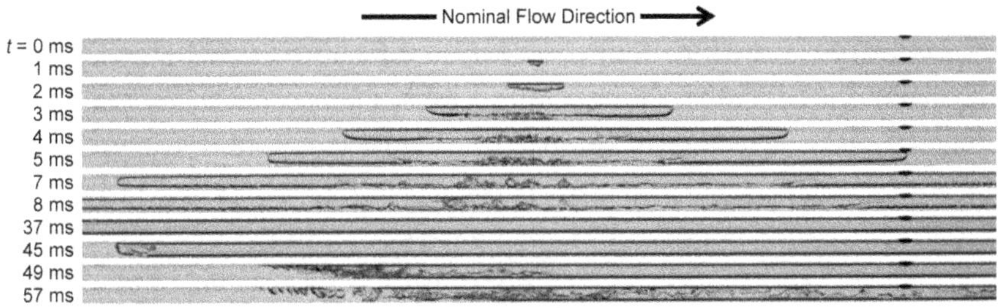

Figure 13. Selected images of the two-phase morphology during the rapid-bubble-growth instability at the onset of boiling. The nominal flow direction is left to right. Local flow reversal is observed and results in a large local pressure spike.

Source: Adapted from Ref. [59].

spikes have the potential to be catastrophic and can lead to premature CHF. It was also shown that higher sub-cooling levels resulted in shorter durations of flow reversal because, as the vapor bubble moved upstream in the microchannel, it encountered cooler fluid temperatures, which halted rapid evaporation.

3.3.2. *Influence of operating conditions on flow boiling instabilities*

Time-periodic dynamic flow boiling instabilities were also studied by Kingston *et al.* [60]. To identify the characteristic frequencies of the time-periodic oscillations, a spectral analysis by means of a fast Fourier transform (FFT) was performed on the time-periodic oscillatory data. Two different types of dynamic instabilities were observed under different operating conditions: (i) time-periodic transitions between single-phase and two-phase flows, which lead to a time-periodic series of rapid-bubble-growth instabilities, and (ii) pressure drop instability. For high levels of inlet liquid sub-cooling and low heat fluxes (circular data points in Figure 14(a)), a time-periodic series of rapid-bubble-growth instabilities was observed, resulting in large-amplitude (~30°C), low-frequency (~1 Hz) wall temperature oscillations. These flow regime transitions were attributed to the relatively large inlet liquid sub-cooling, causing the cool incoming liquid to quench the heated microchannel wall and stop the nucleation of vapor bubbles during depressurization of the upstream compressible volume. Above these heat flux levels, for each sub-cooling level, the time-periodic series of rapid-bubble-growth instabilities gave way to the pressure drop instability (denoted by diamond symbols in Figure 14(a)) because larger heat fluxes allowed for vapor bubbles to continue nucleating during instances of high mass flux. For low levels of inlet liquid sub-cooling (5°C), the pressure drop instability was exclusively observed as boiling was always occurring, which resulted in low-amplitude (~3°C), high-frequency (~12 Hz) wall temperature oscillations. The incoming liquid at this smaller sub-cooling level was much closer to the saturation temperature, which enabled continuous bubble nucleation, regardless of the heat flux level.

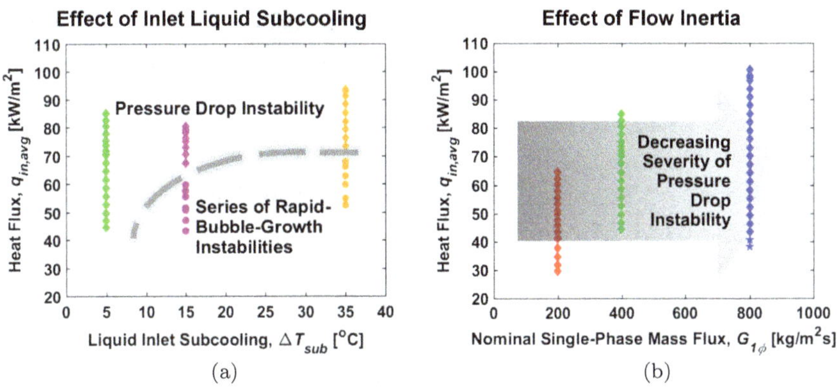

Figure 14. The effect of heat flux and (a) inlet liquid sub-cooling or (b) flow inertia on the type or severity of flow boiling instabilities: The five different combinations of inlet liquid sub-cooling and flow inertia are distinguished by color. Diamond symbols (♦) denote flow boiling operating conditions where pressure drop instability was observed, whereas circles (•) denote flow boiling operating conditions where a time-periodic series of rapid-bubble-growth instabilities was observed. The dashed line in (a) demarcates the regions in which each of the two types of flow boiling instabilities are observed.

Source: Adapted from Ref. [60].

The effect of flow inertia was also quantified, and it was shown that reducing it exacerbated the pressure drop instability and resulted in more severe hydrodynamic and thermal oscillations (Figure 14(b)) [60]. For example, at the lowest time-averaged flow inertia level (a mass flux of 200 kg/m^{2}s), the instantaneous mass flux through the channel oscillated by as much as 100% of the time-averaged value. In comparison, at the highest level of flow inertia (mass flux of 800 kg/m^{2}s) and low heat fluxes ($\leq$40.5 kW/m^2), the pressure drop instability was completely suppressed. These findings suggest that the ratio of flow inertia to upstream compressibility plays a critical role in the pressure drop instability.

In summary, these studies [59, 60] provide critical insight into the behavior and occurrence of dynamic flow boiling instabilities under hydrodynamic boundary conditions similar to those experienced by individual microchannels in a parallel-channel heat sink. While flow boiling instabilities are known to generally reduce the heat transfer and pressure drop performance of microchannel heat sinks, the effect of these instabilities on the overall heat transfer coefficient and CHF has not been fully quantified and is an important topic for future investigation.

4. Effects of Transient Heating on Flow Boiling in Embedded Microchannels

Many electronic components, such as central processing units (CPUs), radio-frequency power amplifiers, laser diodes, and avionics, experience transient operation resulting in time-varying heat generation [61]. These thermal transients can have repercussions on device operation that cause poor performance, reduced reliability, and even device failure. Embedded-cooling strategies — such as those demonstrated in Section 2 — that locate coolant channels within the device substrate, very near the source of heat generation, can be expected to exacerbate the effects of transient heating conditions on the thermal management system [62]. In particular, these heating transients can lead to flow boiling instabilities and oscillations, which have been shown to reduce heat transfer coefficients [63–65] and CHF [66]. To address these concerns, researchers have begun to study the transient temperature response at the onset of boiling when subjected to a single heat pulse [67–69] and during time-periodic heating conditions [68, 70, 71].

The dynamic response of microchannel flow boiling to transient heating conditions and the coupling between flow boiling instabilities and transient heating conditions were recently studied by Kingston *et al.* [72, 73]. The custom-built experimental facility used in their previous studies [54, 59, 60] was modified to enable transient tests. By using a microchannel with an inside diameter of 500 μm and only a 100 μm thick wall, transient heating conditions applied using a 100 nm thick ITO layer could easily propagate to the coolant, as may occur in embedded two-phase heat sinks. A solid-state switch was used to actively switch between two different power levels and apply a square-wave heating profile. Three heat flux levels resulting in vastly different flow behaviors under constant heating conditions were used to conduct transient heating tests: a low heat flux corresponding to single-phase flow (15 kW/m^2), an intermediate heat flux corresponding to continuous flow boiling (75 kW/m^2), and a very high heat flux which exceeds CHF and would cause dry-out if applied continuously (150 kW/m^2). Transient tests were first conducted using a single heat flux pulse between these three heat flux levels, with pulse durations ranging from 0 s (i.e., constant heating) to 10 s. Next, the effect of heating pulse frequency on microchannel flow boiling was explored using a time-periodic series of pulses in the form of a square-wave heating profile with frequencies ranging from 0.1 to 100 Hz while maintaining a constant duty cycle (i.e., the ratio of time at the high heating level to the total time for each period) for each operating scenario. High-frequency measurements (2,500 Hz) of heat flux, wall temperature, pressure drop, and mass flux were synchronized to high-speed flow visualizations to characterize the boiling dynamics during transient heating tests. Spectral analysis of the sensor measurements was used to characterize the dynamic response and flow instability frequencies.

At the onset of boiling for single heat flux pulse tests, the dynamic response of the flow boiling system resembled that of an underdamped mass–spring–damper system subjected to a unit step input, as shown by the oscillatory behavior of the pressure drop and mass flux in Figure 15(b) and the two-phase boiling boundary in Figure 15(c) between 0.171 and 0.302 s. This response was attributed to the complex interaction between the inertia of the flow, upstream compressibility, and vapor within the channel. The time needed to reach time-periodic boiling conditions (~3 s) was an order of magnitude larger than the experimentally characterized thermal time constant of the test section microchannel (0.43 s). This suggests that the dynamic response to boiling is dictated by the thermal-fluidic behavior of the system and cannot be simply attributed to the dynamic response of the test section. When transitioning between single-phase flow and time-periodic flow boiling, the wall temperature would temporarily over-/undershoot the eventual steady operating temperature by up to 20°C; this remarkable finding demonstrates the potential severe repercussions of transient heating conditions on flow boiling performance. It was shown that longer-duration high-heat-flux pulses

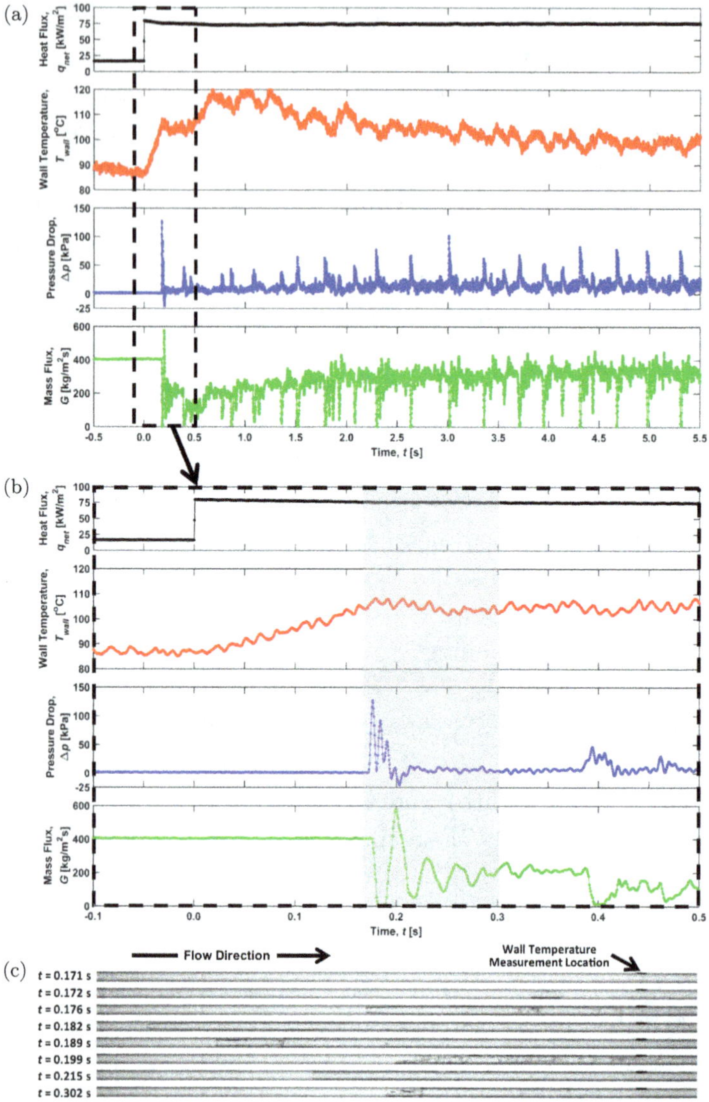

Figure 15. Synchronized measurements of heat flux, wall temperature, pressure drop, and mass flux before and after a single 10 s heat flux pulse from 15 to 75 kW/m^2. A temperature overshoot of 20°C during the transition from single-phase to two-phase flow is shown in (a), while the oscillatory dynamic response of the flow boiling system at the onset of boiling, which resembles that of an underdamped mass–spring–damper system, is shown in (b) and (c).

Source: Reprinted from Ref. [72].

(up to ~50% longer in some cases) can be withstood when the fluid in the microchannel is initially boiling, compared to if it is initially in the single-phase flow regime and goes through a transition to boiling during the pulse. The high-heat-flux pulse was also shown to increase the characteristic frequency of pressure drop instability by an order of magnitude during the pulse.

When subjected to time-periodic heat flux pulses, Kingston *et al.* [73] observed three different time-periodic flow boiling fluctuations: flow regime transitions, pressure drop oscillations, and heating pulse propagation. The types of fluctuations observed were mapped to different operating conditions (see Figure 2 in Ref. [73]). For heat flux pulses between 15 and 75 kW/m^2 and heating pulse frequencies >1 Hz, time-periodic flow regime transitions between single-phase and two-phase flows were observed and attributed to low-frequency (~1 Hz) compressibility effects within the flow loop, which cause relatively cool incoming liquid to quench the channel wall and arrest boiling. For low heating frequencies ($f \leq 1$ Hz), it was observed that the dynamic response resembles multiple independent step changes between different heat flux levels that each reach a new steady condition. For all operating conditions tested, time-periodic pressure drop oscillations resulting from the pressure drop instability were observed; however, these oscillations became significantly coupled to the transient heating profile only at intermediate heating frequencies ($1 < f < 10$ Hz) near the timescale of the thermal time constant of the test section. To illustrate a condition with coupling between the transient heating profile and the dynamic response of the flow boiling system, Figure 16 shows synchronized measurements for transient heat flux pulses between 15 and 150 kW/m^2 at a frequency of 5 Hz. As shown, the transient heat flux induces thermal and hydrodynamic fluctuations during the flow boiling process. A spectral analysis of the sensor signals (Figure 16(b)) was used to verify that the heating pulse frequency of 5 Hz forces the pressure drop instability to occur at exactly this frequency, instead of the intrinsic frequency of the pressure drop oscillations, thus illustrating the coupling of the transient heating conditions to the flow boiling performance and instabilities. At high heating frequencies ($f \geq 10$ Hz), the

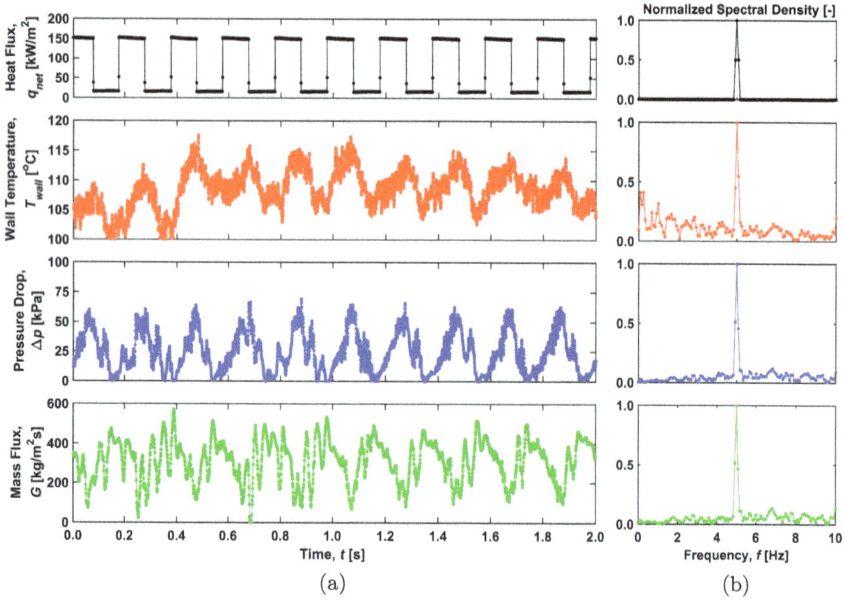

Figure 16. (a) Time-resolved and synchronized measurements of heat flux, wall temperature, pressure drop, and mass flux as a function of time for flow boiling of HFE-7100 in a microchannel during transient heat flux pulses between 15 and 150 kW/m^2 at a frequency of 5 Hz. (b) Normalized spectral density of the sensor measurements used to identify the characteristic oscillation frequency of the pressure drop instability and confirm the coupling between transient heating conditions and flow boiling performance and instabilities.

Source: Reprinted from Ref. [60].

transient heat flux pulses were attenuated by the thermal capacity of the microchannel wall and became effectively a constant heat flux. This observed dependence of flow instabilities and thermal performance on time-varying heating conditions indicates that flow boiling dynamics have a significant influence on overall performance in transient applications.

5. High-Fidelity Modeling of Microchannel Flow Boiling

Experimental investigations of two-phase cooling in microchannel heat sinks (e.g., Refs. 33, 34, 36, 37, as discussed in Section 2) over the past several decades have provided significant advancements; however, heat sink design is heavily reliant on empiricism and correlation-based models, which cannot readily be used for more generalized performance prediction. High-fidelity simulation of flow boiling in microchannels is a promising approach that has recently sparked significant interest because of advances in computing power. However, accurate numerical prediction of two-phase performance remains a challenging problem due in part to the high computational cost of tracking the vapor–liquid interface and accurately quantifying interfacial transport phenomena. Various numerical techniques, such as volume of fluid (VOF) [74–76] and coupled level-set volume of fluid (CLSVOF) [77, 78], have recently been combined with phase-change models that are implemented through user-defined functions, to simulate the flow boiling process.

5.1. *Numerical simulations of microchannel flow boiling*

Pan *et al.* [79] developed a saturated-interface-volume phase-change model using computational expense-saving features that enabled accurate simulation of microchannel flow boiling at a low numerical cost. Mass, momentum, and energy equations were solved using a VOF approach to resolve the two phases, with interfacial surface tension treated as a volumetric body force using the continuum surface force (CSF) model. This surface tension force is ideally calculated normal to the vapor–liquid interface. However, because VOF simulations are discretized over a fixed mesh and the volume fraction is not continuous, errors in the estimation of the surface tension force can produce unphysical spurious velocities near the interface. Pan *et al.* [80] applied an artificial moving reference frame to suppress these spurious currents while maintaining physical accuracy. This technique enabled the use of non-uniform mesh elements. In particular, it allowed the inclusion of high-aspect-ratio elements near the microchannel wall that are required to resolve the thin liquid film and thermal boundary layer, helping reduce computational expense.

The novel, saturated-interface-volume phase-change model developed by Pan *et al.* [79] accounts for evaporative heat and mass transfer at the vapor–liquid interface. This phase-change model anchors the interfacial temperature at the saturation value, which acts as a robust constant-temperature boundary condition. This approach enables the local temperature to be explicitly coupled to the source terms, thereby eliminating the need for iteration and significantly reducing the computational cost without sacrificing physical accuracy. To assess the accuracy of the phase-change model, the simulations were compared to the analytical solution for a one-dimensional Stefan problem. The Stefan problem features a layer of vapor between a superheated, isothermal wall and the surrounding saturated liquid. Heat is transferred from the superheated wall to the vapor–liquid interface, causing evaporation of the liquid and a growing vapor layer. The simulations showed excellent agreement with the analytical solution for all three working fluids: R113, HFE-7100, and water.

Pan *et al.* [79] then used their phase-change model to simulate the growth of a vapor bubble flowing in a heated, 500 μm diameter microchannel. A two-dimensional (2D) axisymmetric and a three-dimensional (3D) simulation were both performed. The 2D and 3D domains were both separated into two different thermal regions along the axial direction: an adiabatic upstream section where the vapor bubble was initialized, and a heated downstream region subjected to a uniform heat flux where evaporation takes place. For both simulations, high-aspect-ratio mesh elements were used near the microchannel wall to resolve the thin liquid layer. These computational expense-saving techniques enabled the 2D simulation to run in $\sim$14 h without any parallelization and the 3D simulation in $\sim$16 h with six-core parallelization in a lab computer (Intel Xeon Processor E3-1245 v3, Windows 7 64bit).

For the 2D simulation, shortly after the vapor bubble traveled into the heated region, it encountered the thermal boundary layer, thereby causing evaporation and bubble growth with time (Figure 17). The bubble position, bubble growth rate, liquid film thickness, and local heat transfer coefficient were quantitatively compared against another numerical simulation in the literature under identical conditions to provide another layer of verification [81].

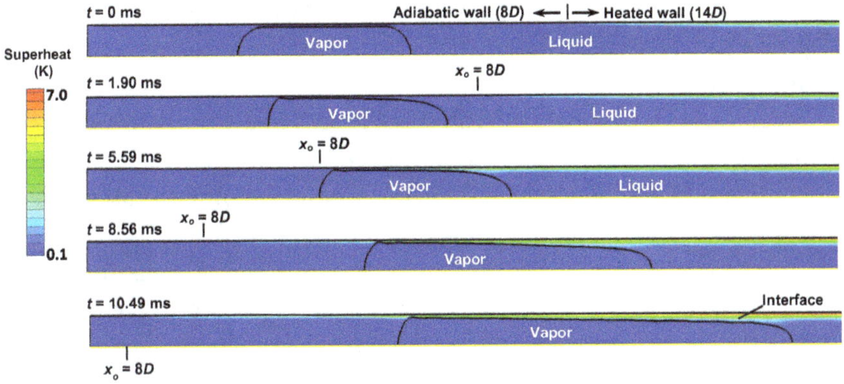

Figure 17. High-fidelity 2D axisymmetric simulation of microchannel slug flow boiling using a saturated-interface-volume phase-change model. The vapor bubble interface profile and temperature field contours for a vapor bubble flowing through a circular cross-section microchannel at different time instants are shown. Evaporation of the liquid surrounding the bubble causes the elongation of the bubble. *Source*: Reprinted from Ref. [79].

The 3D flow boiling simulation performed by Pan *et al.* [79] was used to demonstrate the cost-effectiveness of the model and highlight the heat and mass transport mechanisms during microchannel flow boiling. It was found that the highest heat transfer enhancement relative to the single-phase case was achieved at a location downstream of the bubble tail, rather than along the liquid thin film region, because of the transient heat convection effect and a larger axial velocity in this region; this agrees with other simulations in the literature [81]. The local evaporation rate around the cross-section of the bubble was largely dependent on the local liquid film thickness. In regions near the center of the square cross-section microchannel, where the liquid film was thinner, the evaporation rate was higher. The evaporative rate was also larger near the trailing edge of the vapor bubble, as it was in the 2D simulation.

5.2. *Experimental validation of flow boiling models*

Two-phase flows are typically generated in flow boiling experiments by vapor bubble nucleation from a heated surface. This incipience-based approach causes a streamwise progression of flow regimes, typically beginning with bubbly flow and transitioning to slug flow and then annular flow. Moreover, it leads to stochastic hydrodynamic variations, flow instabilities, and a close proximity between successive vapor bubbles. These phenomena prevent the acquisition of a comprehensive database of well-conditioned experimental results that can be used for the validation of high-fidelity flow boiling models. Thus, the accuracy of state-of-the-art modeling techniques is still assessed using test problems for which simplistic analytical solutions are available [79, 82–85], by comparison to temporally and spatially averaged quantities that can be easily measured experimentally [77, 78, 81, 86–88], or against different numerical modeling approaches [77, 79]. There is a clear need for high-fidelity benchmark experimental data that can be used to validate sophisticated flow boiling models.

To address this need, Kingston *et al.* [89] developed a novel experimental test facility that generates an archetypal microchannel slug flow boiling regime suitable for high-fidelity characterization of key hydrodynamic and heat transfer parameters that can be used for direct experimental validation of flow boiling models. Single-phase vapor and single-phase liquid HFE-7100 were extracted from a constant-pressure reservoir that contained a stratified mixture of saturated liquid and saturated vapor through two separate lines and independently injected into a microscale T-junction. This approach produced a saturated two-phase slug flow consisting of alternating liquid slugs and vapor bubbles in a downstream 500 μm diameter borosilicate glass microchannel (Figure 18). Because this approach does not rely on nucleation from the superheated microchannel wall to generate the vapor phase, flow instabilities, flow regime transitions, and the undesirable close proximity of successive vapor bubbles are not present. A thin ITO layer on the microchannel provided a uniform heat flux ranging from 30 to 5,160 W/m^2 via Joule heating while allowing simultaneous flow visualization. The flow was visualized using a high-speed camera and two different lenses: a macro lens and a high-magnification lens. Images obtained with the macro lens were used to quantify the uniformity of the vapor bubbles and liquid slugs generated, as well as the growth of vapor bubbles under the

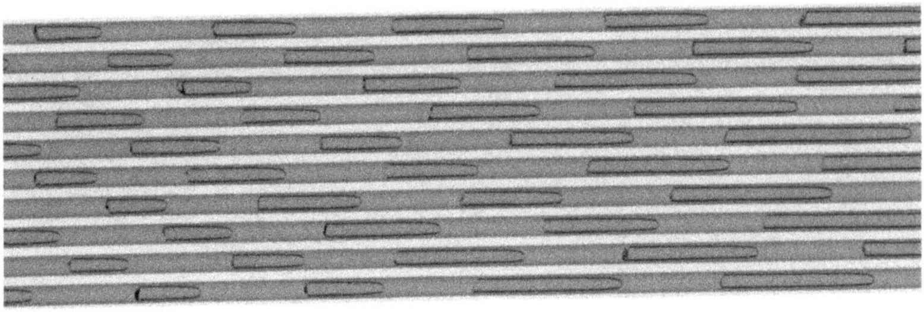

Figure 18. Selected images of an archetypal slug flow boiling regime in a 500 μm diameter microchannel at a heat flux of 3,520 W/m^2. The working fluid is degassed HFE-7100, and the flow direction is from left to right.

Source: Adapted from Ref. [89].

various heat flux levels using a custom image-processing algorithm, whereas images obtained with the high-magnification zoom lens were used to quantify the liquid film thickness.

Selected images of the slug flow boiling regime generated by Kingston *et al.* [89] in the test section microchannel at a heat flux of 3,520 W/m^2 are shown in Figure 18; the flow direction is from left to right. A quantitative analysis confirmed a high degree of uniformity in the initial dimensionless vapor bubble and liquid slug lengths, which for this operating condition were $L/D_0 = 4.6\pm1.1$ and 6.8 ± 0.9, respectively; the variation represents one standard deviation. The growth of individual vapor bubbles was also temporally resolved as a means to assess the evaporation rate. At the highest heat flux of 5,160 W/m^2, vapor bubbles would elongate from a dimensionless length of $L/D = 5$ to $L/D \approx 11$ in 18 ms. The generation of this archetypal slug flow boiling regime and the characterization of key hydrodynamic and heat transfer parameters provide a unique platform for the investigation of microchannel slug flow boiling under controlled, stable conditions and serve as a means to provide a direct comparison to numerical models. Shown in Figure 19 is a direct comparison of experimentally obtained slug flow boiling data generated using the facility developed by Kingston *et al.* [89] with a high-fidelity simulation of the slug flow boiling process modeled using the approach developed by Pan *et al.* [79]. The boundary (e.g., heat flux) and operating (e.g., initial bubble size, bubble frequency, and

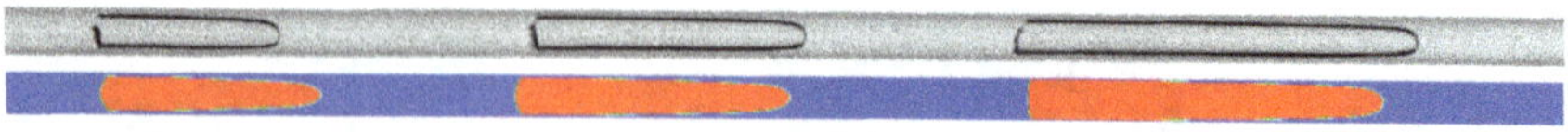

(a)

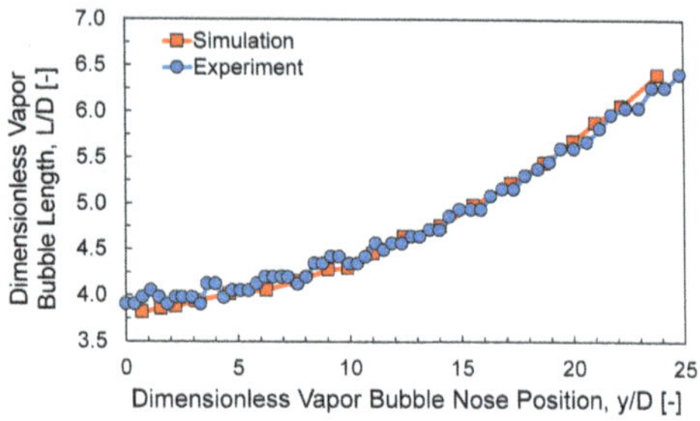

(b)

Figure 19. (a) Qualitative and (b) quantitative comparison of slug flow boiling experiments performed in the unique test facility developed by Kingston *et al.* [89] to a high-fidelity slug flow boiling simulation modeled using the saturated-interface-volume phase-change model developed by Pan *et al.* [79].

flow rate) conditions measured during the experiments were used as inputs to the simulation. As shown in Figure 19(b), the vapor bubbles initially start with a dimensionless length of approximately $L/D = 4$ and grow due to the evaporation of liquid surrounding the bubble, until they reach a dimensionless length of approximately $L/D = 6.5$ at $y/D = 25$. The excellent agreement between the experimental data and model predictions for this operating condition provides an initial experimental validation of the saturated-interface-volume phase-change model developed by Pan *et al.* [79]. Additional thermal and hydrodynamic comparisons between the experimental data and model predictions at more boundary and operating conditions are needed to provide robust validation of the model and enable accurate two-phase heat transfer predictions to be made.

6. Outlook

Next-generation high-power electronic devices will require innovative thermal management strategies in order to provide enhanced functionality. Two-phase embedded microfluidic cooling is an enabling technology that offers significant thermal performance improvement relative to traditional remote-cooling strategies. For example, the total thermal resistance of a two-phase embedded-cooling system is often less than that of a single high-performance thermal interface material [33], several of which are needed in a remote-cooling stack. However, the design, performance prediction, fabrication, and integration of two-phase embedded-cooling systems all require further, non-trivial advances. Notably, a paradigm shift in the electrothermal co-design and fabrication of electronic devices would be required to incorporate cooling channels within the substrate or package, which would require coupled thermal, electrical, and mechanical design approaches, with extensive evaluation of their reliability. Scalable cooling architectures that utilize microfluidic manifolds to route fluid to many discretized heat sinks also need to be realized. With continual advancements in multi-chip stacking and heterogeneous packaging of various devices on the same substrate, these solutions need to be densely integrated such that they can provide high areal and volumetric heat dissipation. Additionally, they need to accommodate thermal transients and localized hotspots, both of which become prominent with densely integrated systems and have been shown to be coupled to the two-phase cooling characteristics. The increasing number of parallel flow paths in these microscale systems via a reduction in channel dimensions causes flow boiling instabilities to become increasingly prevalent. While decades of research on flow boiling instabilities in macroscale systems is available and should be leveraged, embedded microfluidic cooling presents its own unique set of challenges (and opportunities) that need to be considered to minimize the adverse effects of these instabilities on CHF and heat transfer coefficients. Finally, experimentally validated physics-based models that provide more accurate and generalized prediction of thermal and hydrodynamic performance will need to replace current empirical design strategies. With the significant progress toward these required advancements outlined herein, two-phase embedded microfluidic cooling promises to revolutionize the electronics cooling industry and enable a new wave of increasingly capable devices that will have far-reaching industrial and societal implications.

References

[1] K.P. Bloschock, A. Bar-Cohen, Advanced thermal management technologies for defense electronics, *Proc. Spie*, 8405, 2012, pp. 157–168.

[2] A. Bar-Cohen, J.J. Maurer, J.G. Felbinger, DARPA's intra/interchip enhanced cooling (ICECool) program, in: *Compound Semiconductor Manufacturing Technology Conference (CS MANTECH)*, 2013, pp. 171–174.

[3] J.R. Wasniewski, D.H. Altman, S.L. Hodson, T.S. Fisher, A. Bulusu, S. Graham, B.A. Cola, Characterization of metallically bonded carbon nanotube-based thermal interface materials using a high accuracy 1D steady-state technique, *J. Electron. Packag.*, 134 (020901), 2012.

[4] L. Guo, S.L. Hodson, T.S. Fisher, X. Xu, Heat transfer across metal-dielectric interfaces during ultrafast-laser heating, *J. Heat Transfer*, 134 (042402), 2012.

[5] D. Shaddock, S. Weaver, I. Chasiotis, B. Shah, D. Zhong, Development of a compliant nanothermal interface material, in: *ASME International Technical Conference and Exhibition on Packaging and Integration of Electronic and Photonic Microsystems*, 2011, pp. 13–17.

[6] Y. Zhao, D. Strauss, T. Liao, Y.C. Chen, C.L. Chen, Development of a high performance thermal interface material with vertically aligned graphite platelets, in: *ASME/JSME Thermal Engineering Joint Conference*, 2011.

[7] A. Bar-Cohen, K. Matin, S. Narumanchi, Nanothermal interface materials: Technology review and recent results, *J. Electron. Packag.*, 137 (040803), 2015.

[8] J.A. Weibel, S.V. Garimella, Recent advances in vapor chamber transport characterization for high-heat-flux applications, in: E.M. Sparrow, Y.I. Cho, J.P. Abraham, J.M. Gorman (eds.), *Advances in Heat Transfer*, San Diego: Elsevier, 2013, pp. 209–301.

[9] D.H. Altman, J.R. Wasniewski, M.T. North, S.S. Kim, T.S. Fisher, Development of micro/nano engineered wick-based passive heat spreaders for thermal management of high power electronic devices, in: *ASME International Technical Conference and Exhibition on Packaging and Integration of Electronic and Photonic Microsystems*, 2011, pp. 213–220.

[10] H.P.J.D. Bock, S. Chauhan, P. Chamarthy, S.E. Weaver, T. Deng, F.M. Gerner, M.T. Ababneh, K. Varanasi, On the charging and thermal characterization of a micro/nano structured thermal ground plane, in: *IEEE Intersociety Conference on Thermal and Thermomechanical Phenomena in Electronic Systems*, 2010, pp. 1–6.

[11] Q. Cai, Y.-C. Chen, Investigations of biporous wick structure dryout, *J. Heat Transfer*, 134 (021503), 2011.

[12] C. Ding, G. Soni, P. Bozorgi, B.D. Piorek, C.D. Meinhart, N.C. MacDonald, A flat heat pipe architecture based on nanostructured titania, *J. Microelectromech. Syst.*, 19 (4), 2010, 878–884.

[13] S.M. Thompson, B.S. Tessler, H. Ma, D.E. Smith, A. Sobel, Ultrahigh thermal conductivity of three-dimensional flat-plate oscillating heat pipes for

electromagnetic launcher cooling, *IEEE Trans. Plasma Sci.*, 41 (5), 2013, 1326–1331.

[14] C. Oshman, Q. Li, L.-A. Liew, R. Yang, V.M. Bright, Y.C. Lee, Flat flexible polymer heat pipes, *J. Micromech. Microeng.*, 23 (1), 2012, 015001.

[15] Y.S. Ju, M. Kaviany, Y. Nam, S. Sharratt, G.S. Hwang, I. Catton, E. Fleming, P. Dussinger, Planar vapor chamber with hybrid evaporator wicks for the thermal management of high-heat-flux and high-power optoelectronic devices, *Int. J. Heat Mass Transfer*, 60, 2013, 163–169.

[16] A. Bar-Cohen, K. Matin, N. Jankowski, D. Sharar, Two-phase thermal ground planes: Technology development and parametric results, *J. Electron. Packag.*, 137 (1), 2014, 010801–010809.

[17] D. Guo, J. Gao, A.J.H. McGaughey, M. Moran, S. Santhanam, G.K. Fedder, B. Anderson, S.-C. Yao, Design and evaluation of MEMS-based Stirling cycle micro-refrigeration system, in: *ASME International Mechanical Engineering Congress and Exposition*, 2011, pp. 83–90.

[18] J.M. Allison, W.L. Staats, M. McCarthy, D. Jenicek, A.K. Edoh, J.H. Lang, E.N. Wang, J.G. Brisson, Enhancement of convective heat transfer in an air-cooled heat exchanger using interdigitated impeller blades, *Int. J. Heat Mass Transfer*, 54 (21), 2011, 4549–4559.

[19] T. Yeom, T.W. Simon, M. Zhang, M.T. North, T. Cui, High frequency, large displacement, and low power consumption piezoelectric translational actuator based on an oval loop shell, *Sens. Actuat. A Phys.*, 176, 2012, 99–109.

[20] F. Zhou, I. Catton, Numerical evaluation of flow and heat transfer in plate-pin fin heat sinks with various pin cross-sections, *Numer. Heat Transfer Part A Appl.*, 60 (2), (2011), 107–128.

[21] D.B. Tuckerman, R.F.W. Pease, High-performance heat sinking for VLSI, *Electron. Dev. Lett.*, 2 (5), 1981, 126–129.

[22] M. Schultz, F. Yang, E. Colgan, R. Polastre, B. Dang, C. Tsang, M. Gaynes, P. Parida, J. Knickerbocker, T. Chainer, Embedded two-phase cooling of large three-dimensional compatible chips with radial channels, *J. Electron. Packag.*, 138 (021005), 2016.

[23] M.D. Schultz, M. Gaynes, P.R. Parida, F. Yang, G. McVicker, O. Ozsun, T. Chainer, Two-phase flow characteristics in radial expanding channels with embedded pin fins, in: *IEEE Intersociety Conference on Thermal and Thermomechanical Phenomena in Electronic Systems*, 2018, pp. 281–287.

[24] J.W. Palko, H. Lee, C. Zhang, T.J. Dusseault, T. Maitra, Y. Won, D.D. Agonafer, J. Moss, F. Houshmand, G. Rong, J.D. Wilbur, D. Rockosi, I. Mykyta, D. Resler, D. Altman, M. Asheghi, J.G. Santiago, K.E. Goodson, Extreme two-phase cooling from laser-etched diamond and conformal, templatefabricated microporous copper, *Adv. Funct. Mater.*, 27 (45), 2017, 1703265.

[25] R.K. Mandel, D.G. Bae, M.M. Ohadi, embedded two-phase cooling of high flux electronics via press-fit and bonded FEEDS coolers, *J. Electron. Packag.*, 140 (031003), 2018.

[26] C. Woodcock, C. Ng'oma, M. Sweet, Y. Wang, Y. Peles, J. Plawsky, Ultra-high heat flux dissipation with piranha pin fins, *Int. J. Heat Mass Transfer*, 128, 2019, 504–515.

[27] M.H. Nasr, C.E. Green, P.A. Kottke, X. Zhang, T.E. Sarvey, Y.K. Joshi, M.S. Bakir, A.G. Fedorov, Flow regimes and convective heat transfer of refrigerant flow boiling in ultra-small clearance microgaps, *Int. J. Heat Mass Transfer*, 108, 2017, 1702–1713.

[28] M.H. Nasr, C.E. Green, P.A. Kottke, X. Zhang, T.E. Sarvey, Y.K. Joshi, M.S. Bakir, A.G. Fedorov, Hotspot thermal management with flow boiling of refrigerant in ultrasmall microgaps, *J. Electron. Packag.*, 139 (011006), 2017.

[29] S.N. Ritchey, J.A. Weibel, S.V. Garimella, Local measurement of flow boiling heat transfer in an array of non-uniformly heated microchannels, *Int. J. Heat Mass Transfer*, 71, 2014, 206–216.

[30] T. Harirchian, S.V. Garimella, Boiling heat transfer and flow regimes in microchannels — A comprehensive understanding, *J. Electron. Packag.*, 133 (1), 2011, 011001.

[31] P.-S. Lee, S.V. Garimella, Saturated flow boiling heat transfer and pressure drop in silicon microchannel arrays, *Int. J. Heat Mass Transfer*, 51 (3), 2008, 789–806.

[32] T. Chen, S.V. Garimella, Measurements and high-speed visualizations of flow boiling of a dielectric fluid in a silicon microchannel heat sink, *Int. J. Multiphase Flow*, 32 (8), 2006, 957–971.

[33] K.P. Drummond, D. Back, M.D. Sinanis, D.B. Janes, D. Peroulis, J.A. Weibel, S.V. Garimella, A hierarchical manifold microchannel heat sink array for high-heat-flux two-phase cooling of electronics, *Int. J. Heat Mass Transfer*, 117, 2018, 319–330.

[34] K.P. Drummond, D. Back, M.D. Sinanis, D.B. Janes, D. Peroulis, J.A. Weibel, S.V. Garimella, Characterization of hierarchical manifold microchannel heat sink arrays under simultaneous background and hotspot heating conditions, *Int. J. Heat Mass Transfer*, 126, 2018, 1289–1301.

[35] K.M. Razeeb, E. Dalton, G.L.W. Cross, A.J. Robinson, Present and future thermal interface materials for electronic devices, *Int. Mater. Rev.*, 63 (1), 2018, 1–21.

[36] K.P. Drummond, J.A. Weibel, S.V. Garimella, Two-phase flow morophology and local wall temperatures in high-aspect-ratio manifold microchannels, *Int. J. Heat Mass Transfer*, 153 (119551), 2020.

[37] D. Back, K.P. Drummond, M.D. Sinanis, J.A. Weibel, S.V. Garimella, D. Peroulis, D.B. Janes, Design, fabrication, and characterization of a compact hierarchical manifold microchannel heat sink array for two-phase cooling, *IEEE Trans. Compon. Packag. Manuf. Technol.*, 9 (7), 2019, 1291–1300.

[38] A.E. Bergles, S.G. Kandlikar, On the nature of critical heat flux in microchannels, *J. Heat Transfer*, 127 (1), 2005, 101–107.

[39] T. Van Oevelen, J.A. Weibel, S.V. Garimella, Predicting two-phase flow distribution and stability in systems with many parallel heated channels, *Int. J. Heat Mass Transfer*, 107, 2017, 557–571.

[40] T. Zhang, T. Tong, J.-Y. Chang, Y. Peles, R. Prasher, M.K. Jensen, J.T. Wen, P. Phelan, Ledinegg instability in microchannels, *Int. J. Heat Mass Transfer*, 52 (25–26), 2009, 5661–5674.

[41] L.C. Ruspini, C.A. Dorao, M. Fernandino, Dynamic simulation of Ledinegg instability, *J. Nat. Gas Sci. Eng.*, 2 (5), 2010, 211–216.

[42] Y.W. Kuang, W. Wang, J.Y. Miao, X.G. Yu, R. Zhuan, Theoretical analysis and modeling of flow instability in a mini-channel evaporator, *Int. J. Heat Mass Transfer*, 104, 2017, 149–162.

[43] J.A. Boure, A.E. Bergles, L.S. Tong, Review of two-phase flow instability, *Nucl. Eng. Des.*, 25 (2), 1973, 165–192.

[44] S. Kakac, B. Bon, A review of two-phase flow dynamic instabilities in tube boiling systems, *Int. J. Heat Mass Transfer*, 51 (3–4), 2008, 399–433.

[45] L.C. Ruspini, C.P. Marcel, A. Clausse, Two-phase flow instabilities: A review, *Int. J. Heat Mass Transfer*, 71, 2014, 521–548.

[46] Y.K. Prajapati, P. Bhandari, Flow boiling instabilities in microchannels and their promising solutions — A review, *Exp. Thermal Fluid Sci.*, 88, 2017, 576–593.

[47] K. Akagawa, M. Kono, T. Sakaguchi, M. Nishimura, Study on distribution of flow rates and flow stabilities in parallel long evaporators, *Bull. Jpn. Soc. Mech. Eng.*, 14 (74), 1971, 837–848.

[48] A. Koşar, C.J. Kuo, Y. Peles, Suppression of boiling flow oscillations in parallel microchannels by inlet restrictors, *J. Heat Transfer*, 128 (3), 2006, 251–260.

[49] U. Minzer, D. Barnea, Y. Taitel, Flow rate distribution in evaporating parallel pipes — Modeling and experimental, *Chem. Eng. Sci.*, 61 (22), 2006, 7249–7259.

[50] M. Baikin, Y. Taitel, D. Barnea, Flow rate distribution in parallel heated pipes, *Int. J. Heat Mass Transfer*, 54 (19–20), 2011, 4448–4457.

[51] T. Van Oevelen, J.A. Weibel, S.V. Garimella, The effect of lateral thermal coupling between parallel microchannels on two-phase flow distribution, *Int. J. Heat Mass Transfer*, 124, 2018, 769–781.

[52] G. Patankar, T.R. Salamon, Maldistribution of two-phase flow in parallel channel heat sinks: effects of thermal connection between channels, in: *IEEE Intersociety Conference on Thermal and Thermomechanical Phenomena in Electronic Systems*, 2018, pp. 653–663.

[53] R. Flynn, C.-H. Cheng, K. Goodson, Decoupled thermal and fluidic effects on hotspot cooling in a boiling flow microchannel heat sink, in: *ASME International Technical Conference and Exhibition on Packaging and Integration of Electronic and Photonic Microsystems*, 2007, pp. 179–184.

[54] T.A. Kingston, J.A. Weibel, S.V. Garimella, Ledinegg instability-induced temperature excursion between thermally isolated, heated parallel microchannels, *Int. J. Heat Mass Transfer*, 132, 2019, 550–556.

[55] A. Miglani, J.A. Weibel, S.V. Garimella, Measurement of flow maldistribution induced by the Ledinegg instability during boiling in thermally isolated parallel microchannels, *Int. J. Multiphase Flow*, 139, 2021, 103644.

[56] A. Miglani, J.A. Weibel, S.V. Garimella, An experimental investigation of the effect of thermal coupling between parallel microchannels undergoing boiling on the Ledinegg instability-induced flow maldistribution, *Int. J. Multiphase Flow*, 139, 2020, 103536.

[57] C.J. Kuo, Y. Peles, Flow boiling instabilities in microchannels and means for mitigation by reentrant cavities, *J. Heat Transfer*, 130 (7), 2008, 072402.

[58] W. Qu, I. Mudawar, Measurement and prediction of pressure drop in two-phase micro-channel heat sinks, *Int. J. Heat Mass Transfer*, 46 (15), 2003, 2737–2753.

[59] T.A. Kingston, J.A. Weibel, S.V. Garimella, High-frequency thermal-fluidic characterization of dynamic microchannel flow boiling instabilities: Part 1 – Rapid-bubble-growth instability at the onset of boiling, *Int. J. Multiphase Flow*, 106, 2018, 179–188.

[60] T.A. Kingston, J.A. Weibel, S.V. Garimella, High-frequency thermal-fluidic characterization of dynamic microchannel flow boiling instabilities: Part 2 — Impact of operating conditions on instability type and severity, *Int. J. Multiphase Flow*, 106, 2018, 189–201.

[61] T. David, D. Mendler, A. Mosyak, A. Bar-Cohen, G. Hetsroni, Thermal management of time-varying high heat flux electronic devices, *J. Electron. Packag.*, 136 (2), 2014, 021003.

[62] A. Bar-Cohen, Gen-3 thermal management technology: Role of microchannels and nanostructures in an embedded cooling paradigm, *J. Nanotechnol. Eng. Med.*, 4 (2), 2013, 020907.

[63] L.J. Guo, Z.P. Feng, X.J. Chen, Transient convective heat transfer of steam-water two-phase flow in a helical tube under pressure drop type oscillations, *Int. J. Heat Mass Transfer*, 45 (3), 2002, 533–542.

[64] I.W. Park, J. Ryu, M. Fernandino, C.A. Dorao, Can flow oscillations during flow boiling deteriorate the heat transfer coefficient?, *Appl. Phys. Lett.*, 113 (15), 2018, 154102.

[65] M. Sorum, C.A. Dorao, Experimental study of the heat transfer coefficient deterioration during density wave oscillations, *Chem. Eng. Sci.*, 132, 2015, 194–201.

[66] M. Ozawa, H. Umekawa, K. Mishima, T. Hibiki, Y. Saito, CHF in oscillatory flow boiling channels, *Chem. Eng. Res. Design*, 79 (A4), 2001, 389–401.

[67] S. Basu, B. Werneke, Y. Peles, M.K. Jensen, Transient microscale flow boiling heat transfer characteristics of HFE-7000, *Int. J. Heat Mass Transfer*, 90, 2015, 396–405.

[68] H. Huang, N. Borhani, J. Thome, Thermal response of multi-microchannel evaporators during flow boiling of refrigerants under transient heat loads with flow visualization, *J. Electron. Packag.*, 138 (3), 2016, 031004.

[69] G. Chen, P. Cheng, Nucleate and film boiling on a microheater under pulse heating in a microchannel, *Int. Commun. Heat Mass Transfer*, 36 (5), 2009, 391–396.

[70] J.L. Miler, R. Flynn, G. Refai-Ahmed, M. Touzelbaev, M. David, J. Steinbrenner, K.E. Goodson, Effects of transient heating on two-phase flow response in microchannel heat exchangers, in: *ASME International*

Technical Conference and Exhibition on Packaging and Integration of Electronic and Photonic Microsystems, 2010, pp. 563–569.

[71] G.M. Aguiar, G. Ribatski, The effect of transient power hotspots on the heat transfer coefficient during flow boiling inside single microscale channels, *Heat. Transfer Eng.*, 2018, 1–12.

[72] T.A. Kingston, J.A. Weibel, S.V. Garimella, Time-resolved characterization of microchannel flow boiling during transient heating: Part 1 — Dynamic response to a single heat flux pulse, *Int. J. Heat Mass Transfer*, 154 (119643), 2020.

[73] T.A. Kingston, J.A. Weibel, S.V. Garimella, Time-resolved characterization of microchannel flow boiling during transient heating: Part 2 — Dynamic response to time-periodic heat flux pulses, *Int. J. Heat Mass Transfer*, 154 (119686), 2020.

[74] D. Lorenzini, Y.K. Joshi, Computational fluid dynamics modeling of flow boiling in microchannels with nonuniform heat flux, *J. Heat Transfer*, 140 (1), 2018, 011501.

[75] M. Magnini, J.R. Thome, Computational study of saturated flow boiling within a microchannel in the slug flow regime, *J. Heat Transfer*, 138 (2), 2015, 021502.

[76] M. Magnini, J.R. Thome, A CFD study of the parameters influencing heat transfer in microchannel slug flow boiling, *Int. J. Thermal Sci.*, 110, 2016, 119–136.

[77] D. Lorenzini, Y. Joshi, Comparison of the volume of fluid and CLSVOF methods for the assessment of flow boiling in silicon microgaps, *J. Heat Transfer*, 139 (11), 2017, 111506.

[78] D. Lorenzini, Y. Joshi, Numerical modeling and experimental validation of two-phase microfluidic cooling in silicon devices for vertical integration of microelectronics, *Int. J. Heat Mass Transfer*, 138, 2019, 194–207.

[79] Z. Pan, J.A. Weibel, S.V. Garimella, A saturated-interface-volume phase change model for simulating flow boiling, *Int. J. Heat Mass Transfer*, 93, 2016, 945–956.

[80] Z. Pan, J.A. Weibel, S.V. Garimella, Spurious current suppression in VOF-CSF simulation of slug flow through small channels, *Num. Heat Transfer Part A Appl.*, 67 (1), 2015, 1–12.

[81] M. Magnini, B. Pulvirenti, J.R. Thome, Numerical investigation of hydrodynamics and heat transfer of elongated bubbles during flow boiling in a microchannel, *Int. J. Heat Mass Transfer*, 59, 2013, 451–471.

[82] D. Sun, J. Xu, Q. Chen, Modeling of the evaporation and condensation phase-change problems with FLUENT, *Num. Heat Transfer Part B Fundam.*, 66 (4), 2014, 326–342.

[83] C. Kunkelmann, P. Stephan, CFD simulation of boiling flows using the volume-of-fluid method within OpenFOAM, *Num. Heat Transfer Part A Appl.*, 56 (8), 2009, 631–646.

[84] S. Hardt, F. Wondra, Evaporation model for interfacial flows based on a continuum-field representation of the source terms, *J. Comput. Phys.*, 227 (11), 2008, 5871–5895.

[85] Y. Tsui, S. Lin, Y. Lai, F. Wu, Phase change calculations for film boiling flows, *Int. J. Heat Mass Transfer*, 70, 2014, 745–757.
[86] P.R. Parida, T. Chainer, Eulerian multiphase conjugate model development and validation for flow boiling in micro-pin field, in: *IEEE Intersociety Conference on Thermal and Thermomechanical Phenomena in Electronic Systems*, 2016, pp. 1398–1408.
[87] P.R. Parida, A. Sridhar, M. Schultz, F. Yang, M. Gaynes, E. Colgan, B. Dang, G. McVicker, T. Brunschwiler, J. Knickerbocker, T. Chainer, Modeling embedded two-phase liquid cooled high power 3D compatible electronic devices, in: *SEMI-THERM*, 2017, pp. 130–138.
[88] C. Gorlé, H. Lee, F. Houshmand, M. Asheghi, K. Goodson, P.R. Parida, Validation study for VOF simulations of boiling in a microchannel, in: *ASME International Technical Conference and Exhibition on Packaging and Integration of Electronic and Photonic Microsystems*, 2015.
[89] T.A. Kingston, J.A. Weibel, S.V. Garimella, An experimental method for controlled generation and characterization of microchannel slug flow boiling, *Int. J. Heat Mass Transfer*, 106, 2017, 619–628.

https://doi.org/10.1142/9789811279379_0010

Chapter 10

Scaling Limits, Challenges, and Opportunities in Embedded Cooling

Srikanth Rangarajan*, Scott N. Schiffres*, and Bahgat Sammakia

Binghamton University, Department of Mechanical Engineering, Binghamton, NY 13902, USA

Abstract

Many packaging experts believe Moore's law will hold for another decade due to recent technological efforts to merge multiple integrated circuit dies horizontally and vertically into a single package, commonly referred to as heterogeneous integration. With simultaneous shrinkage in chip size and larger heat dissipation, the flux values have increased exponentially. To tackle the skyrocketing thermal management challenges, the electronics industry and academic research community have examined various thermal management solutions. Embedded cooling eliminates most of the sequential conduction resistances from the chip to ambient, unlike the separable cold plates/heat sinks. Though there is a high heat transfer potential by embedding the cooling solution onto an electronic chip, there are still technological risks associated with the implementation of these technologies and uncertainty about which technologies should be adopted. This chapter discusses the thermal spreading

*These authors contributed equally to the work.

limits, fluid selection considerations, as well as conventional and additive manufacturing-based embedded-cooling technologies.

Keywords: Embedded cooling technologies, scaling limits, heat spreading resistance, etched substrates, additive manufacturing

1. Introduction

According to the Merriam-Webster dictionary, embed is "to make something an integral part of" [1], so "embedded cooling" broadly encompasses technologies that cool directly on or inside an active electronic device (e.g., microprocessor, power electronic device). Embedded-cooling strategies can employ purely conduction heat transfer [2–4], single-phase forced convection [5–7], two-phase forced convection [8], natural convection (single or two phase) [9], immersion cooling [10, 11], heat pipes/vapor chambers [12–14], active refrigeration [15, 16], and phase-change cooling [17, 18]. Bringing the cooling closer to the active device is becoming ever more necessary due to the increased heat fluxes in microprocessors and power electronics, as well as the increasing packing densities of devices and their shrinking form factors. In terms of these trends, microprocessor computational power is hitting thermal limits due to the exponential growth in heat dissipation that coincides with Moore's law [19]. As a consequence, microprocessor cores throttle due to thermal considerations. There is also a developing trend to co-locate multiple devices into a package, the so-called heterogeneous integration (HI) and 3D integrated circuits (3DICs) [20]. A tighter packing of electronics means faster communication and computation, which is especially important in high-performance computing. These trends are pushing high-heat-flux packages closer together, meaning that conventional lateral spreading is increasingly difficult due to the tight spatial integration of circuitry [21]. In power electronics, extreme heat fluxes are possible, which require new strategies (e.g., 30 kW/cm^2 at the transistor and at least 1 kW/cm^2 at the die) [22]. Embedded cooling imparts an advantage at these high heat fluxes by eliminating thermal interface

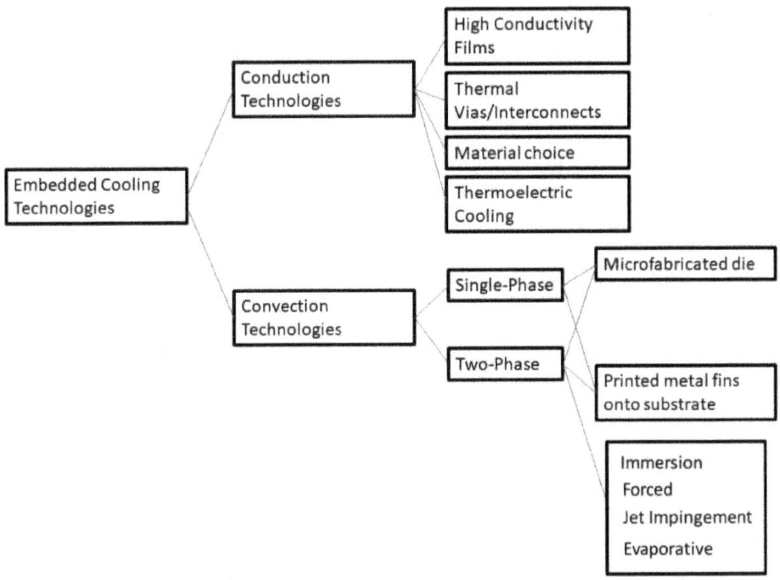

Figure 1. An overview of the wide variety of embedded-cooling technologies.

materials that would otherwise increase the resistance and hurt the thermal budget.

The general architectures of embedded-cooling technologies take a plethora of forms, as shown in Figure 1. The general hierarchy includes technologies that are based on conduction and convection. The conduction technologies encompass films with high thermal conductivities (e.g., CVD diamond) for near-junction heat spreading, thermal vias made of copper to transport heat vertically through a chip, and thermoelectric devices to actively cool devices. Convection-focused embedded-cooling technologies typically employ fluid-directing structures on top of or inside the active electronic device (i.e., inter-chip or intra-chip cooling) and extended surfaces. These convection-based embedded-cooling devices have primarily been made from etched microchannels produced on microfabricated dies. More recently, printed metal fins are also being explored to create freeform metal cooling structures directly onto the active electronic devices [23–26, 109, 110]. These strategies are introduced and discussed in depth in the following section.

Passive embedded-cooling strategies enhance conduction heat transfer internal to the chip by employing materials with high thermal conductivities that help lower the thermal resistance. One method involves incorporating vertical thermal vias and lateral thermal interconnects, which can serve as both thermal pathways and potential electrical signal pathways. As copper has a much greater conductivity than traditional dielectrics used in the microprocessor (0.1–1 W/m-K), increasing the density of copper in the interconnect layer, beyond the electrical signal requirements, increases thermal performance. Conduction-based thermal management strategies include through-silicon vias that conduct heat vertically over hotspots [4], coating active layers in materials with high thermal conductivity, such as diamond [2, 27] or pyrolytic graphite [28].

Convective embedded-cooling technologies directly attach to the active device, as in etched silicon microchannels [29, 30] and metal 3D-printed cooling fins [23–26, 109, 110]. Etched silicon microchannel-based technologies have been widely investigated, although alternative technologies, such as printing metal fins onto silicon, are also being examined. The choice of coolant fluid and whether that fluid operates in single- or two-phase convection are also important considerations. Two-phase cooling can operate with small temperature differences, even when cooling multiple devices in series, owing to the large latent heat of many fluids. Designers must operate below critical heat fluxes, as abrupt failures occur if they are exceeded. Water has excellent heat capacity and latent heat and has been used for safe electronic cooling with robust plumbing solutions, such as the IBM mainframe ZEnterprise 196. However, water cannot contact active electronic circuits and is still viewed as risky by some end users, so various dielectric fluids have also been explored, such as Novec and mineral oil. These dielectric fluids are more expensive than water and often possess less desirable heat-dissipating properties (thermal conductivity, specific heat, latent heat, and critical boiling heat flux), but are electrically insulating. A full comparison of fluids for forced convection, natural convection, and two-phase cooling is provided in Section 3.

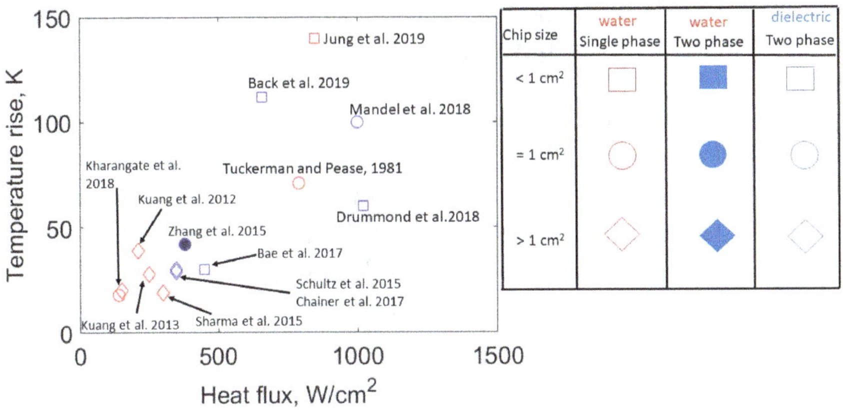

Figure 2. Regime map of the recent advancements in embedded cooling trends.

The general challenges that have so far limited widespread adoption of many embedded-cooling technologies are their perceived complexity, high costs, high levels of coordination required between electrical circuitry and thermal engineering, new manufacturing processes, reliability risks, yield reduction, reworkability, and supply chain risks. These challenges, whether real or only perceived, have limited the adoption of many embedded-cooling technologies; however, there is growing consensus that these technologies are on the critical path to advancing electrical performance. The following chapter examines the scaling limits of embedded cooling for different cooling, heat spreading, and fluid combinations and examines the challenges and opportunities for convective cooling technologies in microfabricated etched die channels and 3D-printed fins.

Figure 2 shows the regime map for different chip sizes, microchannel design concepts, and coolants from the recent literature on embedded cooling. Please note that Figure 2 pertains only to chips with uniform heat fluxes. Future ideas have to concentrate on technologies operable at lower temperatures to manage higher heat fluxes through a smart choice of coolant, microchannel design, and shorter fluid paths to lower caloric resistance. A shorter fluid path enables local heat removal, thereby eliminating the caloric resistance of the cold plate, as demonstrated by Drummond *et al.* [31].

2. Scaling Limits for Convective Embedded Cooling vs. Non-Embedded Cooling

The trade-off between employing embedded cooling and bulky non-embedded cooling (such as a heat spreader/lid combined with a heat sink) must be studied to understand the challenges of moving to direct embedded cooling. The overall thermal resistance can be deconstructed into resistances due to the spreader and the heat sink. These two resistances exhibit conflicting trends in area ratio and the ratio of spreader area to chip area. The area ratio AR is defined as

$$\mathrm{AR} = A_s/A_c$$

The convection resistance of the heat sink decreases with area ratio, while the spreading resistance increases. We first examine the total thermal resistance for air cooling and then for water cooling. Subsequently, we study how a thicker, high-conductivity electronic die can improve heat dissipation from hotspots.

The analytical solutions provided by Shah and London [32] and Zhang *et al.* [6] were employed to calculate the heat sink resistance for developing-to-developed flow through a parallel microchannel. The spreading resistance was modeled using the approximate analytical solutions given by Song *et al.* [33]:

$$\begin{aligned}\text{Total resistance of heat sink} &= \text{Conduction resistance} + \text{Convection}\\ &\quad\ \text{resistance} + \text{Caloric resistance}\\[1em] \text{Total resistance of the spreader} &= \text{Direct 1D conduction resistance}\\ &\quad + \text{Spreading resistance} \qquad (1)\end{aligned}$$

In the forthcoming discussions, the spreading resistance refers to only the spreading component of the total spreading resistance, excluding the direct 1D conduction resistance.

Air cooling

Chip size (A_c)	1	cm^2
Air velocity	10	m/s
Channel width	1	mm

Fin/channel height	25	
Spreader thickness	10	mm
Spreader Conductivity	150	W/m-K

Figure 3 makes it evident that the enhancement in area reduces the convection resistance while increasing the spreading resistance within the considered resistance decomposition framework. The general vapor chamber resistance range is indicated by the shaded gray region on the plot [13]. Another factor is that whenever non-embedded cooling is used, an additional conduction resistance from TIM1 (the thermal interface material between chip and lid) is added but not accounted for in these values. With AR > 1, the junction temperature is determined by the sum of the heat sink resistance, spreader resistance, and TIM1 resistance. A properly designed air-cooled heat sink can have a thermal resistance as low as 0.25 K/W for $AR > 30$ and an active area of 1 cm^2, excluding the spreading and TIM resistances. The total external resistance, including the

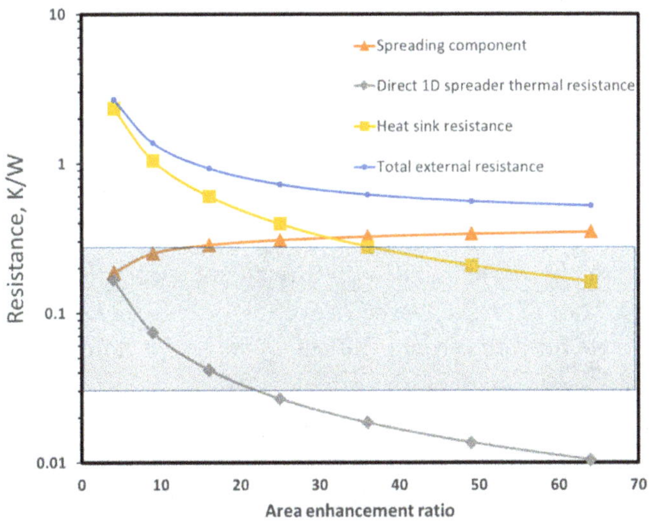

Figure 3. Trade-off analysis for air cooling with variable area enhancement ratios, assuming the chip size $A_c = 1$ cm^2, metal conductivity $k = 150$ W/m-K (heat sink and spreader), spreader thickness $t = 10$ mm, $w_{ch} = 1$ mm, $h_{ch} = 25$ mm, and $V_{air} = 10$ m/s. This figure excludes the TIM resistance. The gray region indicates the thermal resistance of the vapor chamber heat spreader [13].

heat sink and spreading resistances, is estimated to be 0.5 K/W. The total external resistance could be further minimized by using a vapor chamber heat spreader. On the other hand, a directly air-cooled heat sink with $AR = 1$ has a resistance of above 2.5 K/W, making it clear that direct air cooling is not a feasible solution even for modest heat loads.

2.1. *Liquid cooling*

Chip size	1	cm^2
Liquid water flow rate	0.5	lpm
Channel width	0.05	mm
Fin/ channel height	0.3	mm
Spreader thickness	10	mm
Metal conductivity	150	W/m-K

The resistance of a liquid-cooled cold plate versus area enhancement ratio is shown in Figure 4. The liquid-cooled cold plates take advantage of the high heat transfer coefficient in the microchannel, and hence, enhancing the area leads to a marginal performance benefit with the area ratio beyond about 4. The liquid cooled cold plates with no area enhancement can have a thermal resistance of 0.095 K/W. The total thermal resistance would just be the heat sink resistance without the spreading resistance. The total external thermal resistance is only plotted for AR $>$ 1. For AR $=$ 1, the total external resistance is the same as the heat sink resistance shown in the plot. On the other hand, non-direct liquid cooling with an area ratio of 4 would have a lower heat sink resistance of 0.025 K/W; however, the total external resistance would be dominated by the sum of the resistances of the spreader (0.35 K/W) and the TIM. *This shows the benefit of embedded cooling without any spreading for liquid cooling compared to that with conventional spreading, which would have overall higher resistances.* Printing the heat sinks directly onto silicon is an especially appealing technology because it allows for increased convective surface area with minimum interfacial resistance [25, 26, 109, 110]. If an unconventional spreader or spreaders utilizing phase change, such as a vapor chamber, were used, the

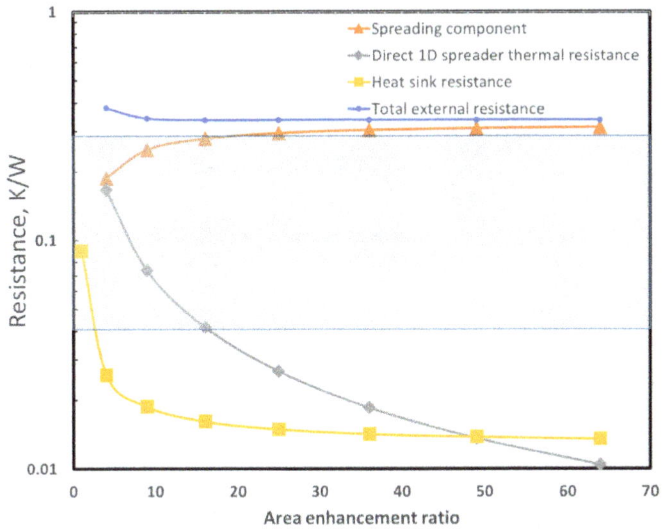

Figure 4. Trade-off analysis for a liquid-cooling scenario for chip size $A_c = 1$ cm^2, metal conductivity $k = 150$ W/m-K, spreader thickness $t = 10$ mm, $w_{ch} = 0.05$ mm, $h_{ch} = 0.3$ mm, $Q = 0.5$ lpm. The plot excludes the TIM resistance. The gray region indicates the thermal resistance of the vapor chamber heat spreader [13].

calculations would show a smaller overall resistance than a conduction spreader, as its effective conductivity is larger. Vapor chambers have thermal resistances in the range of 0.35–0.03 cm^{2}K/W [13]. Technologies, such as direct printing, could enable interfacing vapor chambers to chips without the use of thermal interface materials.

2.2. *Substrate thickness and spreading number (Sp)*

Another important scenario to consider is an embedded-cooling solution that uses the thickness of the electronic substrate as a spreader for a local hotspot. Figure 5 shows the variation in spreading resistance versus spreader thickness and heat transfer coefficient. Beyond a particular heat transfer coefficient, known as the transitional heat transfer coefficient, additional silicon thickness actually adds more total resistance, so it is very important to know the right value of this transitional heat transfer coefficient when making decisions

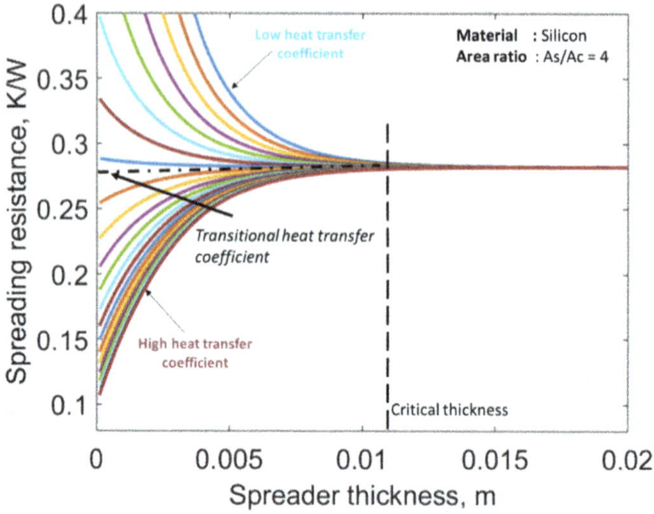

Figure 5. Variation of the spreading component of the spreading resistance versus spreader thickness for different heat transfer coefficients. For high enough heat transfer coefficients, the minimum spreader/lid thickness is optimal.

on the thickness of the electronic substrate for the purposes of thermal spreading. This design rule is elucidated in the concluding part of this section.

We propose a dimensionless number, which we call the spreading number, to capture this trade-off between the length of the spreader, the heat transfer coefficient, and the thermal conductivity of the spreader. The spreading number S_p in Eq. (2) has a form identical to the Biot number, but uses the characteristic length of the spreader as the characteristic length. *Sp* will later be used to solve for the critical thickness of a conduction spreader:

$$S_p = \frac{h_t L_s}{k_t} \tag{2}$$

where h_t: transitional heat transfer coefficient, W/m^2-K; L_s: spreader length, m, ($\sqrt{A_s}$); k_t: spreader conductivity, W/m-K.

The threshold spreading number [6] is solved for a range of materials and chip sizes by means of a genetic algorithm. The correlation for the threshold spreading number is shown in Eq. (3):

$$S_{p,t} = 9.56 + 0.081 \times \frac{A_s}{A_c} - \frac{31.93}{9.66 + \frac{A_s}{A_c}} - 0.0001894 \times \left(\frac{A_s}{A_c}\right)^2 \quad (3)$$

which is valid for $50 < k < 5,000$ W/m-K and $0.25\ \text{cm}^2 < A_c < 2\ \text{cm}^2$.

The threshold spreading number can be used to derive the transitional heat transfer coefficient, h_t, by setting Eq. (2) equal to our transitional spreading number correlation given in Eq. (3).

The critical thickness is defined as the thickness of the spreader beyond which the spreading component of the resistance is independent of the thickness and the heat transfer coefficient (within 1% of the asymptotic value). The critical thickness is marked in Figure 5.

As shown in Figure 6, the critical thickness increases with increasing area ratio. Knowing the right value of critical thickness helps designers optimize the spreaders. (Beyond the critical thickness, there is no gain in performance with additional thickness.)

We have developed an empirical correlation for the critical thickness:

$$\boldsymbol{t_c = 17.9 + 0.12 A_c A_R - \frac{65.1}{4.44 + 0.461\, A_c A_R} - 8.14 \times 10^{-4} k - 0.078 A_R} \quad (4)$$

which is valid for $50 < k < 5{,}000$ W/m-K and $0.25\ \text{cm}^2 < A_c < 2\ \text{cm}^2$.

Designers can use the correlation in Eq. (4) to determine the critical thickness. The physical implications of this correlation are as follows:

1. If the available spreader thickness is greater than the critical thickness, the only way to optimize performance is by selecting a suitable material and area ratio for a given chip size. Often, there is a constraint imposed on the available area such that material conductivity will remain the only option. From Eq. (4), it is also evident that t_c has a weak dependency on the material conductivity k. Therefore, beyond the critical thickness, there are limited opportunities for achieving optimized heat transfer performance.

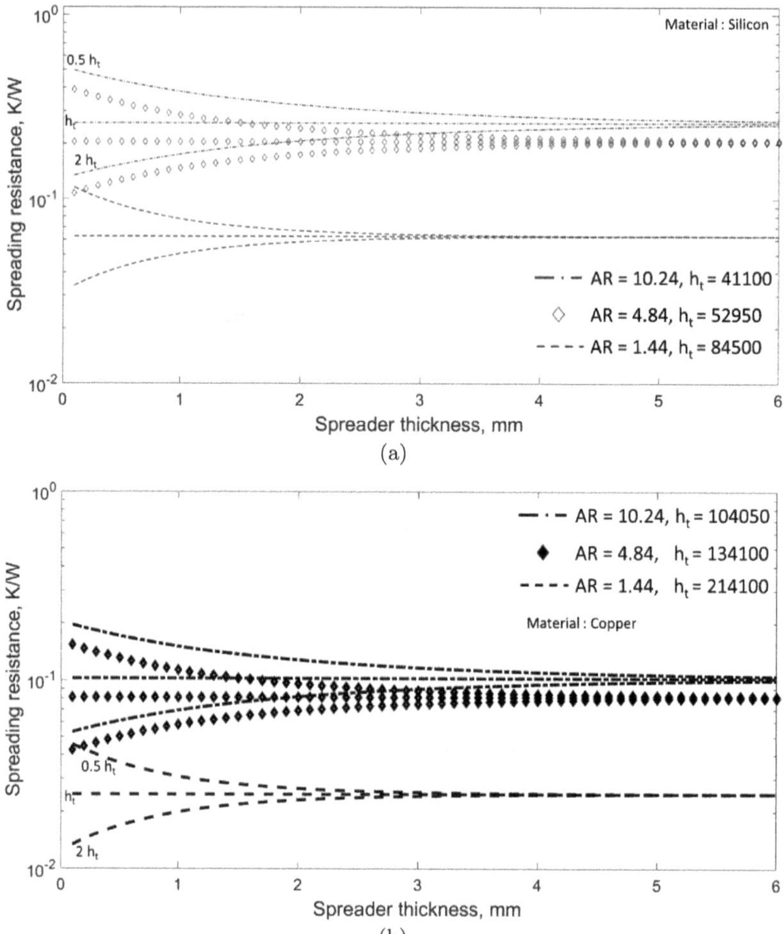

Figure 6. Spreading resistance versus spreader thickness for various area spreading ratios and heat transfer coefficients for cold plates made of (a) silicon and (b) copper.

2. If the available spreader thickness is less than the critical thickness, there are more opportunities to optimize the system performance, such as by thickening the spreader, increasing the heat transfer coefficient, and selecting a material with a higher thermal conductivity.

Figure 6 shows the spreading resistances for silicon and copper for different area ratios, thicknesses, and heat transfer coefficients. This chart enables the designer to make a quick decision on whether to employ embedded cooling with the available chip area or implement measures to spread the heat.

Figure 6 is a useful chart in the context of understanding the opportunities to augment the limits of embedded-cooling heat transfer by increasing spreader area (by using a chip lid or an integrated heat spreader) or by increasing the thickness of the spreader for different cooling scenarios (high or low heat transfer coefficients). It is noteworthy that the gain in the spreader area enhances the convection heat transfer (usually estimated from numerical analysis or empirical correlations) but also trades off with increased spreading resistance, as seen in Figure 6. This chart allows the designer to choose a particular area ratio and thickness of the spreader that can reach the required external resistance. Furthermore, as seen in Figures 3 and 4, given a high enough heat transfer coefficient, the optimal spreader thickness goes to zero.

Figure 7 shows the design flowchart for employing the spreading number in real-time scenarios, allowing some design guidelines to be outlined. First, the chip thickness can be compared to its critical thickness, beyond which it is never beneficial to add material. Next, the threshold convection coefficient can be calculated from the threshold spreading number. If the convection coefficient is less than this critical convection coefficient, a certain thickness becomes

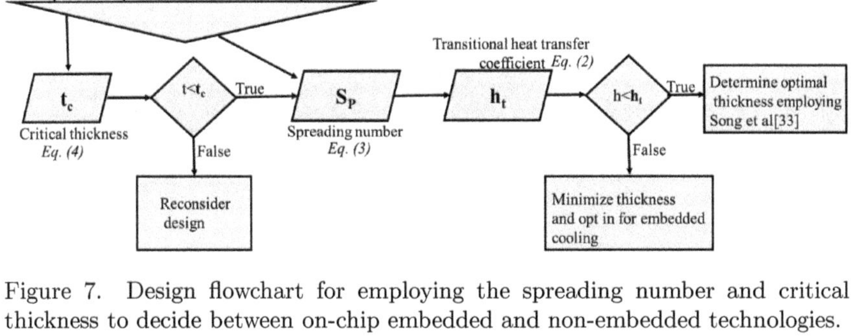

Figure 7. Design flowchart for employing the spreading number and critical thickness to decide between on-chip embedded and non-embedded technologies.

optimal (e.g., increasing thickness reduces spreading resistance while increasing direct 1-D conduction resistance, leading to an optimal thickness). If the convection coefficient is greater than this critical convection coefficient, no thickness of spreader material is beneficial and it is better to embed the cooling solution with no spreader thickness (e.g., increasing thickness increases spreading resistance and also increases direct 1-D conduction resistance, making no thickness of material optimal for that given area ratio).

3. Comparison of Embedded Cooling Fluids

For indirect cooling with cold plates, chemical and electrical compatibility between the liquid and electronic components is not an issue; therefore, water, a liquid with excellent heat transfer properties, can be employed. In the case of indirect cooling, where the fluid does not directly contact active devices, stringent constraints on the electrical properties of the fluid are not required (class "non-active contacting"). However, for embedded cooling where the fluid contacts powered active devices, as shown in Figure 8, stringent chemical and electrical requirements are imposed on the liquid (class "active contacting"). The opportunities to maximize heat transfer in different classes of embedded cooling are shown in Figure 8. The constraints for optimal fluid selection depend on the class of embedded cooling and the chemical and electrical properties of the liquid. Compatibility with chip substrate and printed circuit board materials, such

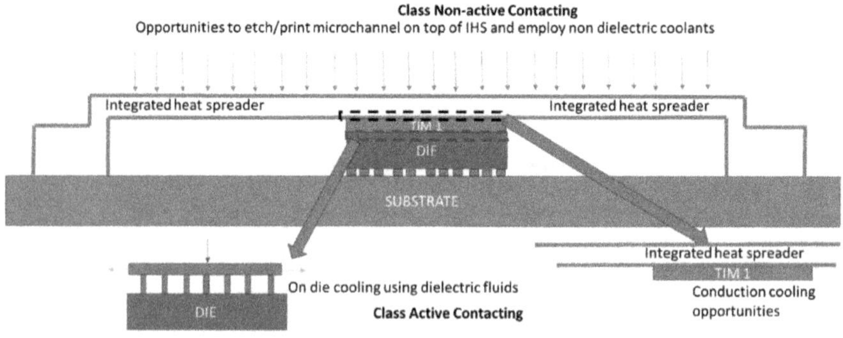

Figure 8. Opportunities in embedded cooling of a die/chip module.

Table 1. Different classes of embedded cooling and their details.

Parameter	Class active contacting	Class non-active contacting
Choice of coolant	Dielectric	Non-dielectric permissible
Thermal resistance	Cold plate resistance	Sum of the resistances of TIM, spreader, and cold plate

as printed wiring boards, silicone, and plastic materials, requires the liquid to be chemically inert and non-absorbing by these materials. Furthermore, it must provide electrical isolation between closely spaced conductors.

The desired properties of the fluid for embedded cooling are stability, non-toxicity, non-flammability, inertness, high dielectric strength, and low environmental impact (non-ozone depleting, low global warming potential, and capable of eventually decomposing after accidental release).

Fluid selection for embedded cooling can be posed as an optimization problem with the objective of maximizing heat transfer for the minimum junction temperature rise, subject to maximum reliability and good material compatibility. The constraints for the embedded technology to be implemented for practical applications are shown in Figure 9.

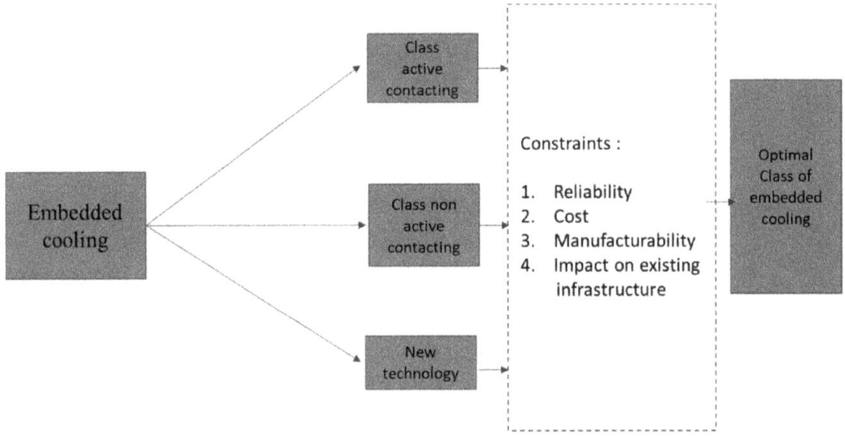

Figure 9. Constraints on embedded cooling technology.

Table 2. Comparison of thermophysical properties of three most commonly used coolants for embedded cooling.

Property	FC-72	HFE-7100	Water
Boiling point 1 atm (°)	56	61	100
ρ, Density $\times 10^{-3}$ (kg/m^3)	1.680	1.51	0.997
Cp, Specific heat $\times 10^{-3}$ (J/kg-K)	1.088	1.183	4.179
k, Thermal conductivity (W/m-K)	0.057	0.069	0.613
Kinematic viscosity (cSt)	0.38	0.38	8.55
h_{fg}, Heat of vaporization (kJ/kg)	88	112	243.8
σ, Surface tension $\times 10^3$ (N/m)	10	13.6	72
β, Thermal coefficient of expansion (K^{-1})	0.0016	0.0018	0.003
Dielectric constant	1.72	7.4	78.0

Source: 3M Company.

There may be numerous coolants which can provide adequate cooling, but only a few will be chemically compatible. Water is an example of a liquid which has excellent heat transfer characteristics but is generally unsuitable for direct immersion cooling on account of its chemical characteristics. Fluorocarbon liquids (e.g., FC-72, FC-86, FC-77) are generally considered the most suitable liquids for direct chip cooling, despite their poor thermo-physical properties. Although FCs, such as the family of Freons, have serious environmental problems that make them unsuitable due to their global warming potential and ozone depletion capacity, we include them in the discussion for the purpose of comparison with more environmentally benign coolants, such as HFEs.

Figure 10 illustrates the trade-offs in the thermal and electrical properties of the three main classes of coolants employed in embedded cooling. Liquid water is highly desired for heat transfer; however, the risks associated with circuit contact make it undesirable for active contacting embedded cooling. A recent study by Patrick *et al.* [10] demonstrated a proof of concept that employed liquid water for direct chip cooling by efficiently isolating the electrical circuit conformal layers (~1 um thick) of Parylene C. This may, however, pose some serious practical difficulties once deployed on a large scale due to manufacturing challenges.

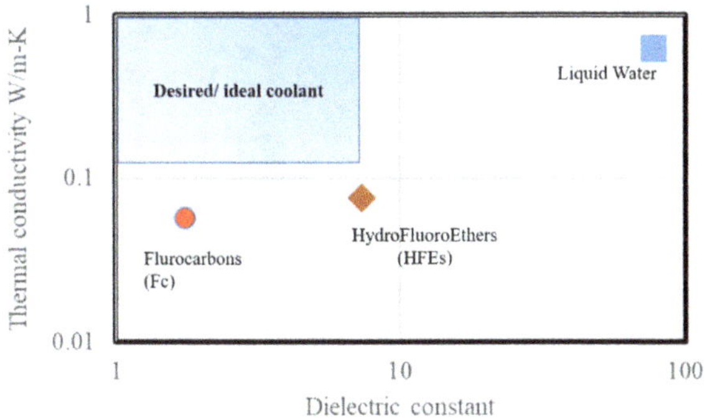

Figure 10. Three classes of coolants preferred for embedded cooling of electronics displaying the trade-off in heat transfer and electrical properties.

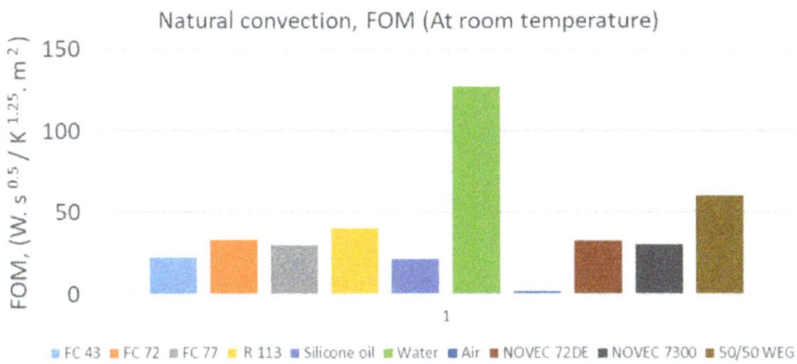

Figure 11. Figures of merit (FOMs) for coolants employed in natural convection.

Saylor *et al.* [34] developed figures of merit (FOMs) to compare the effectiveness of different fluids for different modes of cooling, which we summarize in Figures 11–14.

3.1. *Natural convection*

$$\mathrm{FOM} = \left(\frac{\beta\rho^2 C_p k^3}{\mu}\right)^{0.25} \tag{5}$$

Figure 11 makes it apparent that water is the best coolant for natural convection, whereas HFEs and FCs have significantly lower FOMs.

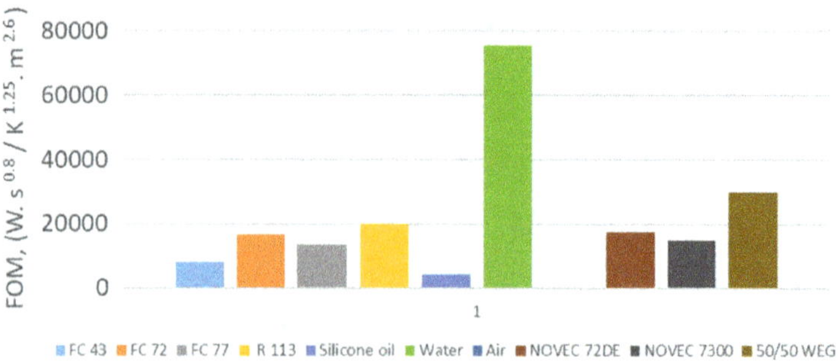

Figure 12. Figures of merit (FOMs) for coolants employed in forced convection (at room temperature).

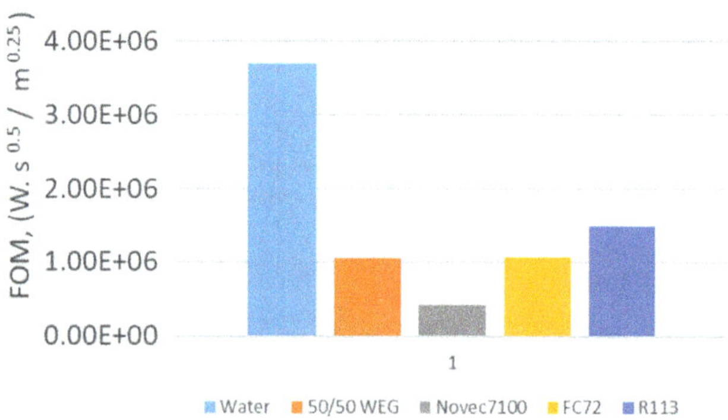

Figure 13. Figures of merit (FOMs) for coolants employed in immersion pool boiling (@1atm).

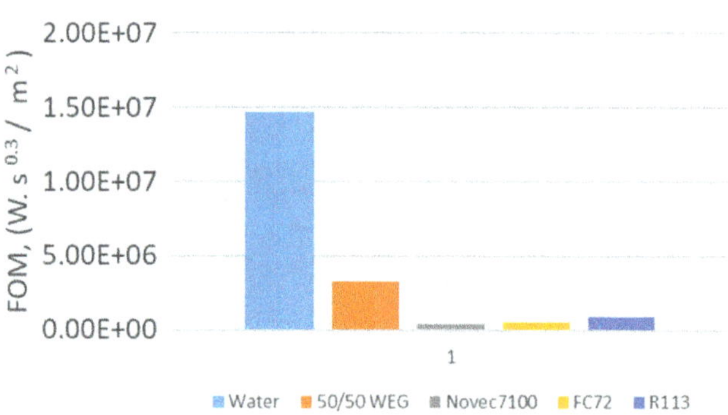

Figure 14. Figures of merit (FOMs) for coolants employed in embedded flow boiling (@1atm).

3.2. *Forced convection*

$$\mathrm{FOM} = \left(\frac{\rho^m C_p^n k^{1-n}}{\mu^{m-n}}\right)^1 \tag{6}$$

3.3. *Pool boiling*

$$\mathrm{FOM} = \sqrt{\rho_g} h_{fg} (\sigma\, g\, (\rho_f - \rho_g))^{0.25} \tag{7}$$

3.4. *Flow boiling*

$$\mathrm{FOM} = \rho_g^{0.239} \times \rho_f^{0.396} \times \sigma^{0.365} \times h_{fg} \tag{8}$$

The FOMs shown in Figures 11–14 calculated from Eqs. (5)–(8) make it evident that water exhibits significantly better thermal performance than FCs, HFEs, and other dielectric liquids in all the heat transfer regimes and is clearly the coolant of choice, though its dielectric properties require it to be used in a non-active circuit contacting solution. The chlorinated fluorocarbon, R-113, is almost always found to offer better heat removal capability than the other FC and HFE liquids. Electrical and chemical considerations have, however, compelled the electronic industry to focus its attention on the inert FC and HFE coolants. Despite their chemical and electrical attractiveness, the inherently poor heat removal capability of FC and HFE coolants offers a significant challenge and opportunity to chemists involved in the development or application of new liquids.

4. Embedded-Cooling Solutions

4.1. *Embedded-cooling solutions based on etched electronic substrate*

Etched electronic technologies can utilize single- or two-phase systems. Various techniques have been explored, including parallel microchannels [35], radial channels with fins [30, 36], impinging jets [37], and nanomembrane evaporation [8]. The first published discussions of etched silicon microchannels date back to the late 1970s and early 1980s [38, 39].

The seminal work by Tuckerman and Pease [39] demonstrated a thermal resistance of 0.09 cm^2K/W over a 1 cm^2 surface area with fins 50 μm in separation and width and 300 μm in height when operating under uniform heat fluxes of up to 790 W/cm^2 and with water as the working fluid (a maximum pressure drop of 31 psi or 213 kPa). The fluid entered and exited manifolds in the etched silicon fin array via holes in the glass cover.

Alternative fluids and designs have been studied since the work of Tuckerman and Pease. IBM demonstrated a two-phase radial microchannel solution using the refrigerant R1234ze [36]. This system had a critical heat flux of 340 W/cm^2 and a low standard deviation between measured temperature values in the chip (4.5°). The thermal resistance just below this critical heat flux was 0.09 cm^2K/W at 120 kPa [30]. Notably, the critical heat flux for this two-phase device was less than the heat flux demonstrated for single-phase water by Tuckerman and Pease, while the thermal resistance was approximately the same. Many cryo-refrigerants have relatively low critical heat fluxes compared to water, owing to their lower heats of vaporization and weaker surface tension. The flow boiling critical heat flux can be increased by increasing the refrigerant flow rate.

Two-phase jet-impingement cooling sprays a fine mist of fluid impinging on a hot surface, which creates a thin fluid film that cools via a combination of convection and evaporation at low superheats ($T_w - T_{sat} < 15°$) and via boiling at higher superheats [37]. The benefits of spray cooling include low superheats, the ability to carry high heat fluxes before exceeding the critical heat flux, and the ability to multiplex many spray nozzles over a single surface. The downsides of spray cooling include relatively high pressure drops across nozzles (e.g., 70–300 kPa, typically), the fouling potential of small nozzle openings (e.g., $\o = 20\mu$m), and diminished performance when non-condensable gases enter the system. The thermal resistances and critical heat fluxes are, respectively, 0.10–0.12 cm^2K/W and 525–945 W/cm^2 for water [40], 0.16–0.21 cm^2K/W and 357–490 W/cm^2 for methanol, 0.45–0.60 cm^2K/W and 65–83 W/cm^2 for FC-72, and 0.49–0.63 cm^2K/W and 79–90 W/cm^2 for FC-87, where the range represents variation due to the operating pressure difference and other conditions (such as nozzle separation to surface, flow rate, nozzle design, and mean spray diameter) [37].

Evaporative cooling with nanoporous membranes (600 nm thick, pore ø < 140 nm) was recently demonstrated to provide device-level thermal resistances and critical heat fluxes of 0.15 cm^2K/W and 144 W/cm^2 for water and 0.055 cm^2K/W and 550 W/cm^2 for pentane [8, 41]. In this study, a network of channels was created to support fluid replenishment in the nanoscale pores, through which evaporative cooling proceeded. It also demonstrated ultra-high heat fluxes through individual pores (up to 2.7 kW/cm^2), which is consistent with the findings of the study on individual nanopores [42]. The reliability of this new kind of device needs further study, as the accumulation of non-volatile contaminants accumulated in the pores degrades performance after a relatively short period of time [41].

Making silicon microchannels with high aspect ratios can be achieved in several ways; for a detailed discussion on the manufacturing techniques, we refer readers to Kandlikar *et al.* [43]. Tuckerman and Pease [39] used anisotropic wet chemical etching (KOH). This process proceeds at an etch rate of about a micron per minute, so the etching time for heat removal devices will be on the order of hours. Moreover, the anisotropic etchants must consider the substrate crystal orientation, which acts as an etch stop. Anisotropic etching can also be done via deep reactive-ion etching (DRIE), which uses alternating cycles of etching and polymeric passivation. Typical DRIE etch rates up to 40 $\mu m/min$ are possible with specialized tools, though 2–8 $\mu m/min$ are more typical. In industries, there is a reluctance to etch onto active devices because of concerns about reduced device yields, so an additional step of bonding onto the active device is considered necessary [36].

4.2. *Additive-manufacturing-based embedded-cooling solutions*

While embedded cooling via silicon etching has demonstrated noteworthy cooling capabilities [30, 36], alternative 3D-printed embedded-cooling strategies have begun to be explored in the literature. The techniques for printing onto electronic substrates to build cooling devices have assumed several architectures to date: selective laser melting using an interlayer alloy [23–26, 109, 110], melting metal onto a physical vapor-deposited film [44–49], and printing thermoelectric cooling devices onto hot spots [15, 16].

Silicon has many advantages as a heat sink material, including the ability to process fine features and the use of direct silicon-to-silicon. However, there are also challenges: the backside of an actual silicon device is unlikely to be micromachined due to poor yield, other materials with higher thermal conductivities than silicon are available, and building high-aspect-ratio features with silicon microfabrication is challenging and time-consuming. This is why a prototype microchannel cooling device is usually made on a separate chip and then solder-bonded to the device [36]. These challenges are motivating factors that have led to recent interest in printed embedded-cooling features (e.g., fins, channels, thermoelectrics).

Direct fabrication of cooling features with metal printing is appealing due to the freeform nature of the manufacturing process. As has been seen in the nascent literature on additively manufactured heat sinks, the relatively unconstrained nature of printing can offer advantages in performance that optimize cooling for hot spots. In terms of electronics cooling, removing the interface to cooling can reduce the significant temperature drops across typical thermal interface materials. Even relatively high-performing solid metal TIMs can have resistances on the order of 0.2 cm^2K/W. When this is installed directly over next-generation microprocessor hotspots, which will be roughly 1,000 W/cm^2, the thermal budget is exceeded just in the first chip-to-lid TIM (TIM 1). Even the highest-performing liquid metal TIMs with datasheets showing day-zero resistances of 0.025 cm^2K/W would consume 25° of the thermal budget just in the chip-to-lid TIM for intense hotspots without spreading, which leaves precious little thermal budget from the lid to the liquid coolant, considering a typical overall thermal budget of 40°C–60°C in the case of chiller-less liquid or two-phase cooling.

One prior research demonstrated that microfins can be laser printed onto a silicon device. The manufacturing process involves printing a low-melting-point brazing alloy that rapidly forms silicide bonds with silicon and subsequently printing a high-thermal-conductivity metal on top [50, 109, 110]. While the brazing bonding layer has a thermal conductivity comparable to that of solder, albeit potentially thinner, subsequent layers can be made of high-thermal-conductivity metals, such as Cu, Ag, or Al. The manufacturing technique of direct metal printing has the potential to create very thin intermetallic-like bonding layers (e.g., titanium silicides of <100 nm) due to the rapid laser processing conditions that limit diffusion times. The printing has been demonstrated in silicon and graphite (Figure 15) [23, 24, 109, 110]. Another method involves depositing a thick metalized interlayer using physical vapor deposition [44–47].

Conventional metal laser printers have resolution limits on the order of 100 μm due to laser sizes and metal powder dimensions. However, research has demonstrated that metals can be printed via selective laser sintering at 1–5 μm resolutions [51, 52]. The materials developed in macro-printing onto silicon could be applied to these finer-resolution printers that deposit thin layers of nanoparticle inks (1.5 μm wet film), dry the inks (500 nm dry film), and then laser expose the desired pattern, where nanoparticle inks are composed of appropriate alloys. The film is then laser sintered with a beam-expanded laser that selectively heats the surface by way of a digital micromirror array. A continuous wave or nanosecond laser could alternatively be used. The substrates in these experiments were aluminum and glass. The ability to build structures as tall as 7 μm with a thickness of 50 μm was demonstrated.

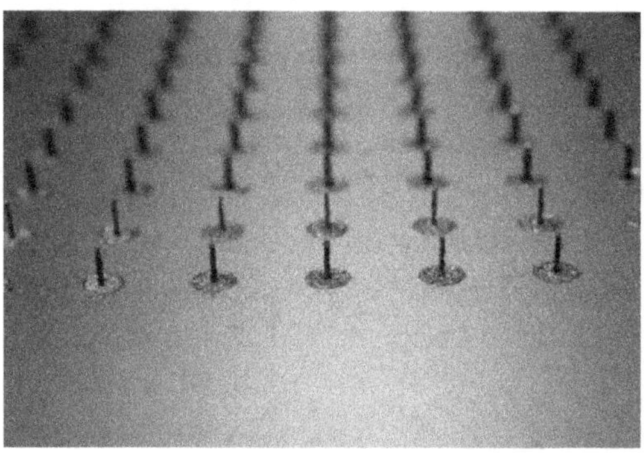

Figure 15. Printing high-aspect-ratio pin fins onto silicon (diameter 100 μm) [23, 24].

While conventional metal microchannel designs [53, 54] could be made via additive manufacturing, they would not capture the freeform features unique to additive manufacturing. Simulations of potential designs made via direct metal printing indicate pin-fin-type structures used in single phase with water can have thermal resistances less than 0.03 K/W (calculated based on maximum die temperature and a die size of 4 cm^2) [25]. This design is not yet fully optimized, and significant reductions in thermal resistance are anticipated in future design iterations. Another application of additive manufacturing can arise from its integration with a manifold [55].

While these technologies are appealing due to their higher performance, customizability, and thermal conductivity than silicon-etched fins, more research is required to demonstrate the practicality of the technology, determine a scalable manufacturing strategy, and demonstrate lifetime reliability.

4.3. *Material selection/processing*

The thermal conductivity of materials commonly used in electronics varies by four orders of magnitude, from a relatively low thermal conductivity of around 0.1 W/m-K for many polymers and glasses to 400 W/m-K for Ag. Exceptional thermal conductors, such as natural diamond and pyrolytic graphite with thermal conductivities of 2,200 W/m-K [56] and 1,700 W/m-K in-plane, respectively, are also of great interest. The use of materials with high thermal conductivities increases the surface area for convection via lateral spreading and lowers the conduction resistances of internal semiconductor components. Material choices for internal features, such as the substrate, dielectric films, and thermal vias/interconnects, are often based on their electrical performance, with thermal performance being a secondary consideration.

The thermal conductivity of electronic substrates varies greatly. Silicon has a thermal conductivity similar to that of aluminum, 140 W/m-K for undoped silicon. The thermal conductivities of silicon and other semiconductors are reduced with doping [57]. GaAs, InP, GaN, Si, and SiC have thermal conductivities of 50, 68, 140, and 450 W/m-K, respectively [22]. Material properties may be well known in bulk; however, the performance of these materials *in situ* can be greatly reduced due to size effects and processing conditions caused by defects, cracks, dislocations, and other microstructural changes. For instance, diamond thin films have thermal conductivities lower than those of natural bulk diamond (30–600 W/m-K). The interface between diamond grains adds an interfacial resistance of 0.0001–0.1 cm^2K/W [58–61], which contributes to thermal conductivity reduction.

While an extreme example, the reduction in thermal performance of thin films requires that material properties be measured experimentally for accurate thermal modeling of embedded-cooling systems. Interfacial resistance can also play an important role at interfaces between materials, especially at the high heat fluxes seen at hotspots. Interfacial resistance exists even on perfectly contacting surfaces that have no interfacial roughness due to the scattering of thermal carriers at the interfaces between the two materials. For detailed discussions of thermal interfacial resistances in general and for various systems, we refer readers to Refs. [58–63] (Figure 16).

Technologies that enhance lateral spreading can include growing electronic substrates with lower thermal conductivities on higher-conductivity substrates and introducing higher-conductivity materials into electronic substrates, such as copper thermal vias and thermal interconnects [4, 20]. More recently, diamond thermal vias and GaN on diamond have even been demonstrated [9, 64, 67, 68, 70]. For a detailed summary of these technologies, we refer readers to Ref. [22].

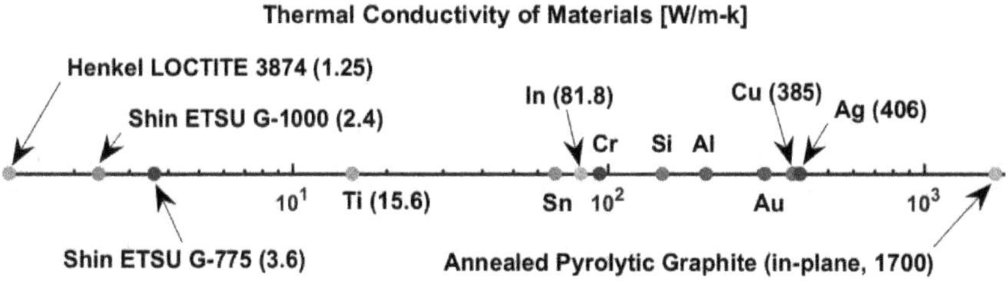

Figure 16. Plot of thermal conductivity for different electronic packaging materials.

5. History and Future of Embedded Cooling

5.1. *IP adoption history of embedded cooling*

Currently, high-performance computing and demanding power electronic applications have increased the need for adopting leading-edge technologies, as these applications prize performance the most and are among the least cost-sensitive. High-performance computing includes supercomputers, servers, mainframes, and advanced workstations. All such systems share the common characteristics of advanced technology, superior performance, fast processing, and large-capacity memory and storage. High performance requires very dense packaging for short signal flight times, which results in high power dissipation, at the processor level as well as at the module and system levels. High-power dissipation often requires complex thermal management schemes, whether air or a liquid is used as the heat transfer medium. The thermal management requirements often introduce additional mechanical and reliability concerns, and in many cases, the reliability requirements for high-performance systems are much more stringent than those for mobile personal computing and other systems. All of these considerations necessitate special considerations in packaging technologies.

Patents related to embedded cooling may be categorized into the following:

- **Passive conductive approaches:** These include the addition of metal layers and thermally plated through holes (PTHs) that are there solely for thermal management. These may be in the device itself or in the substrate that carries the device [12, 71–77]. It is noteworthy that conductive thermal PTHs and vias, as well as conductive copper layers, are commonly used in commercial products at both the device and package levels.
- **Passive vapor chamber and heat pipes:** These are attached directly to the device or are part of the package that is in immediate contact with the device [14, 78]. Passive vapor chambers and heat pipes are very commonly used in air-cooled heat sinks and have become very inexpensive to manufacture. However, such devices are not commonly embedded directly in the device.

- **Forced convection cooling:** Here, the coolant makes direct contact with the device. These would include channels for carrying cooling fluid that are integrated into the device or come into direct contact with the device [50, 52, 65, 67, 79–86]. Forced convection cooling has only been commonly used in heat sinks and cold plates at the package level. Immersion cooling with pool boiling or forced convection boiling has been used in supercomputers (e.g., CRAY), wherein the entire system is immersed in a dielectric fluid.
- **Active structures:** They are embedded into the device, such as thermoelectric cooling systems [87–94].

Embedded cooling, as defined here, would eliminate several thermal resistances but, at the same time, would often introduce new and rarely practiced approaches, such as exposing the devices to direct contact with liquids or two-phase systems. Another major consideration is that embedded solutions would often require changes to the device-manufacturing infrastructure, which is prohibitively expensive. These reasons explain why disclosed and patented solutions in this field are rarely practiced in mainstream manufacturing. Instead, most practices focused on trying to reduce the thermal resistances between the device and the cooling medium without direct contact between them. Early examples of the non-embedded minimization of the internal resistances between the device and the cooling medium (air, liquid) include the air-cooled IBM 4381, the water-cooled IBM Thermal Conduction Module (TCM) series, Hitachi RAM, Mitsubishi HTCM, Honeywell SLIC, NECSX LCM, NTT, VAX 9000, and Fujitsu VP2000. A detailed review of many of those systems is given in Refs. [95, 96], and detailed analyses and designs are found in Refs. [97–108], amongst many others.

5.2. *Future adoption of embedded cooling*

What types of embedded cooling strategies are adopted and when will depend on the pull of the industry (e.g., the need for higher heat fluxes to meet performance targets, requirements for compact package sizes, and closer integration of components) and the capabilities of technologies in solving these problems with the least perceived disruption, risk, and expense. Some combination of passive and active cooling solutions is likely given the current electronic and packaging trends.

Acknowledgments

We acknowledge the support of NSF-1941181 (SNS, BGS), NSF-1846157 (SNS), and SRC CHIRP (BGS, SNS, SR).

References

[1] Embed | Definition of embed by Merriam-Webster. https://www.merriam-webster.com/dictionary/embed (accessed June 24, 2020).

[2] A. Bar-Cohen, J.J. Maurer, A. Sivananthan, Near-junction microfluidic cooling for wide bandgap devices, *MRS Adv.*, 1 (2), 2016, 181–195.

[3] S. Bhopte, S.B. Desu, B. Sammakia, An integrated nano-structured heat spreader for high heat flux electronic systems, in *2010 14th International Heat Transfer Conference*, Washington, DC, USA, January 2010, Vol. 3, pp. 629–635. doi: 10.1115/IHTC14-22918.

[4] Hu Xu, V.F. Pavlidis, G. De Micheli, Analytical heat transfer model for thermal through-silicon vias, in *2011 Design, Automation & Test in Europe*, Grenoble, March 2011, pp. 1–6. doi: 10.1109/DATE.2015763069.

[5] C.R. Kharangate *et al.*, Experimental investigation of embedded micropin-fins for single-phase heat transfer and pressure drop, *J. Electron. Packag.*, 140 (2), June 2018. doi: 10.1115/4039475.

[6] H.Y. Zhang, D. Pinjala, T.N. Wong, K.C. Toh, Y.K. Joshi, Single-phase liquid cooled microchannel heat sink for electronic packages, *Appl. Thermal Eng.*, 25 (10), July 2005, 1472–1487.

[7] K.W. Jung *et al.*, Embedded cooling with 3D manifold for vehicle power electronics application: Single-phase thermal-fluid performance, *Int. J. Heat Mass Transfer*, 130, March 2019, 1108–1119.

[8] D.F. Hanks *et al.*, Nanoporous evaporative device for advanced electronics thermal management, in *Fourteenth Intersociety Conference on Thermal and Thermomechanical Phenomena in Electronic Systems (ITherm)*, May 2014, pp. 290–295. doi: 10.1109/ITHERM.2014.6892295.

[9] A. Bar-Cohen, J.J. Maurer, A. Sivananthan, Near-Junction Microfluidic Cooling for Wide Bandgap Devices, *MRS Adv.*, 1 (2), 2016, 181–195.

[10] P. Birbarah *et al.*, Water immersion cooling of high power density electronics, *Int. J. Heat Mass Transfer*, 147, February. 2020, 118918.

[11] S.R.C. Jr, Immersion cooled high density electronic assembly, US4590538A, May 20, 1986. Available at: https://patents.google.com/patent/US4590538A/en (accessed August 6, 2020).

[12] Z. Stefanoski, Embedded heat pipe in a hybrid cooling system, US7551442B2, June 23, 2009. Available at: https://patents.google.com/patent/US7551442B2/en (accessed August 6, 2020).

[13] Semi-Therm, Celsia, G. Meyer, https://semi-therm.org/wp-content/uploads/2017/04/Heat-Pipe-Vapor-Chamber-Heat-Sink-Design-Guidelines-Thermal-Live-2016.pptx.pdf.

[14] A.S. Conte, Cooling multi-chip modules using embedded heat pipes, US5355942A, October 18, 1994. Available at: https://patents.google.com/patent/US5355942A/en. (accessed August 6, 2020)

[15] H. Zhang, D. Hobbis, G.S. Nolas, S. LeBlanc, Laser additive manufacturing of powdered bismuth telluride, *J. Mater. Res.*, 33 (23), December 2018, 4031–4039.

[16] M.J. Carter, A. El-Desouky, M.A. Andre, P. Bardet, S. LeBlanc, Pulsed laser melting of bismuth telluride thermoelectric materials, *J. Manuf. Proc.*, 43, July 2019, 35–46.

[17] A.S. Fleischer, *Thermal Energy Storage Using Phase Change Materials: Fundamentals and Applications.* Springer, 2015.

[18] M. Schultz *et al.*, Embedded Two-Phase Cooling of Large 3D Compatible Chips With Radial Channels, p. V003T10A007, July 2015, doi: 10.1115/IPACK2015-48348.

[19] E. Pop, S. Sinha, K.E. Goodson, Heat generation and transport in nanometer-scale transistors, *Proceedings of the IEEE*, 94 (5), August 2006, 1587–1601, doi: 10.1109/JPROC.2006.879794.

[20] V. Venkatadri, B. Sammakia, K. Srihari, D. Santos, A review of recent advances in thermal management in three dimensional chip stacks in electronic systems, *J. Electron. Packag*, 133 (4), December 2011, doi: 10.1115/1.4005298.

[21] G. Hariharan *et al.*, Reliability Evaluations on 3D IC Package beyond JEDEC, in *2017 IEEE 67th Electronic Components and Technology Conference (ECTC)*, May 2017, pp. 1517–1522. doi: 10.1109/ECTC.2017.298.

[22] A. Bar-Cohen, J.J. Maurer, D.H. Altman, Embedded cooling for wide bandgap power amplifiers: A review, *J. Electron. Packag.*, 141 (4), December 2019, doi: 10.1115/1.4043404.

[23] A. Azizi, S.N. Schiffres, Laser metal additive manufacturing on graphite, in *Solid Freeform Fabrication Symposium*, Austin, TX, 2018, p. 1.

[24] A. Azizi, M.A. Daeumer, S.N. Schiffres, Additive laser metal deposition onto silicon, *Addit. Manuf.*, January 25, 2019, 390–398.

[25] V. Radmard *et al.*, Direct micro-pin jet impingement cooling for high heat flux applications, presented at the *Thermal Technologies Workshop: Semi-Therm 2020*, Redmond, WA, Submitted.

[26] A. Azizi, M.A. Daeumer, J.C. Simmons, B.G. Sammakia, B.T. Murray, S.N. Schiffres, Additive laser metal deposition onto silicon for enhanced microelectronics cooling, in *2019 IEEE 69th Electronic Components and Technology Conference (ECTC)*, May 2019, pp. 1970–1976. doi: 10.1109/ECTC.2019.00302.

[27] D. Setiadi, H. Liu, Chip having thermal vias and spreaders of cvd diamond, US20100140790A1, June 10, 2010. Available at: https://patents.google.com/patent/US20100140790A1/en. (accessed June 24, 2020).

[28] M.A. Ali, C.W. Peterson, K.M. McNab, Electronic structure having an embedded pyrolytic graphite heat sink material, US6075701A, June 13, 2000. Available at: https://patents.google.com/patent/US6075701A/en (accessed June 24, 2020).

[29] B. Dang *et al.*, Integration and packaging of embedded radial micro-channels for 3d chip cooling, in *2016 IEEE 66th Electronic Components and Technology Conference (ECTC)*, Las Vegas, NV, May 2016, pp. 1271–1277. doi: 10.1109/ECTC.2016.60.

[30] M. Schultz *et al.*, Embedded two-phase cooling of large 3D compatible chips with radial channels, presented at the ASME 2015 International Technical Conference and Exhibition on Packaging and Integration of Electronic and Photonic Microsystems collocated with the ASME 2015 13th International Conference on Nanochannels, Microchannels, and Minichannels, November 2015. doi: 10.1115/IPACK2015-48348.

[31] K.P. Drummond *et al.*, Characterization of hierarchical manifold microchannel heat sink arrays under simultaneous background and hotspot heating conditions, *Int. J. Heat Mass Transfer*, 126, November 2018, pp. 1289–1301.

[32] R.K. Shah, A.L. London, *Laminar Flow Forced Convection in Ducts: A Source Book for Compact Heat Exchanger Analytical Data.* Academic Press, 2014.

[33] L.S. Song, S.V. Au, K.P. Moran. Constriction/Spreading Resistance Model for Electronics Packaging. *Proceedings of the 4th ASME/JSME Thermal Engineering Joint Conference*, Hawaii, March 19–24, 1995, 4, 199–206.

[34] J.R. Saylor, A. Bar-Cohen, Lee Tien-Yu, T.W. Simon, Tong Wei, Wu Pey-Shey, Fluid selection and property effects in single- and two-phase immersion cooling (of electronic components), *IEEE Trans. Compon. Hybrids Manuf. Technol.*, 11 (4), December 1988, 557–565.

[35] V. Narayan, S.-C. Yao, Modeling and optimization of micro-channel heat sinks for the cooling of 3D stacked integrated circuits. *ASME 2011 International Mechanical Engineering Congress and Exposition (IMECE)*, 2012, 999–1011.

[36] B. Dang *et al.*, Integration and packaging of embedded radial micro-channels for 3D chip cooling, in *2016 IEEE 66th Electronic Components and Technology Conference (ECTC)*, May 2016, pp. 1271–1277. doi: 10.1109/ECTC.2016.60.

[37] L. Lin, R. Ponnappan, Heat transfer characteristics of spray cooling in a closed loop, *Int. J. Heat Mass Transfer*, 46 (20), September. 2003, 3737–3746.

[38] W. Anacker, Liquid cooling of integrated circuit chips, *IBM Tech. Disclosure Bulletin*, 20, 1978, 3742–3743.

[39] D.B. Tuckerman, R.F.W. Pease, High-performance heat sinking for VLSI, *IEEE Electron. Dev. Lett.*, 2 (5), May 1981, 126–129.

[40] R.-H. Chen, L.C. Chow, J.E. Navedo, Effects of spray characteristics on critical heat flux in subcooled water spray cooling, *Int. J. Heat Mass Transfer*, 45 (19), September. 2002, 4033–4043.

[41] D.F. Hanks, Z. Lu, J. Sircar, I. Kinefuchi, K.R. Bagnall, T.R. Salamon, D.S. Antao, B. Barabadi, E.N. Wang, High heat flux evaporation of low surface tension liquids from nanoporous membranes, *ACS Appl. Mater. Interfaces*, 12 (6), 2020, 7232–7238.

[42] Y. Li *et al.*, Ultrafast diameter-dependent water evaporation from nanopores, *ACS Nano*, 13 (3), March 2019, 3363–3372.

[43] S.G. Kandlikar, W.J. Grande, Evolution of microchannel flow passages—Thermohydraulic performance and fabrication technology, *Heat Transfer Eng.*, 24 (1), January 2003, 3–17.

[44] M. Conrad, *et al.* Packaging for power semiconductors based on the 3D printing technology selective laser melting, in *2014 16th European Conference on Power Electronics and Applications*; IEEE: Lappeenranta, Finland, 2014, pp. 1–7.

[45] M. Conrad, *et al.* Implementation aspects of on-chip printed micro heat sinks for power semiconductors, in *2015 IEEE Energy Conversion Congress and Exposition (ECCE)*, IEEE: Montreal, QC, Canada, 2015, pp. 5716–5723.

[46] M. Conrad, *et al.* Purpose, potential and realization of chip-attached micro-pin fin heat sinks. *Microelectron. Reliab.* 2015, 55 (9–10), 1992–1996. https://doi.org/10.1016/j.microrel.2015.07.007.

[47] M. Conrad, R.W. De Doncker, M. Schniedenharn, A. Diatlov, Packaging for power semiconductors based on the 3D printing technology selective laser melting, in *2014 16th European Conference on Power Electronics and Applications*, August. 2014, pp. 1–7, doi: 10.1109/EPE.2014.6910965.

[48] N.K. Roy, *et al.*, M. A. A Comprehensive study of the sintering of copper nanoparticles using femtosecond, nanosecond, and continuous wave lasers. *J. Micro Nano-Manuf.* 6 (1), 2017, 010903–010921. https://doi.org/10.1115/1.4038455.

[49] N.K. Roy, *et al.* A novel microscale selective laser sintering (μ-SLS) process for the fabrication of microelectronic parts, *Microsyst Nanoeng.*, 5 (1), 2019, 1–14.

[50] US Patent for Microchannel heat sink Patent (Patent # 7,331,378 issued February 19, 2008) — Justia Patents Search. https://patents.justia.com/patent/7331378 (accessed August 6, 2020).

[51] An index of patents with Donald Tilton listed as inventor. https://radaris.com/f/Donald/Tilton/Inventor (accessed August 6, 2020).

[52] P.S. Andry, E.G. Colgan, L.S. Mok, C.S. Patel, D.E. Seeger, Semiconductor integrated circuit chip packages having integrated microchannel cooling modules, US7230334B2, June 12, 2007. Available at: https://patents.google.com/patent/US7230334B2/en (accessed August 6, 2020).

[53] W. Qu, I. Mudawar, Experimental and numerical study of pressure drop and heat transfer in a single-phase micro-channel heat sink, *Int. J. Heat Mass Transfer*, 45 (12), June 2002, 2549–2565.

[54] P.-S. Lee, S.V. Garimella, D. Liu, Investigation of heat transfer in rectangular microchannels, *Int. J. Heat Mass Transfer*, 48 (9), April 2005, 1688–1704.

[55] C.U. Gonzalez-Valle, S. Samir, B. Ramos-Alvarado, Experimental investigation of the cooling performance of 3-D printed hybrid water-cooled heat sinks, *Appl. Thermal Eng.*, 168, March 2020, 114823.

[56] J.R. Olson, R.O. Pohl, J.W. Vandersande, A. Zoltan, T.R. Anthony, W.F. Banholzer, Thermal conductivity of diamond between 170 and 1200 K and the isotope effect, *Phys. Rev. B*, 47 (22), June 1993, 14850–14856.

[57] A.D. McConnell, S. Uma, K.E. Goodson, Thermal conductivity of doped polysilicon layers, *J. Microelectromech. Syst.*, 10 (3), September 2001, 360–369.

[58] C. Monachon, L. Weber, C. Dames, Thermal boundary conductance: A materials science perspective, *Annu. Rev. Mater. Res.*, 46 (1), 2016, 433–463.

[59] A. Assy, S. Gomès, Heat transfer at nanoscale contacts investigated with scanning thermal microscopy, *Appl. Phys. Lett.*, 107 (4), 2015, 043105.

[60] E.T. Swartz, R.O. Pohl, Thermal boundary resistance, *Rev. Mod. Phys.*, 61 (3), July 1989, 605–668.

[61] P.E. Hopkins, L.M. Phinney, J.R. Serrano, T.E. Beechem, Effects of Surface Roughness and Oxide Layer on the Thermal Boundary Conductance at Aluminum/Silicon Interfaces. *14th International Heat Transfer Conference (IHTC)*, March 2011, 313–319. https://doi.org/10.1115/IHTC14-22268.

[62] S.N. Schiffres, S. Harish, S. Maruyama, J. Shiomi, J.A. Malen, Tunable electrical and thermal transport in ice-templated multilayer graphene nanocomposites through freezing rate control, *ACS Nano*, 7 (12), December 2013, 11183–11189.

[63] S. Majumdar *et al.*, Vibrational mismatch of metal leads controls thermal conductance of self-assembled monolayer junctions, *Nano Lett.*, 15 (5), May 2015, 2985–2991.

[64] D. Setiadi, H. Liu, Chip having thermal vias and spreaders of CVD diamond, US20100140790A1, June 10, 2010. Available at: https://patents.google.com/patent/US20100140790A1/en (accessed August 6, 2020).

[65] G.M. Chrysler, R. Prasher, Integrated micro channels and manifold/plenum using separate silicon or low-cost polycrystalline silicon, US6992382B2, January 31, 2006. Available at: https://patents.google.com/patent/US6992382B2/en (accessed August 6, 2020).

[66] R.J. Bezama, E.G. Colgan, J.H. Magerlein, R.R. Schmidt, Apparatus and methods for microchannel cooling of semiconductor integrated circuit packages, US7139172B2, November 21, 2006. Available at: https://patents.google.com/patent/US7139172B2/en (accessed August 6, 2020).

[67] M.S. Hodes *et al.*, Techniques for microchannel cooling, US7204298B2, April 17, 2007. Available at: https://patents.google.com/patent/US7204298B2/en (accessed August 6, 2020).

[68] D.H. Altman, A. Gupta, M. Tyhach, Development of a diamond microfluidics-based intra-chip cooling technology for GaN, *ASME International Technical Conference and Exhibition on Packaging and Integration of Electronic and Photonic Microsystems*, July 2015, 1–7.

[69] G. Campbell *et al.*, Advanced cooling designs for GaN-on-diamond MMICs, in *International Electronic Packaging Technical Conference and Exhibition*, 2015, vol. 56901, p. V003T04A007.

[70] K.P. Drummond *et al.*, Evaporative intrachip hotspot cooling with a hierarchical manifold microchannel heat sink array, in *2016 15th IEEE Intersociety Conference on Thermal and Thermomechanical Phenomena in Electronic Systems (ITherm)*, 2016, pp. 307–315.

[71] P.A. Reddy, V.A. Jairazbhoy, R.E.B. Jr, Embedded metal planes for thermal management, US5792677A, August 11, 1998. Available at: https://patents.google.com/patent/US5792677A/en (accessed August 6, 2020).

[72] Oleg Naigertsik Inventions, Patents and Patent Applications — Justia Patents Search. https://patents.justia.com/inventor/oleg-naigertsik (accessed August 6, 2020).

[73] M.J. Glickman, Nano-copper solder for filling thermal vias, US20150114707A1, April 30, 2015. Available at: https://patents.google.com/patent/US20150114707A1/en (accessed August 6, 2020).

[74] P. Loney, R.M. Young, D.R. Queen, A.A. Hathaway, J.X. Przybysz, Cryogenic integrated circuit having a heat sink coupled to separate ground planes through differently sized thermal vias, US10229864B1, March 12, 2019. Available at: https://patents.google.com/patent/US10229864B1/en (accessed August 6, 2020).

[75] P. Loney, R.M. Young, D.R. Queen, A.A. Hathaway, J.X. Przybysz, Multiple ground plane thermal sink, US20190157184A1, May 23, 2019. Available at: https://patents.google.com/patent/US20190157184A1/en (accessed August 6, 2020).

[76] P.A. Loney, A.A. Hathaway, D.R. Queen, J.X. Przybysz, R.M. Young, Thermally isolated ground planes with a superconducting electrical coupler, US10629535B2, April 21, 2020. Available at: https://patents.google.com/patent/US10629535B2/en (accessed August 6, 2020).

[77] J. David, Lima Inventions, Patents and Patent Applications — Justia Patents Search. https://patents.justia.com/inventor/david-j-lima (accessed August 6, 2020).

[78] T.J. Chainer *et al.*, Cooled electronic system with liquid-cooled cold plate and thermal spreader coupled to electronic component, US9307674B2, April 05, 2016. Available at: https://patents.google.com/patent/US9307674B2/en (accessed August 6, 2020).

[79] US Patent for Microchannel cooling using aviation fuels for airborne electronics Patent (Patent # 5,692,558 issued December 2, 1997) — Justia Patents Search. https://patents.justia.com/patent/5692558 (accessed August 6, 2020).

[80] A.H. Monfarad, J.L. Yang, Embedded microchannel cooling package for a central processor unit, US7515415B2, April 7, 2009. Available at: https://patents.google.com/patent/US7515415B2/en (accessed August 6, 2020).

[81] US Patent for Silicon based microchannel cooling and electrical package Patent (Patent # 8,581,392 issued November 12, 2013) — Justia Patents Search. https://patents.justia.com/patent/8581392 (accessed August 6, 2020).
[82] W.J. Benett, R.J. Beach, D.R. Ciarlo, Monolithic microchannel heatsink, US5548605A, August 20, 1996. Available at: https://patents.google.com/patent/US5548605A/en (accessed August 6, 2020).
[83] P.S. Andry, E.G. Colgan, L.S. Mok, C.S. Patel, D.E. Seeger, Electronic module comprising memory and integrated circuit processor chips formed on a microchannel cooling device, US7888786B2, February 15, 2011. Available at: https://patents.google.com/patent/US7888786B2/en (accessed August 6, 2020).
[84] C.H. Jun, C.A. Choi, Y.T. Kim, Method of fabricating microchannel array structure embedded in silicon substrate, US6582987B2, June 24, 2003. Available at: https://patents.google.com/patent/US6582987B2/en (accessed August 6, 2020).
[85] A. Heydari, J.L. Yang, Direct contact cooling liquid embedded package for a central processor unit, US7289326B2, October 30, 2007. Available at: https://patents.google.com/patent/US7289326B2/en (accessed August 6, 2020).
[86] A. Kulik, A. Roemer, Apparatus and method for embedded fluid cooling in printed circuit boards, US6665185B1, December 16, 2003. Available at: https://patents.google.com/patent/US6665185B1/en (accessed August 6, 2020).
[87] US Patent for Integrated thermoelectric cooling Patent (Patent # 5,032,897 issued July 16, 1991) — Justia Patents Search. https://patents.justia.com/patent/5032897 (accessed August 6, 2020).
[88] R. Bhatia, Package with integrated thermoelectric module for cooling of integrated circuits, US6094919A, August 1, 2000. Available at: https://patents.google.com/patent/US6094919A/en (accessed August 6, 2020).
[89] R.C. Chu, M.J.E. Jr, R.E. Simons, Electronic module with integrated programmable thermoelectric cooling assembly and method of fabrication, US6548894B2, April 15, 2003. Available at: https://patents.google.com/patent/US6548894B2/en (accessed August 6, 2020).
[90] Z. Ren, Z. Yu, J.C. Kim, J. Lee, TSV-integrated thermoelectric cooling by holey silicon for hot spot thermal management, *Nanotechnology*, 30 (3), November 2018, 035201.
[91] G. Sui, J. Chen, H. Ni, Y. Ma, X. Gao, N. Wang, Improvement of LED performance with an integrated thermoelectric cooling package, *IEEE Access*, 8, 2020, 116535–116543.
[92] R. Mittal, H.J. Park, Y.H. Kang, Systems and methods for reducing leakage power of a system on chip with integrated thermoelectric cooling, US10101756B2, October 16, 2018. Available at: https://patents.google.com/patent/US10101756B2/en (accessed August 6, 2020).

[93] W. Huang, Integrated thermoelectric cooler for three-dimensional stacked Dram and temperature-inverted cores, US10553258B2, February 4, 2020. Available at: https://patents.google.com/patent/US10553258B2/en (accessed August 6, 2020).

[94] A.G. Fedorov, C. Green, Y. Joshi, Devices including composite thermal capacitors, US8378453B2, February 19, 2013. Available at: https://patents.google.com/patent/US8378453B2/en?oq=52.Fedorov%2c +Andrei+G.%2c+Craig+Green%2c+and+Yogendra+Joshi.+%22Devices +including+composite+thermal+capacitors.%22+U.S.+Patent+8%2c378 %2c453%2c+issued+February+19%2c+2013 (accessed August 6, 2020).

[95] A. Bar-Cohen, An addendum and correction to thermal management of air — and liquid — cooled multi — chip modules, *IEEE Trans. Compon. Hybrids Manuf. Technol.*, 11 (2), 1988, 333–334.

[96] A. Bar-Cohen, Thermal management of air- and liquid-cooled multichip modules, *IEEE Trans. Compon. Hybrids Manuf. Technol.*, 10 (2), June 1987, 159–175.

[97] Allen, E., Donohue, B. and Pei, H., Cooling the VAX 8000 family of computers: An overview, *Proc. of the Technical Conf., Seventh Annual Intl. Electronic Packaging Society Conf.*, Boston, MA, vol. 2, 1987, pp. 712–717.

[98] R.C. Chu, U.P. Hwang, R.E. Simons, Conduction cooling for an LSI package: A one-dimensional approach, *IBM J. Res. Develop.*, 26 (1), January 1982, 45–54.

[99] Y. Emoto, M. Tsuchiya, S. Ogiwara, T. Sasaki, F. Kobayashi, K. Otsuka, A high performance package approach on 1st level packaging for main frame, *Proc. of the 36th Electronic Components Conf., Organized by the IEEE Components, Hybrids, and Manufacturing Technology*, Seattle, WA, May 5–7, 1986, pp. 564–570.

[100] S. Hopper, D.L. Edwards, S.P. Young, The thermal management of IBM's ES/9000 advanced thermal conduction module, *Proc. of the 42nd Electronic Components and Technogy Conf., Organized by IEEE Soc.* on Components, Hybrids and Manufacturing Technology, San Diego, CA, May 18–20, 1992, pp. 997–1001.

[101] K. Kaneko, K. Seyama, M. Suzuki, LSI packaging and cooling technologies for Fujitsu VP2000 series, *Fujitsu Sci. Technol. J.*, 41 (1), 1990, 12–19.

[102] B. Kalita, W. English, Cooling the VAX 8600 processor, *Digital Tech. J.*, 1(1), pp. 60–70. August 1985.

[103] F. Kobayashi, A. Anzai, M. Yamada, A. Takahashi, S. Yamazaki, G. Toda, Packaging technologies for the ultrahigh-speed processor Hitachi M-680 H/M-682 H, *Proc. of the 36th Electronic Components Conf., Organized by the IEEE Components, Hybrids, and Manufacturing Technology*, Seattle, WA, May 5–7, 1986, pp. 571–577.

[104] M. McPhee, T.S. O'Toole, M. Yedvabny, Cooling the VAX 9000, *Electro/90, Conference Record*, Boston, MA, May 9–11, 1990, pp. 288–292.

[105] K. Okutani, K. Otsuka, K. Sahara, K. Satoh, Packaging design of a SiC ceramic multi-chip RAM module, *Proc. of the Technical Conf., Fourth Annual Intl. Electronic Packaging Society Conf.*, Baltimore, MD, October 29–31, 1984, pp. 299–304.

[106] T. Mizuno, M. Okano, Y. Matsuo, T. Watari, Cooling technology for the NEC SX supercomputers, *Proc. of the Intl. Symposium on Cooling Technology for Electronic Equipment*, Hawaii, 1987, pp. 110–125.

[107] T. Watari, H. Murano, Technology for the NEC SX Supercomputer, *IEEE Trans. Compon. Hybrids Manuf. Technol.*, 8 (4), December 1985, 462–467.

[108] R.E. Simons, The evolution of IBM high performance cooling technology, in *Proceedings of 1995 IEEE/CPMT 11th Semiconductor Thermal Measurement and Management Symposium (SEMI-THERM)*, February 1995, pp. 102–114.

[109] S.N. Schiffres, A. Azizi. Additive manufacturing processes and additively manufactured products. United States US20200047288A1, filed August 9, 2019, and issued February 13, 2020. https://patents.google.com/patent/US20200047288A1/en.

[110] S.N. Scott, A. Azizi. Additive manufacturing processes and additively manufactured products. United States US20200049415A1, filed August 9, 2019, and issued February 13, 2020. https://patents.google.com/patent/US20200049415A1/en.

https://doi.org/10.1142/9789811279379_0011

Chapter 11

Co-Design of Thermal and Electromagnetics for Development of Light and Ultrahigh Efficiency Electric Motors

Ziaur Rahman[*,†], Michael Ohadi[†,‡,§], and David Deisenroth[‡]

[*]*Booz Allen Hamilton, Washington, DC, USA*
[†]*Advanced Research Project Agency — Energy, Department of Energy, Washington, DC, USA*
[‡]*Department of Mechanical Engineering, University of Maryland, College Park, MD, USA*
[§]*michael.ohadi@hq.doe.gov*

Abstract

Enhanced cooling, coupled with novel designs and packaging of semiconductors, has revolutionized communications, computing, lighting, and electric power conversion. It is time for a similar revolution that will unleash the potential of electrified propulsion technologies to drive improvements in fuel-to-propulsion efficiency, emission reduction, and increased power and torque densities, which are essential for the decarbonization and efficiency enhancement of aviation and other sectors. To improve the cooling of emerging electrical machines and realize performance and cost metrics of next-generation electric motors, it is crucial to enable electromagnetic and thermo-mechanical co-design through innovative design topologies, materials, and manufacturing techniques. This chapter focuses on the most recent progress in thermal management of electric motors, with particular focus on the future electrification of aviation propulsion, where ultrahigh-efficiency and high-specific-power (kW/kg) electric motors and power electronics play a key role. The chapter also includes a co-design-based simulation case study that analyzes the comparative impacts of selected technological areas, which, among others, can enable cost-effective efficiency enhancements and weight reductions and address the limitations of existing materials and cooling technologies.

Keywords: Embedded cooling of electric motors, thermal and electromagnetic codesign, inter-/intra-winding cooling, advanced potting materials

1. Introduction

An electric motor is an energy conversion device that converts electrical input (e.g., voltage and current) to mechanical output (e.g., torque and rotation). It operates through the interaction between the electric current in a wire winding and a motor's magnetic field. Typically, the winding sits in the slots of the stationary part of the motor, known as the stator, while the magnetic field is generated from magnets that sit on the rotating part of the motor, known as the rotor. The torque is generated through the interaction between the stator current and the rotor magnetic field and provides rotational motion to the shaft of the motor. Figure 1 illustrates the main components (also known as the "active components") of a typical permanent-magnet-type motor structure. As the winding spirals through the stator slots, some portions of its total length hang outside the stator and do not contribute to the production of torque. Those portions of the winding are known as end windings. The total mass of these active components is often termed as the active mass of the motor and is taken into account in the calculation of the active power (or torque) density of a motor. Besides these, there are other essential components of the motor, such as housing and bearings, that provide structural and functional support. These components, although

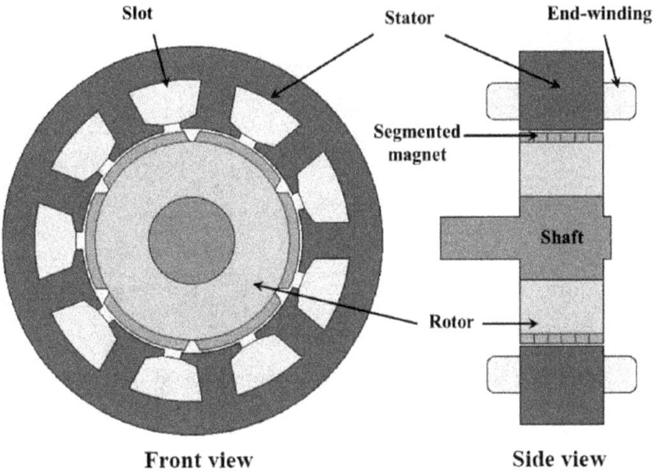

Figure 1. The front and side views of a typical permanent-magnet motor structure showing the "active" components.

not considered in the active mass calculation, contribute to the total motor mass.

Light and ultraefficient motors are needed in many industrial sectors, particularly in the next-generation electric/hybrid electric propulsion systems for aviation. A possible electric propulsion architecture is schematically shown in Figure 2, in which the all-electric powertrain is represented by the green box. It comprises high-efficiency, high-specific-power electric motor(s) with, if needed, a torque amplifier (e.g., a gearbox), its power electronics and controller, and a thermal management system (TMS) for both the power electronics and the electric motor(s). The electric power generated will drive a set of next-generation, high-efficiency propulsors (at least two, or more in the case of distributed propulsion), each driven by its own all-electric powertrain. As shown in Figure 2, the motor is run by an electronic device, commonly known as the motor drive, that harnesses and controls the electrical energy sent to the electric motor using power electronics and associated control logic. The drive can feed electricity into the motor in varying amounts and at varying frequencies, thereby enabling control of the motor's speed and torque. The electric motor and its drive also have a TMS to dissipate the

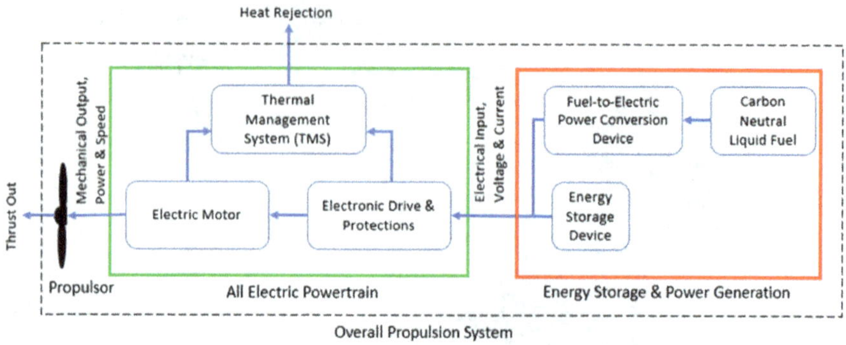

Figure 2. Block diagram of overall electric propulsion system: The electric powertrain system and its individual components (the green box) are the focus of this chapter.

heat generated from the associated conversion losses. The TMS for the electric motor and the drive can be a common, shared unit or can be two separate and independently operated TMS units. More often than not, applications demand complete powertrain solutions of high efficiency and high specific power rather than the optimization of individual components.

In the power electronics area, advances in wide-bandgap semiconductors are bringing a fully packaged motor drive unit with a high specific power of 20 kW/kg and a high efficiency of 98.5% within reach [1]. Wide-bandgap semiconductors, such as SiC and GaN, have comparatively higher power densities than more traditional Si devices. The benefits of wide-bandgap devices also include higher switching speeds and higher temperature operation that somewhat eases the need for highly effective thermal management [2]. Electric machines, on the other hand, suffer from relatively low power densities compared to their respective drives. The three key inhibitors, among others, are as follows: (1) the soft magnetic steel used in the stator and rotor saturates around 1.8–2.2 T; (2) materials used for electrically insulating the wires from the stator have very low thermal conductivities (~1 W/mK), thus making heat rejection difficult; and (3) a lack of sufficient innovation in the winding shapes and their associated thermal management schemes to support the boosting of current density in the windings without affecting the lifecycle, durability, and reliability of the motor. The best-in-class commercially available electric motor today is believed to have a continuous power density of 5 kW/kg, with a demonstrated peak power output of 10 kW/kg for about 20 s [3–5], and this is often without factoring in the entire weight of the supporting TMS for the motor.

To achieve high power density and efficiency simultaneously, the two most commonly considered approaches are (i) to reduce motor losses and (ii) to improve the heat rejection from the motor. Motor losses can be divided into five major categories: core losses, windage and friction losses, stator resistance (I^2R) losses, rotor resistance (I^2R) heating losses, and stray load losses. Figure 3 shows the typical fraction of losses in a four-pole, 50 Hz induction motor (IM) of different power ratings [6]. The absolute dominance of I^2R losses (the sum of stator and rotor losses) is evident in small (<1 HP) to medium (<500 HP) IMs. Core losses are load-independent "fixed"

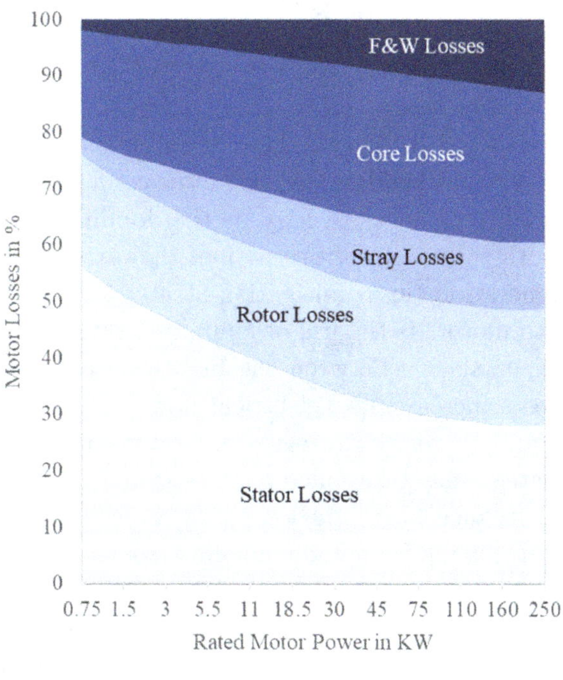

Figure 3. Typical fraction of losses in 50 Hz, four-pole IMs.
Source: Adopted from Ref. 6.

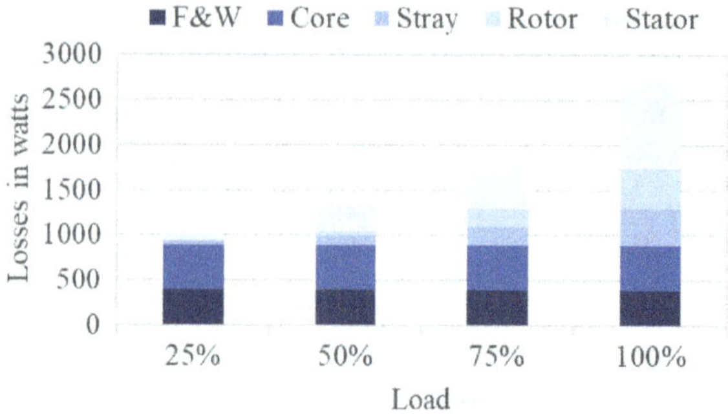

Figure 4. Losses vs. load for a standard IM.
Source: Adopted from Ref. 7.

losses that occur whenever the motor is energized and, therefore, become significant at partial load conditions, as seen in Figure 4 [7]. In modern permanent-magnet (PM) motors, the largest contributor to heat generation is also the copper stator windings. For example, in the optimized air-cooled motors for the NASA SCEPTOR program, the copper (resistive) losses contributed 43% of the total thermal load, while the iron losses (primarily due to Joule heating by eddy currents) contributed 37%, and the PM rotor load contributed the balance of 20% [8]. Depending on the motor configuration and operating conditions, copper losses may be the dominant source of heat generation, with values upward of 64% of the total [9].

A heat generation (loss) mitigation strategy that has long been integrated into motors is the use of laminated rotor and stator iron. The electrical insulation between the high magnetic permeance silicon steel sheets increases the electrical flow path resistivity within the material and therefore decreases eddy currents and iron losses. The same strategy may be applied to PMs segmented along the rotational axis to reduce magnetic losses, as seen in the side view in Figure 1 [10]. It should be noted, though, that an effective electrical insulation is often also a good thermal insulation, which limits the heat dissipation of the generated heat load. The bonding material that holds permanent magnets in place also has thermal resistance, but in most cases, heat generation in the PMs is less than that in copper or iron. In the past, significant reductions in motor losses have been achieved by using new materials technologies. Low-loss magnetic circuit steel materials, high-energy-density PM materials, and the selective and increased use of copper and aluminum conductors have collectively provided low-loss excitation for electric machines. However, reducing losses further has proven difficult and will require a combination of advanced manufacturing technologies (e.g., additive manufacturing, AM) enabling novel design topologies, advanced cooling techniques, and co-design of thermal and electromagnetic components.

For the second approach of improving the heat rejection from the motor, opportunities exist in the development of integrated, low-thermal-resistance cooling techniques. Design innovations for such systems should particularly be aimed at achieving compact, conformal systems with high heat removal rates while keeping the associated fluid coolant pumping power low. In most cases, cooling technologies for motors and their associated drives/power electronics have fallen behind the rapid progress in cooling technologies that have been widely adopted in microelectronics and, to some extent, in selected power electronics [11, 12]. Figure 5 provides a graphical comparison of traditional cooling methods that are used in electric

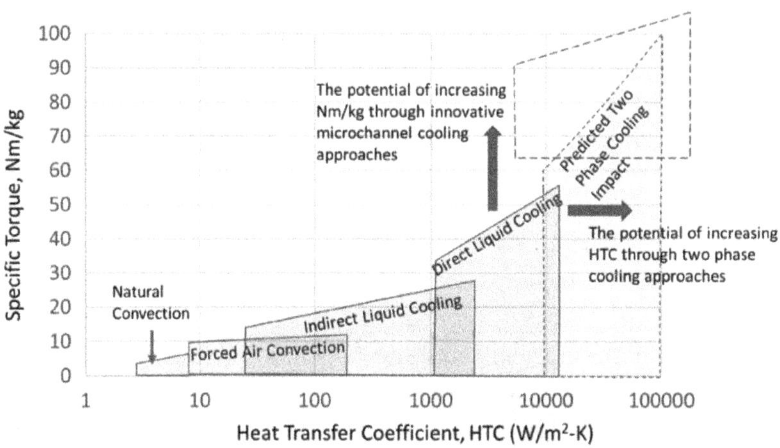

Figure 5. A figurative depiction of the influence of different cooling methods and possible pathways for further increasing specific torque density of an electric motor.

motors [13] with the respective achievable specific torque density and a possible pathway of how innovative cooling approaches, such as two-phase and microchannel cooling, can increase the specific torque density of an electric motor. The trapezoidal box envisioned at the top of the figure can be considered a potential "white space" for impactful research and innovations. However, one must note that increasing the heat transfer coefficient (HTC) must accompany co-design approaches that prevent saturation of the electromagnetic field, while efficient cooling paves the way for improved current density.

The remaining contents of this chapter are organized as follows: A collection of established advanced indirect and direct liquid-cooling practices for high-power-density electric motors is discussed in Section 2. Section 3 describes a few more experimental direct liquid-cooling concepts that have been applied in research environments to further push the envelope of electric motor performance. Section 4 discusses various working fluids that are used in pumped two-phase cooling for electronics and their potential for use in electric motor cooling. Section 5 includes a thermal and electromagnetic co-design-based simulation study that analyzes the comparative impacts of selected technological areas, which, among other benefits, can enable cost-effective efficiency enhancements and weight reductions while addressing the limitations of existing materials and cooling technologies, followed by a summary and conclusion.

2. Established Advanced Cooling Approaches

The focus of this section is on the approaches that have been developed to extract heat from high-power-density motors. Inadequate cooling in the presence of heat generation causes a detrimental temperature rise. Excessive temperatures can cause catastrophic failure in polymer electrical insulation and bonding materials due to loss of electrical and mechanical integrity, ultimately leading to motor failure. Therefore, the temperature threshold of polymer components must not be exceeded.

In addition to the polymer temperature threshold, temperature cycling can lead to fatigue in a variety of motor parts due to mechanical stresses that result from cyclical thermal expansion. Fatigue is exacerbated by the magnitude of the temperature cycle, which drives the need for reduced peak temperatures. Fatigue failure is of particular concern in PM motors because the coefficient of thermal expansion of the magnetic materials is about double that of silicon steel, leading to high cyclical stresses with temperature cycles. PMs are also susceptible to demagnetization at high temperatures, leading to a loss of functionality.

Excessive temperatures can lead to decreased efficiency of electric motors. Heat generation by Joule heating increases due to temperature rises, in part because the electrical resistivity of copper also increases with temperature. Finally, the performance of PMs may decrease with elevated temperatures due to reduced remanence and coercivity.

2.1. *Water jacket cooling*

This falls into the category of indirect liquid cooling and is widely used in electric propulsion applications [10]. Among the highest performance commercial motors are those that power racing aircraft. One example is a motor that can sustain power densities of more than 5 kW/kg and sprint with power densities up to 12.1 kW/kg [5]. The peak power of that 14 kg motor is 170 kW, and the peak torque density is 17.9 Nm/kg. Heat generation in this axial flux PM motor is handled by a water jacket surrounding the stator.

In axial flux motor architectures, a large area of the high-thermal-conductivity stator windings is exposed around the perimeter of the stator, allowing for excellent heat exchange via direct contact with the water jacket. In contrast, the majority of the stator windings in radial flux motors generally cannot be put in direct contact with a circumferential water jacket because of the low-thermal-conductivity steel stator body that surrounds the windings around the exterior circumference of the stator [14].

In any case, water jacket cooling may be configured with circumferential flow paths, helical ducts, axial serpentine channels, or circumferential serpentine channels. Each of these flow designs may increase the water jacket's performance when configured appropriately. Water jacket cooling is a robust design that has achieved quite high performance; however, further improvements to motor cooling are certainly possible, which are discussed in the following.

2.2. *Oil bath cooling*

When combined with water jacket cooling, oil bath cooling increases the heat dissipation ability of the cooling system beyond air cooling and water jacket cooling. It falls in between the categories of indirect and direct liquid cooling. With oil bath cooling, oil is contained within the motor housing and exposed to the stator windings, stator iron, and sometimes the rotor as well. The rotor may be partially or fully submerged in oil, and this leads to two potential issues. First, fluid in the air gap may reduce electromagnetic efficiency. Second, the stirring of the oil by the rotor leads to viscous losses that both reduce available shaft power and cause additional heating of the cooling fluid. An additional requirement for oil bath cooling is that the fluid must be chemically stable and non-flammable and have high magnetic permeability and high electrical resistivity due to the conditions within the motor housing [13]. Automatic transmission fluid (ATF) is a common choice for this application. Passive or active pumping may be used to transport the ATF to an external heat exchanger.

Production traction motors have used partial filling of the motor housing with oil. In this configuration, oil becomes entrained with the rotor and may be sprayed about within the enclosure, ultimately generating an enhanced cooling flow similar to jet-impingement or non-evaporative spray cooling [13]. The motor enclosure may be entirely submerged due to other considerations [9]. Although oil bath cooling is an improvement over air cooling and commercial motors with water jacket cooling, its integration with the electric motor can be complex.

2.3. *Heat pipe cooling*

Portable electronics increasingly implement “heat pipes” to cool high-heat-flux devices, such as processors. A capillary wick is generally used to circulate the flow, transporting liquid from the condenser back to the evaporator with surface tension as the driving force. "Pulsating heat pipes" are a rapidly growing area of study because these devices are not limited by relatively weak capillary forces driving standard heat pipes but can still utilize passive (pumped by internal fluid forces) phase-change cooling. In pulsating heat pipes, the flow is driven by differential bubble expansion and contraction in the evaporator and condenser sections, resulting in flow velocities much higher than those in capillary pumping [15]. Water or alcohol are popular working fluids for both pulsating and capillary wick heat pipes. A recent study used micropillar wicks for passive, capillary-pumped evaporation of water, achieving a peak semiconductor chip-scale heat flux of 45 W/cm^2 at 19°C superheat before becoming limited by the efficacy of capillary pumping [16]. These passive systems may produce HTCs comparable to pool boiling, on the order of 10,000 W/m^2K, which are suitable for many modern applications.

Heat pipe cooling has been applied to electric motors for many decades [17]. It has been shown that heat pipes integrated into the external fins of a motor can decrease the external temperature of the motor from 102°C to 68°C, reducing the thermal resistance from 0.5°C/W to 0.28°C/W, with a heat load of 150 W [18]. Designs for integrating capillary heat pipes into stator windings have also been studied and patented [19]. Higher power-output and power-density motors require rotor cooling as well [20]. As a result, rotating capillaries [21, 22] and rotating pulsating heat pipes [23] have been developed and applied to electric motor cooling. Patents have even been granted for the implementation of heat pipes into the traction motors of modern automobiles [24].

However, as motors become more power dense, the need for higher cooling densities than can be provided by heat pipes is increasing. The direct liquid-cooling approach brings the fluid as close as possible to the heat generation, increasing the manageable heat fluxes and cooling effectiveness. The following section of this chapter focuses on experimental pumped direct liquid-cooling systems.

3. Direct Liquid-Cooling Concepts

The following discussion describes direct liquid cooling concepts that have been applied in research environments to push the envelope of electric motor performance. As discussed previously, the majority of the heat production generally occurs in the stator windings, with the rotor and stator being secondary sources of heat generation. This section describes improvements to the thermal performance of the stator iron as well as several approaches to direct winding liquid cooling, including inter-winding cooling (between windings) and intra-winding cooling (within windings).

3.1. *Stator iron thermal enhancement*

Pyrhonen *et al.* [25] experimentally studied an intra-iron (within stator iron) cooling improvement in an axial flux, dual stator, single rotor configuration. Starting with a commercially available motor, copper rods were embedded in the stator iron to improve heat conductivity from the stator iron to the cooling water heat exchanger. The water heat exchanger was located on the opposite side of the stators from the rotor, as illustrated in Figure 6. This approach to improved thermal management is significant because the high-thermal-conductivity copper bars provided heat flow paths perpendicular to the thermally resistive electrical insulation between the

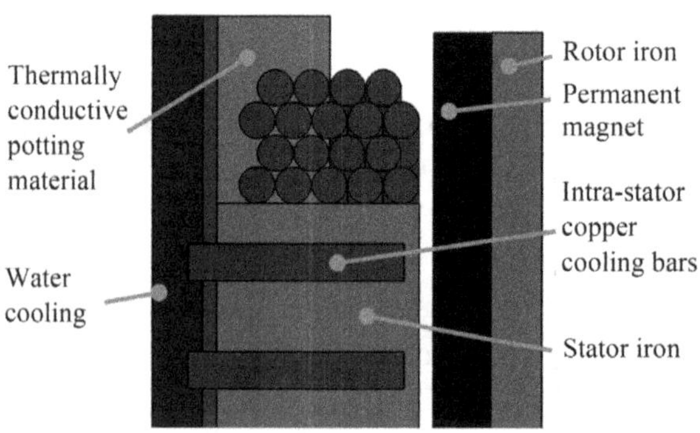

Figure 6. Illustration of intra-iron copper cooling paths and thermally conductive end winding potting material in an axial flux, permanent magnet motor.

stator iron laminates. The copper heat conductors provided a heat flow path from deep within the stator directly into the water flow, where they protruded as pin fins. In addition to the copper heat conductors, aluminum-oxide-based Ceramacast 675-N potting material with a thermal conductivity of 100 W/mK was added between the stator end windings and the water jacket, further improving the thermal performance of the stator.

The study by Pyrhonen *et al.* [25] demonstrated the individual thermal enhancement of one copper bar per stator tooth, three conductive bars per stator tooth, and the potting material on the end windings. A single copper bar per stator tooth reduced the winding temperatures by about 15°C, while three copper bars reduced the winding temperatures by about an additional 10°C. Under the same conditions, the potting material had the effect of cooling the windings by less than 2°C. Ultimately, the improved motor could operate continuously at 75 kW power output with a temperature of less than 125°C in the stator slot, while in the commercial off-the-shelf configuration, it exceeded the temperature limit of 130°C after 70 min of operation at the same power output.

The study by Pyrhonen *et al.* [25] illustrated an important co-design consideration for intra-iron improvements. It was found that a detriment of a 3% increase in stator iron losses was caused by a decreased iron area due to the copper cooling bars occupying that space. Therefore, the addition of copper bars caused a decrease in the electromagnetic efficiency of the stator for an improvement in the overall power density of the motor within the temperature limit of its insulating materials. Thus, this study provides an example of the sometimes-conflicting interaction of thermal and electromagnetic system performance. This provides further motivation for accurate, combined thermal and electromagnetic modeling for the further development of high power- and torque-density motors.

3.2. *Inter-winding cooling*

An effective approach to inter-winding cooling was studied by Rhebergen *et al.* [26]. The approach brought liquid cooling channels into direct contact with the outer perimeter of stator windings (relative

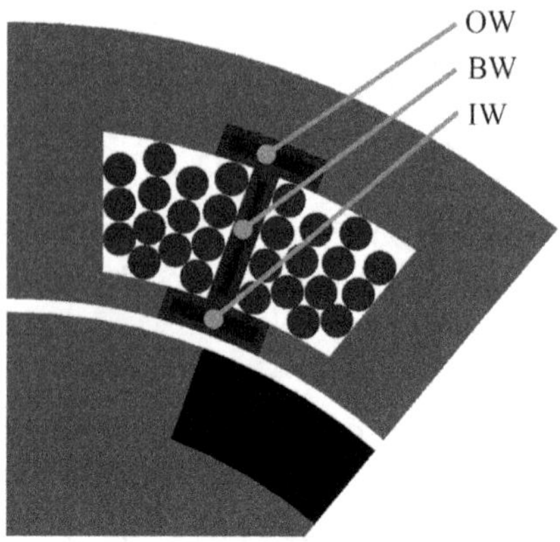

Figure 7. Illustration of three heat exchanger positions in radial flux motor inter-winding positions demonstrated in the literature: outside of windings (OW), between windings (BW), and interior of windings (IW).

to the direction of winding) by placing the channels in the interior of the winding (relative to the stator radius) in the gap between stator teeth at the air gaps. This liquid cooling channel design is illustrated in Figure 7, along with the interior of windings (IW) position.

In the design, the channels ran in a serpentine path, cooling several windings in one fluid pass from the external heat exchanger to the winding. The cooling channels were modeled as a thermally conductive polymer in order to have a relative permeability of near unity. Contact resistance was reduced by bonding the cooling channels to the windings with a thermally enhanced polymer with a thermal conductivity of 2 W/mK. Simulations predicted that with water supplied at room temperature, the peak temperatures of the stator and windings could be reduced by nearly 100°C under a 5 kW heat load to approximately 133°C, when compared to the standard natural convection case. The inter-winding cooled motor, therefore, could operate at much higher powers than it could with its original configuration while maintaining the integrity of its electrical insulators.

Sixel *et al.* [27] additively manufactured polymer cooling channels that were located between the coil windings (BW in Figure 7) and on the outside of the coil windings (OW in Figure 7) simultaneously, illustrating the excellent utility of polymer AM in heat exchanger design [28]. The flow channels were designed as aluminum-filled polycarbonate with a thermal conductivity of 1 W/mK, formed by fused deposition modeling [27]. It was concluded that a 44% decrease in the winding temperature rise could be achieved by this inter-winding water heat exchanger when used as an enhancement in addition to the standard water jacket cooling.

Semidey and Mayor [29] designed, fabricated, and tested an internally micro-featured copper heat exchanger that was placed between the windings (BW position in Figure 7). The heat exchanger was fabricated by micromachining copper plates. The two machined halves were then assembled to form a channel for water flow. The unmodified motor (Golden Motor HPM-10KW) handled a steady-state current density of 8.24 A/mm^2 and a transient current density of 14.7 A/mm^2, and after the inter-winding heat exchanger improvement, the machine was capable of a steady current density in excess of 24.7 A/mm^2 and a transient current density of more than 40 A/mm^2. The motor used class F insulation, which is limited to an operating temperature of 155°C. Although it could not be confirmed due to test bench limitations, it can be estimated that the improved motor produced approximately three times more power than in the stock configuration, resulting in a power density increase by a factor of three.

3.3. *Intra-winding cooling*

Intra-winding cooling shows high potential for enhanced winding cooling. One method is to use the winding for a coaxial flow of fluid and electricity. Lindh *et al.* [30] demonstrated this approach with an axial flux PM synchronous machine, which is illustrated in Figure 8. Stainless steel coolant conduits were tightly wrapped with stranded Litz wire, and the coaxial tubing and electrical conductor served as the stator windings. A shortcoming of this approach is the long fluid flow path generated by the single inlet and outlet of the continuous internal tubing. With polyalphaolefin oil as the cooling fluid, the required pressure drop was nearly 6 bar. Nevertheless, it was found that compared to the stock motor, intra-winding oil cooling decreased the average end-winding temperature by 67°C, while

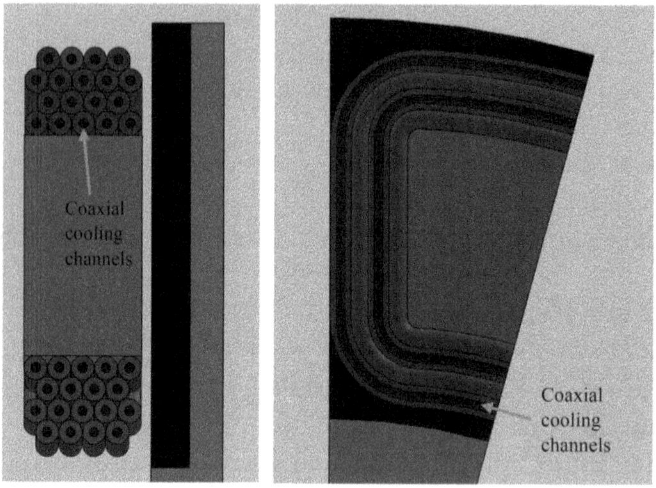

Figure 8. Illustration of intra-winding coaxial cooling paths within stator windings of an axial flux, permanent magnet motor.

intra-winding water cooling decreased the average end-winding temperature by 97°C. The unmodified motor winding current density reached 4 A/mm^2, while the thermally improved motor produced 14 A/mm^2, an improvement that may correspond to more than a factor of three in motor power and torque densities.

Wohlers *et al.* [31] demonstrated intra-winding cooling using AM of the coil windings with integrated cooling channels. The design used a single coil wrapped around the stator tooth, with 11 wraps. Typically, this large electrical cross-section in the winding would cause large cross-sectional current density disparities, or current crowding, leading to reduced efficiency and increased localized heating. The geometrical freedom of AM, though, allowed Wohlers *et al.* [31] to locally tune the winding cross-section to significantly reduce current crowding. Furthermore, the cooling channels were not constrained to a coaxial configuration but were instead routed to have inlets and outlets on either end of the stator, which significantly reduced the pressure drop per cooling capacity of the system. The windings were laser sintered from $AlSi_{10}Mg$ powder, an alloy that has approximately double the electrical resistance of copper and therefore double the heat generation per current density as copper. It was found that the steady-state current density of the coil geometry could reach 58.5 A/mm^2. Extrapolating from those results, it was determined that, with a similar configuration, cast copper could possibly reach a current density of 130 A/mm^2, which is about five times the current state of the art in electric motors. Nevertheless, detailed co-design should be used to evaluate whether or not the more than three times lower mass density of aluminum could offer an overall performance advantage for electric aircraft.

This section summarized the current state of progress in pumped single-phase direct liquid-cooling systems, which — with some modification — may also be compatible with high-heat-flux pumped two-phase cooling. When compared to single-phase cooling, pumped two-phase cooling is advantageous because it can remove more heat per unit pumping power due to much higher HTCs, thus achieving higher heat flux rates and a more uniform temperature distribution for the surface(s) being cooled. However, although much research has focused on two-phase flow cooling over the past four (and more) decades, single-phase cooling still remains the dominant technique in practice when it comes to cooling of high-heat-flux electronics equipment. Major barriers preventing the wide adoption of two-phase flow cooling include the additional complexity associated with phase-change cooling, including possible flow regime instability if not properly designed, gravity-dependence (and vehicle acceleration dependence) of the flow regimes, and a lack of well-established and easy-to-use models and performance predictive tools. However, interest in the application of two-phase flow cooling for high power density electronics continues, with active research being carried out in the field [32–34]. The following section discusses the working fluids used in pumped two-phase cooling for electronics and their potential for use in electric motor cooling.

4. Comparison of Cooling Fluids

Desirable fluid properties for highly effective thermal management include high thermal conductivity, high heat capacity, and high latent heat of vaporization (in two-phase cooling). Low viscosity is also

Table 1. Comparison of common working fluids for embedded two-phase coolers. All properties at 25°C unless otherwise noted.

Fluid	Saturation temp. at atm pressure (°C)	Saturation pressure (kPa)	Specific heat (kJ/kgK)	Latent heat of vaporization [41] (kJ/kg)	Liquid thermal conductivity (W/mK)	Liquid density (kg/m^3)	Vapor density* (kg/m^3)
FC-72 [39]	56	—	1.1	88	0.057	1680	13
HFE-7100 [39]	60	—	1.18	112	0.069	1510	9.9
R236fa [39]	—	272	1.24	145	0.069	1360	18
R245fa [40]	—	148	1.32	190	0.081	1340	8.5
R134a [40]	—	666	1.43	178	0.083	1210	32
ATF [35]	—	—	1.95	—	0.16	870	—
sCO_2^{**} [41]	—	—	4.00	290	0.078	710***	
Water [40]	100	—	4.18	2250	0.59	1000	0.6

Note: *At normal boiling point; **properties taken at 35°C and 10 MPa; ***supercritical fluid.

desirable for low pressure drop and pumping power. Table 1 provides a comparison of common working fluids for two-phase electronics coolers.

As shown in Table 1, water has some excellent heat transfer properties among several common fluids used for two-phase electronic cooling. The latent heat of vaporization and thermal conductivity of water are an order of magnitude greater than those of electronics cooling fluids (FC-72 and HFE-7100) and refrigerants frequently researched for electronics cooling (R236fa, R245fa, and R134a). Furthermore, the specific heat of water is more than three times that of those fluids. Moreover, high-purity water is multiple orders of magnitude less expensive than engineered fluids.

Therefore, water is in many ways a natural choice for high-heat-flux cooling, as is evident by its ubiquitous use in heat transfer applications. However, water has significant drawbacks for electronics cooling. First, unless it is consistently deionized, water can conduct electricity, and the presence of an electric potential can accelerate corrosion. Second, water has a higher freezing temperature than most engineered fluids, which could result in catastrophic failure in low ambient temperature environments due to its expansion upon freezing and/or loss of pumpability (including pump damage) as a solid. Ethylene glycol has the benefit of suppressing the freezing temperature of water, but to the detriment of reducing its thermal conductivity and specific heat.

Water has a relatively high boiling point, which requires channel wall temperatures in excess of 100°C to initiate boiling (at atmospheric pressure). The temperatures away from the channel wall will be higher due to the thermal resistance from the channel wall to the heat-generation source. Therefore, two-phase water cooling may require excessively high temperatures that lead to loss of insulation integrity in some applications, even if a high rate of heat transfer is generated by the potency of water as a cooling fluid. Ethylene glycol increases the already relatively high boiling point of water, which would lead to further increases in temperature at the heat source if ethylene glycol mixtures were relied upon for two-phase cooling. Although the saturation (boiling) temperature can be reduced by operating the cooling loop at sub-atmospheric pressures, this option is generally not preferred due to the risk of air ingress over long time periods.

Engineered fluids can mitigate many of the concerns presented by water. These fluids are quite inert, dielectric, available in a variety of saturation temperatures (at atmospheric pressure), and have a more favorable (lower) liquid–vapor density ratio, which can further improve the feasibility of two-phase configurations. Among the older-generation fluids, FC-72 has been considered by many to be the industry standard fluid for single- and two-phase electronics cooling because of these favorable properties. HFE-7100 (also manufactured by 3M) has a lower global warming potential and 27% greater latent heat but may be more corrosive than FC-72. R134a is an industry standard refrigerant for vapor-compression cycles but may soon be replaced by R1234ze in some applications.

ATF is not typically used in two-phase systems, in part due to its (intentionally) prohibitively high boiling temperature. It is, though, a potent single-phase cooling liquid, which has a specific heat of about 30% greater and a thermal conductivity double that of common refrigerants. In addition to heat transfer, ATF is designed for hydrodynamic power transmission and lubrication [35].

Low-pressure refrigerants are reported more often for embedded semiconductor cooling, with R245fa being most common for its low pressure, high latent heat, and high thermal conductivity compared to R236fa. The saturation pressure of R245fa at 25°C is about half that of R236fa, at 1.5 atm compared to 2.7 atm, respectively. Reduced pressure requirements also reduce the structural strength requirement of the embedded-cooling package, which may also lead to reduced package size and weight. Another advantage is that the latent heat of vaporization of R245fa exceeds that of R236fa by 31% (at 190 kJ/kg compared to 145 kg/kg, respectively), while the liquid thermal conductivity is also 17% greater (at 0.081 W/mK compared to 0.069 W/mK, respectively). These fluid property values are preferable because they both tend to improve the effectiveness of two-phase heat removal.

Among other considerations, the proper working fluid is selected based on the operating and storage temperatures ranges. While high-pressure refrigerants, such as R134a, are the more common working fluids for vapor-compression cycles in automotive and building air-conditioning applications, they are also used in some pumped two-phase flow cooling applications. Specific benefits of R-134a include its favorable thermophysical properties, such as a low freezing temperature, a low viscosity, and a high corrosion resistance.

Among other high-pressure working fluids, CO_2 is an increasingly commonly used fluid for transcritical and supercritical cycles. In contrast to vapor compression, transcritical cooling cycles pressurize the working fluid beyond its critical pressure and temperature. This supercritical fluid is as dense as a liquid but expands and compresses much like a gas. For an R134a cycle rejecting heat at a fluid temperature of 70°C, the highest pressure occurring in the cycle is 21 bar. In contrast, a transcritical CO_2 cycle rejecting heat at a fluid temperature of 70°C requires pressure to exceed 160 bar. This high pressure introduces some engineering challenges but provides significantly higher cooling system power density.

In particular, the size and weight of the heaviest and often bulkiest component of the refrigeration system, the compressor, can be significantly reduced. It has been shown that the compressor displacement can be reduced in excess of 80% when compared to a conventional R134a vapor-compression cycle [36]. In pressure containment devices, including the compressor, thicker walls are required to contain the pressure; however, the reduced internal volume decreases the total amount of material needed, resulting in decreased size and weight.

The benefits of transcritical CO_2 as a refrigerant (also called R744) have been under investigation for two decades for automotive applications, and it has been shown that R744 can reduce the size, weight, and power required by the air-conditioning system [37]. Mercedes-Benz committed to introducing transcritical CO_2 cycles into their luxury automobiles for increased cooling performance and to reduce the refrigerant global warming potential [38]. It should be noted that in order to realize its benefits, special precautions must be taken to mitigate the corrosive nature of supercritical CO_2.

As power electronics increasingly need pumped fluid cooling and drive electronics are increasingly integrated into motors, the symbiosis between high-heat-flux electronics and high-power density motors is increasing. Further, with wide-bandgap semiconductors, the need for pumped single- and two-phase cooling for logic and power devices is increasing [39]. At the same time, logic and power devices are increasingly integrated with electric machines [2]. So, although the thermal management challenges of motors and semiconductors are intercoupled, the industry and research focus on the essential thermal enhancements of motors lags behind that of semiconductors. Highly potent working fluids that offer excellent thermophysical and fluid flow properties, such as select supercritical fluids that offer the density of a liquid and the viscosity of a gas, can also contribute to substantial improvements in the cooling of motors. This can directly enhance power and torque densities and efficiency. However, some supercritical fluids may impose additional operating pressure limitations for the cooling system and thus added design complexity. The following section discusses a co-design-based simulation approach to analyze the comparative impacts of selected advanced technological areas, including the two-phase cooling approach, that can enable cost-effective efficiency enhancements and weight reductions while addressing the limitations of existing materials and cooling technologies.

5. Thermal and Electromagnetic Co-Design Simulation

The motivation to carry out a simulation study was to identify and prioritize impactful research opportunities and compare technological areas that can enable cost-effective efficiency enhancements and weight reductions while addressing the limitations of existing materials and cooling technologies. The objective of this section is to quantify the role of several enabling technologies on two existing motor architectures, induction motors and PM motors, for an initial assessment of the role each technology can play and the total impact on the resulting power density and efficiency of the motor. A standard multiphysics modeling and simulation approach was taken using a commercially available software platform [43].

The modeling task started by selecting state-of-the-art traction motors of two different types (induction and PM) as the baseline, the Tesla S60 and BMW i3, realizing as adequate test data for these two motors are available in the public domain for necessary model validations [44–48]. Figure 9 shows the 2D radial, axial, and 3D models of Tesla S60. Figure 10 shows the same for BMW i3. Both electric motors use state-of-the-art water jacket cooling, as described in Section 2.1. Next, five key enabling technologies were identified that have the potential to significantly improve the power density and efficiency of the baseline designs, as follows: (1) advanced winding topologies using AM, (2) advanced insulating and potting materials, (3) advanced cooling techniques that span both single-phase and phase-change cooling, (4) advanced structural materials, and (5) the co-design of thermal and electromagnetics in the integration of the motor and its drive. These technologies were sequentially incorporated into the two baseline models, as explained in the following, although only the results related to the BMW i3 are reported here. Further details on each of these enabling technologies, as well as additional operational

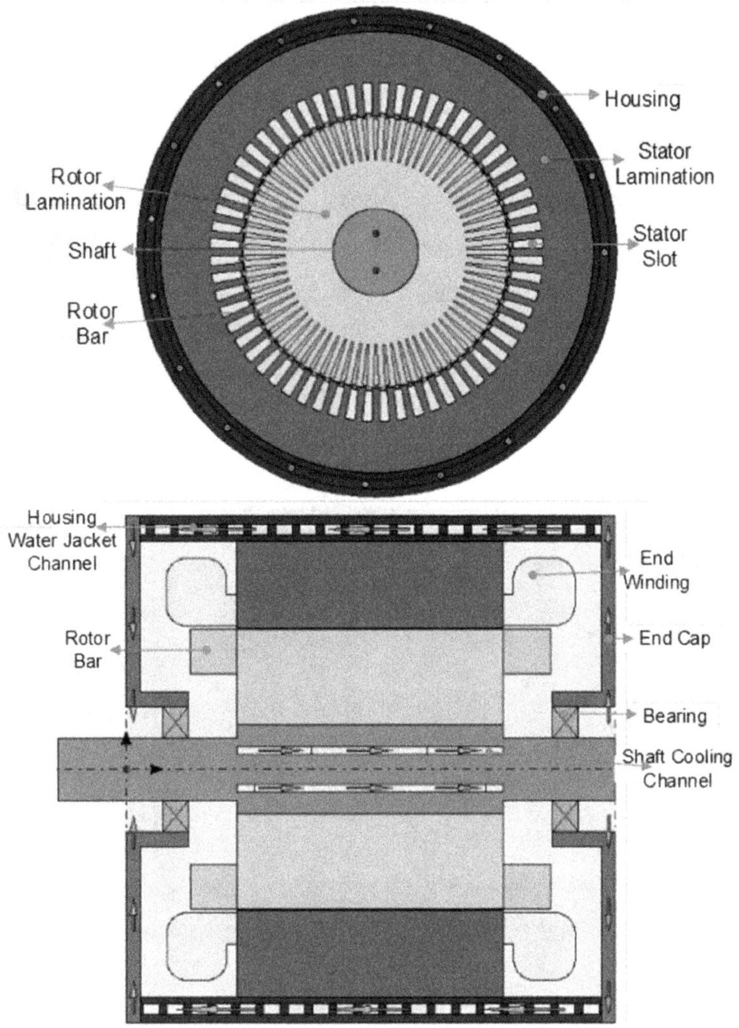

Figure 9. 2D radial (top) and axial cross-section (bottom) of the modeled Tesla S60.

specifications (i.e., bus voltage, maximum speed, ambient temperature, etc.) can be found in the **A**viation-class **S**ynergistically **C**ooled **E**lectric-motors with i**N**tegrated **D**rives (ASCEND) program specifications [47].

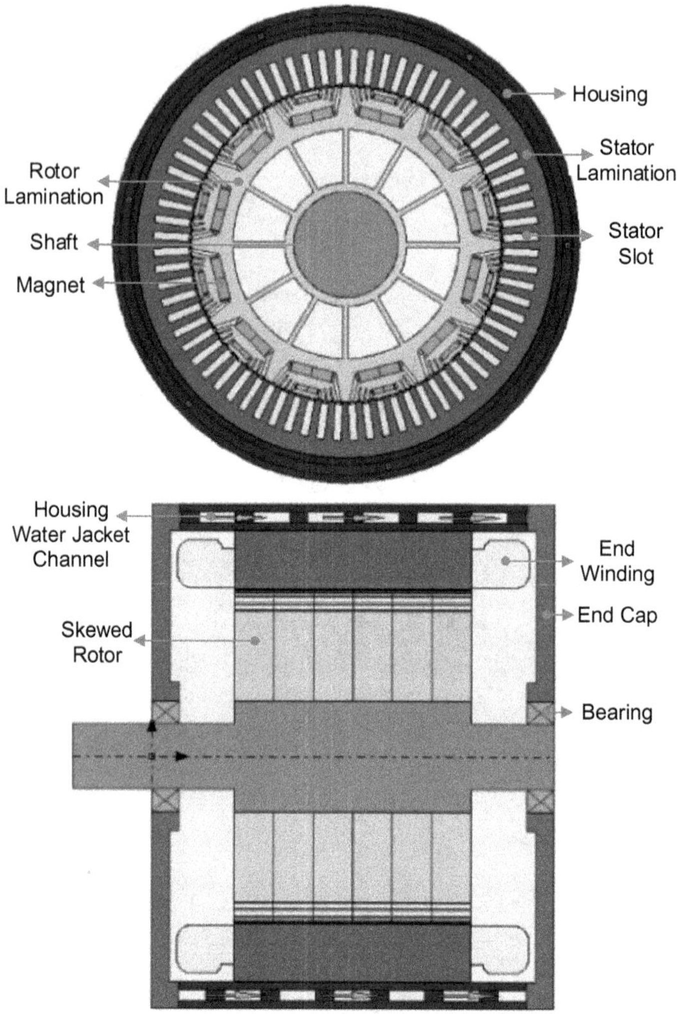

Figure 10. 2D radial (top) and axial cross-section (bottom) of the modeled BMW i3.

ASCEND is a program launched by the **A**dvanced **R**esearch **P**roject **A**gency-**E**nergy (ARPA-E) [48] to support the development of novel lightweight and ultraefficient electric motors, drives, and associated thermal management systems that will facilitate net-zero carbon emissions in the single-aisle, 150–200-passenger commercial aircraft segment. The program requires demanding figures of merit for specific power (≥12 kW/kg) and efficiency (≥93%) for the fully integrated all-electric powertrain system, as shown in Figure 2. For the purpose of this study, it is anticipated that a power density of 30 kW/kg (including TMS) and a 98% efficiency for the motor drive subsystem may be achievable. If so, the electric motor subsystem needs to be at 20 kW/kg and 97% efficiency to meet the system-level metrics. The key metrices used in this study for the electric motor subsystem are given in Table 2. Figure 11 shows the delivered mechanical power as a function of the time flight profile to be used for designing the electric propulsion system.

Table 2. Targets and metrices for the electric motor subsystem.

Ambient condition	Description	Target
At takeoff and climb	Mechanical shaft power output	≥250 kW
	Maximum rotational speed (after gear box)	5,000 RPM
	Average efficiency	≥97%
	Specific power	≥20 kW/kg
At cruise altitude	Nominal mechanical shaft power output	≥83 kW
	Nominal rotational speed	3,500–4,500 RPM
	Nominal efficiency	≥97%
At both	Rail-to-rail bus voltage, electrical input	≥1 kV DC

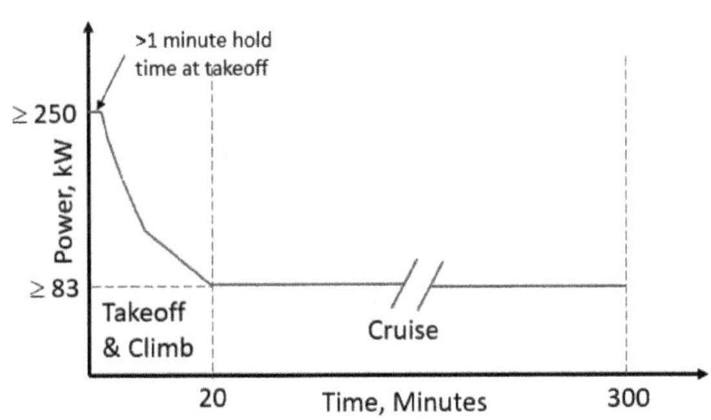

Time, Minutes	Power, kW
0	250
1	250
5	175
10	125
15	100
20	83
300	83

Figure 11. Delivered mechanical power as function of the time flight profile.

5.1. *Advanced winding topologies using additive manufacturing*

Advances in AM make it possible to introduce novel end-winding designs and advanced insulation materials within the windings to facilitate high thermal conductivities while meeting electrical insulation requirements [49]. Such improvements can increase the gravimetric power density of the motor, provided that the conductor conductivity will not degrade with the AM process — a shortfall that some AM technologies may yet need to overcome through innovative winding topologies (e.g., shaping of the conductors). AM techniques may also allow the use of complex geometries that could enhance the air-gap magnetic flux and incorporate highly effective and integrated cooling mechanisms within the motor winding designs [50].

At the first step of this modeling study, it is assumed that the early advancements from AM-enhanced winding will yield a 50% reduction in the end-winding length and double the slot fill factor from the baseline designs. It is also anticipated that AM can shape and/or reduce the wire structures and transpose the wire strands that will diminish the eddy loss effects in the wires. Therefore, in this study, eddy losses were neglected, while also assuming no degradation in copper conductivity due to AM processing. Figure 12 shows the basic modifications made in the electric motor modeling from the baseline designs. The stranded wire winding was changed to rectangular wire hairpin winding with the eddy loss calculation set

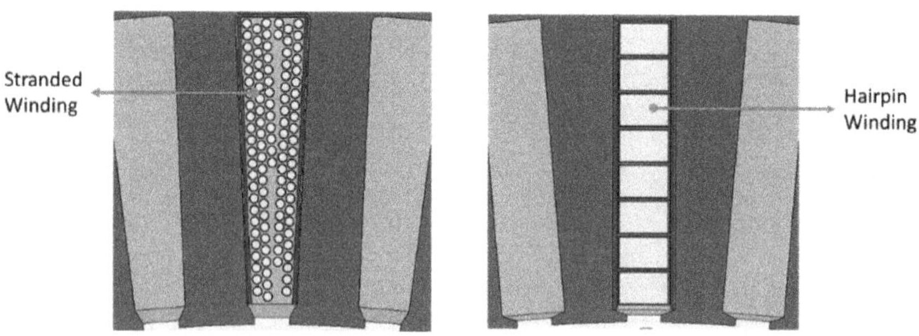

Figure 12. One of few examples of modifications made to the baseline models to emulate AM-based winding advancements; changing from stranded wire winding (left) to rectangular wire hairpin winding (right).

to zero to emulate the envisioned AM-based winding advancements, as explained earlier. With such optimistic assumptions, the analysis showed a modest 25% improvement in specific power from the baseline designs. This, however, did not incorporate the possibilities of AM-enabled slot integrated intra-winding cooling techniques, as described in Section 3.3 (see Figure 8), at this stage. This will be discussed later in Section 5.3.

5.2. *Advanced insulation and potting materials*

A principal bottleneck in dissipating the heat generated from copper losses inside a motor can be the slot insulation material that sits between the winding and stator lamination. Materials utilized as electrical insulators inside the motor inherently have poor thermal properties that limit heat rejection from the windings, thus limiting the achievable motor power density and efficiency. In the power electronics area, the use of boron nitride/aluminum nitride (BN/AlN)-based coating and potting materials filled with graphite/carbon nanotubes has improved the heat rejection process from the switching device to the heat sink. Similar processes have recently been adopted in electric motors as well, and a 10× improvement in thermal conductivity over industry standards has been reported [51]. In this modeling study, the impacts from advanced insulating and potting materials were considered by assuming a 10× improvement in thermal conductivity from the baseline designs. Figure 13 shows the basic modifications made to the baseline designs, with improved insulation and potting materials (in green) introduced at the end-winding areas. Figure 14

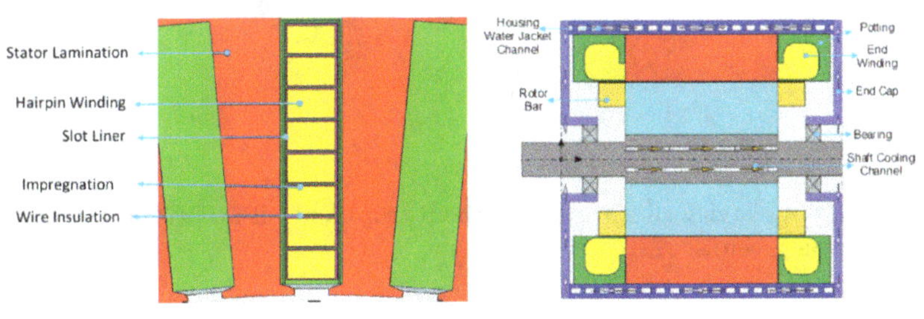

Figure 13. Improved insulation of 2 W/mK slot liners (left) and addition of potting materials (in green) of 10 W/mK (right) to observe the effect of advanced insulation and potting materials.

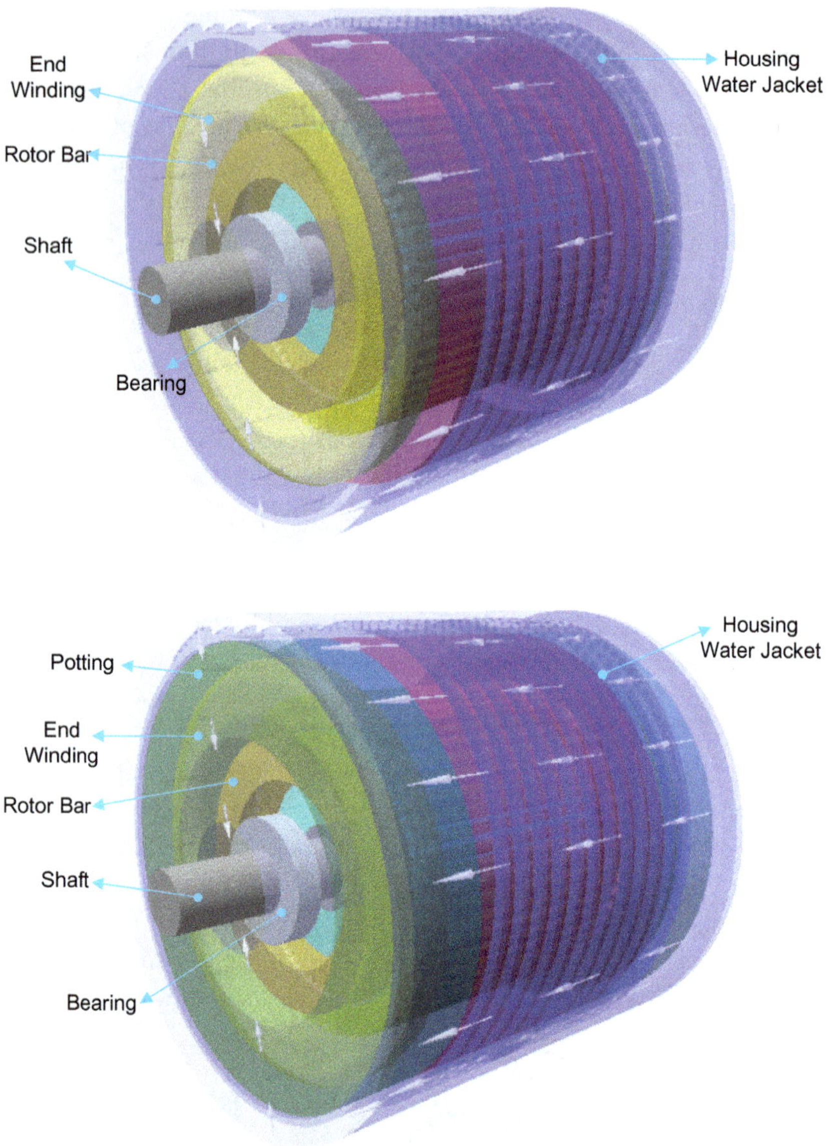

Figure 14. A 3D view of the motor without (top) and with (bottom) the potting materials (in green).

shows a 3D view of the motor without and with the potting materials. Analysis showed a good 55% improvement in specific power from the baseline designs.

In the current simulation study, it was not possible to incorporate further thermal enhancements, as described in Section 3.1, such as introducing copper bars inside the stator tooth (see Figure 6), due to the limitations of the existing electric motor modeling and analysis software. This was left as one of the topics for future analysis.

A sensitivity analysis was also performed to see if further improvements in thermal conductivity will have any significant impact on motor performance, and 10 W/mK ($\sim$10$\times$ from baseline) for the potting impregnation material appeared to be the inflexion point in this study, as seen in Figure 15 (right). A similar analysis was performed for slot liner materials as well, and the inflexion point was observed at 2 W/mK ($\sim$10$\times$ from baseline), as seen in Figure 15 (left). In both cases, it is seen that a minimum improvement is gained past a thermal conductivity of $\sim$2 W/mK for the insulation materials and 10 W/mK for the impregnation materials.

For sensitivity analysis, only one parameter, namely the thermal conductivity in this case, was varied, while other parameters were kept constant. As such, only one thermal resistance (potting impregnation or slot liner) at a time in the thermal circuit was changed in

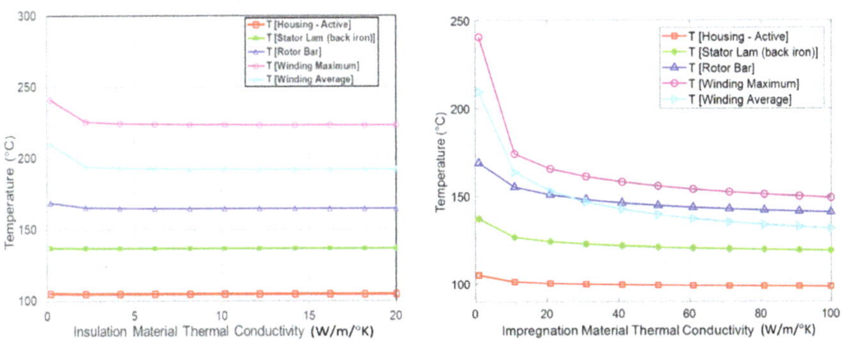

Figure 15. Sensitivity analysis performed on potting material by varying the thermal conductivity from 1 to 100 W/mK (right). Similar analysis performed on slot liner material (left).

the software. However, the final temperature of the motor is determined by several thermal resistances, as can be seen in Figure 16, where the dark-green rectangles represent the thermal resistance of the respective potting impregnation and the yellow rectangles represent the respective thermal resistance of the slot liners. In the parametric analysis for the potting impregnation material, the thermal resistances that are represented by the dark-green rectangles of the thermal circuit were calculated to mimic the effect of increasing the thermal conductivity of the potting impregnation material. At some point, the influence of the decrease in thermal resistance of these dark-green rectangles becomes negligible in the complete thermal circuit, thus suggesting thermal saturation for the potting material. The machine's temperatures will now be dictated by other thermal resistances. This is when the inflexion point occurs for the potting material. A similar study was done for the insulation material represented by the yellow rectangular blocks in the thermal circuit shown in Figure 16.

5.3. *Advanced cooling techniques*

For most commonly known designs, the gravimetric torque density of an electric motor is a strong function of the current density, which is, in turn, directly influenced by the rate of heat rejection from the motor. Commercial state-of-the art electric motors are generally thermally limited at high power ratings. Power and torque density capability, as well as efficiency, can be increased by mitigating the thermal limitations of the motors through a co-design approach. Such an approach would utilize electromagnetics and thermo-mechanical design aspects for the thermal design of the system. Current methods for commercial motor cooling, such as oil bath or water jacket cooling (see Sections 2.1 and 2.2), impose limitations on power density due to a lack of integrated, low-thermal-resistance and highly effective motor cooling practices. This is particularly true for motors that operate at higher voltages, which in most cases require thicker electrical insulation layers that further limit heat dissipation. Researchers have shown that significant potential exists for improved direct (or

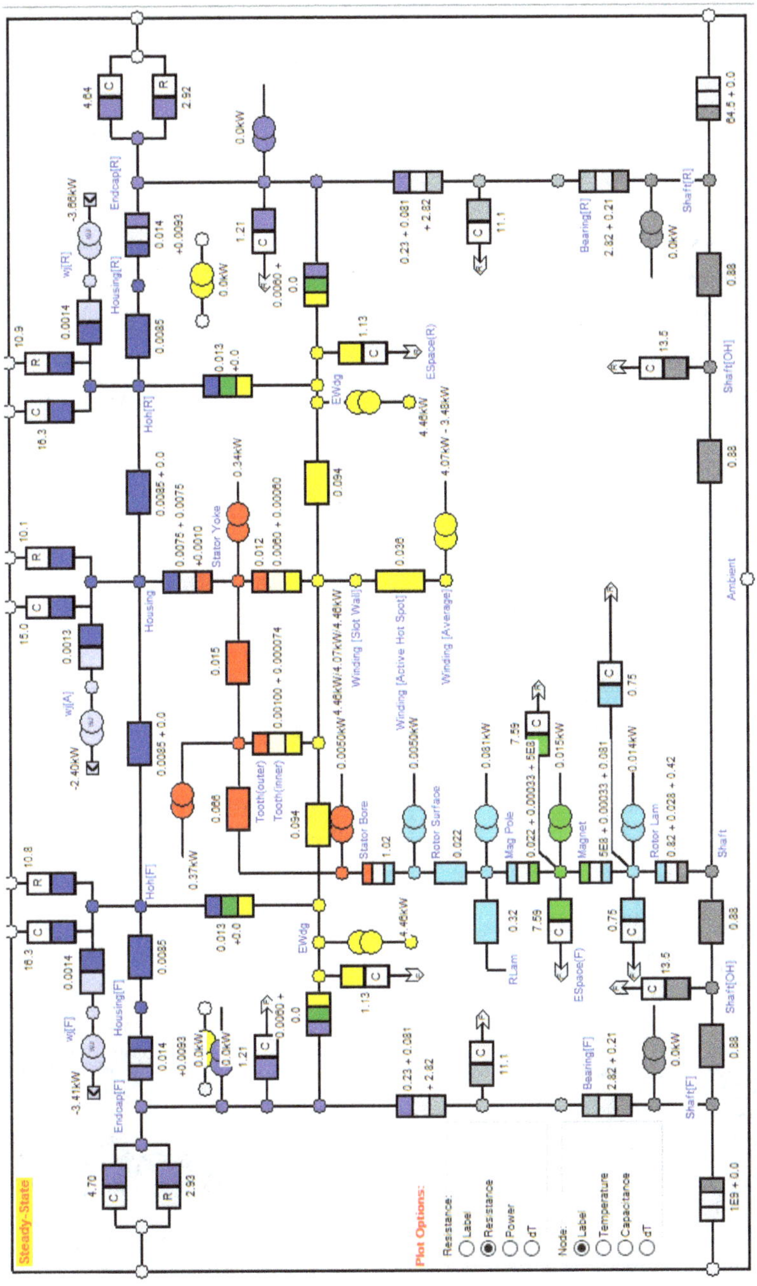

Figure 16. Thermal circuit for the BMW i3 motor model: The dark-green rectangles represent the thermal resistances from the potting material, and the yellow rectangles represent the thermal resistances from the slot liners.

near source) cooling techniques (see Section 3). For instance, the stator windings generate most of the heat generation in electric motors, and case examples show that nearly two-thirds of the total motor heat generated is due to such resistive losses. By employing inter-winding (Section 3.2) and intra-winding (Section 3.3) thermal management, recent studies have demonstrated that the power density of electric motors can be increased substantially. Highly potent working fluids that offer excellent thermo-physical and fluid flow properties, such as select supercritical fluids which offer the density of a liquid and the viscosity of a gas, can also contribute to substantial additional cooling of motors (see Section 4). This can directly enhance power and torque densities and efficiency. However, some supercritical fluids may impose additional operating pressure limitations for the cooling system and thus added design complexity.

In-slot cooling channels were incorporated into the baseline models to investigate the effect of advanced cooling techniques, as shown in Figure 17. The positions of the channels were, however, fixed by the modeling software at the top and bottom of the slot due to a software feature limitation; otherwise, it is understood that moving these channels in the middle of the slots may have provided better heat rejection, as described in Sections 3.2 and 3.3. This was left as one of the topics for future analysis. It is noted that the inclusion of the cooling channels inside the slot will take over spaces available for the copper wires, as can be seen in Figure 17. This will increase

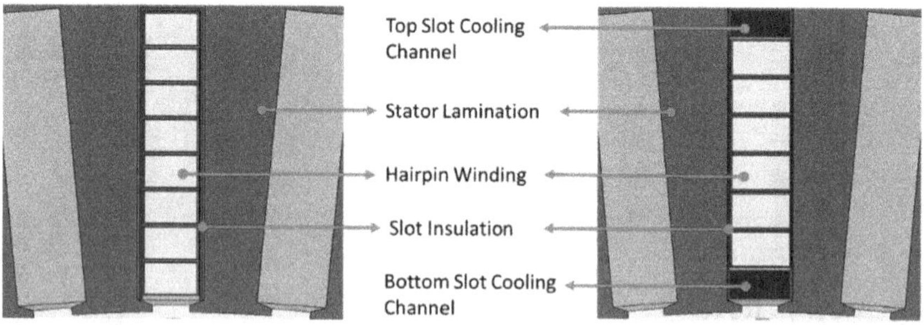

Figure 17. The baseline model of the stator slot with no in-slot cooling (left). The modified model with top and bottom of the slot now occupied by two cooling channels (right).

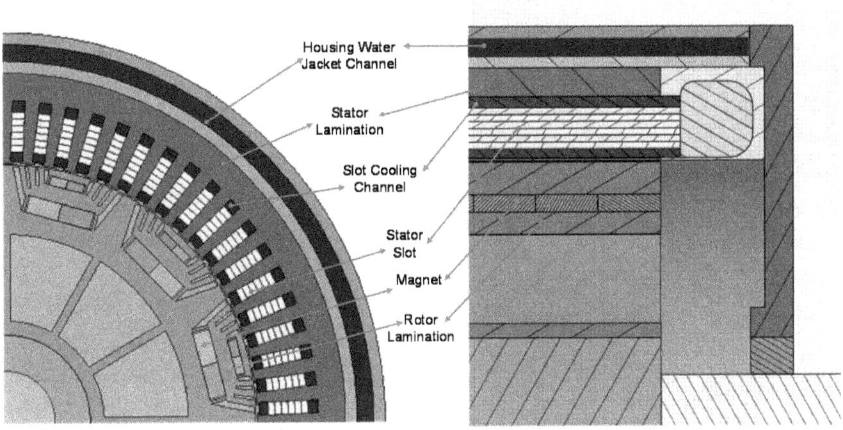

Figure 18. 2D radial (left) and axial cross-section (right) of the modeled BMW i3 showing in-slot cooling channels and the cooling flow inside the slots.

the copper current density (i.e., current per square area of copper) and hence the copper losses. However, improved thermal performance from in-slot cooling should more than compensate for these increased copper losses. Figure 18 provides further details of the in-slot cooling modeled in the baseline BMW i3 motor.

Supercritical CO_2 that can be utilized for enhanced heat rejection was selected as the coolant because of its interesting thermophysical properties, as shown in Figure 19 and also provided in Table 1. A parametric and sensitivity analysis was performed to obtain flowrate and pressure, and it was found that 10 L/min, with the inlet temperature set at 35°C and under 80 atm, provides the best overall result for the limited scope studied here. Figure 20 provides one example of several parametric analyses done in this study.

For the sensitivity analysis, only one parameter, either pressure or flow rate in this case, was varied, while other parameters were kept constant. The change in HTC was estimated from the pressure and/or flow rate changes using Gneilinski's equations for a fully developed turbulent flow. This calculated HTC is used to determine the thermal resistance for cooling. The thermal circuit is updated with the newly calculated thermal resistance, and the final temperature of the motor is determined by solving the complete

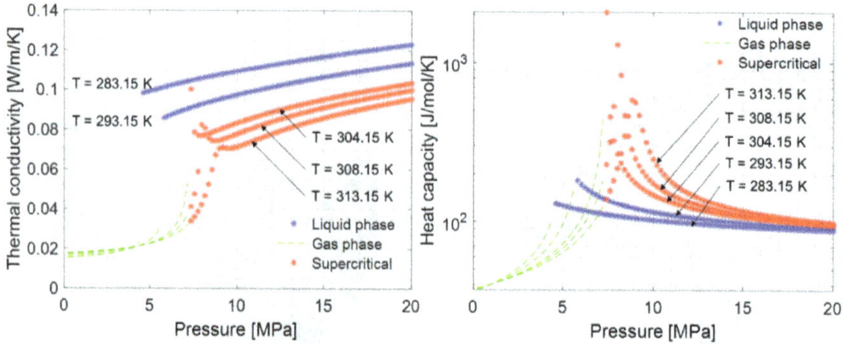

Figure 19. Comparison of thermal conductivity of CO_2 (left) and the heat capacity of CO_2 (right) in gaseous, liquid, and supercritical fluid: Green dashed lines represent the fluid in the gaseous phase. Blue and red asterisks represent the fluid in the liquid phase and supercritical state, respectively [52].

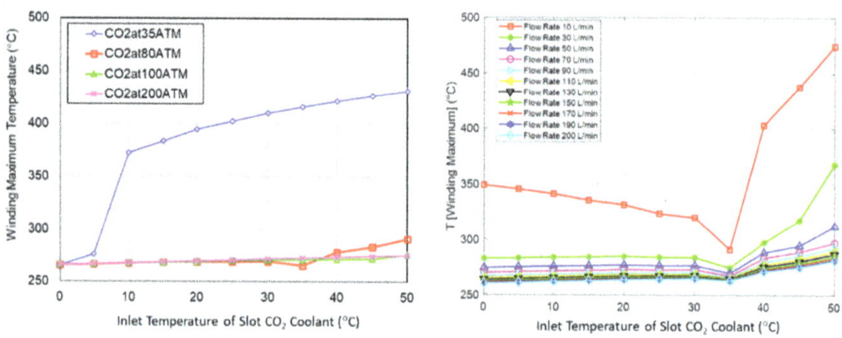

Figure 20. Sensitivity analysis is performed on in-slot cooling by first varying the pressure of CO_2 (left) from 35 to 200 atm and inlet temperature from 0°C to 50°C with flowrate fixed at 100 L/min and then varying flowrate (right) from 10 to 200 L/min and inlet temperature from 0°C to 50°C with CO_2 pressure fixed at 85 atm.

thermal circuit. If one thermal resistance is continuously decreasing (by increasing the pressure and/or flow rate), at some point, the influence of this thermal resistance on the complete thermal circuit becomes negligible, implying thermal saturation for this parameter. The machine's temperatures are now dictated by other thermal resistances. This is when the inflexion point occurs for that parameter.

Several interesting trends were observed in the parametric study. The coolant inlet temperature and flow rate had some impact on the winding maximum temperature, which occurs at the end-winding region, but it never fell below 265°C. With the winding average temperature previously observed in Figure 13 settling ~25°C below the winding maximum temperature, a winding average temperature of 240°C will fall within the operating range of a polyimide-based insulated wire. Nevertheless, the high winding maximum temperature of 265°C at the end-winding region needs to be addressed separately through jet-impingement or non-evaporative spray cooling [13].

Also observed in Figure 20 (left) is that with the coolant inlet temperature at 0°C, the peak winding temperature was minimized and found to be independent of the coolant CO_2 pressure. However, it did not vary much with the coolant inlet temperature when CO_2 pressure was 80 atm or higher. It also showed that the lowest winding maximum temperature when coolant CO_2 pressure and temperature were at their supercritical conditions of 80 atm and 35°C, respectively. Under these conditions, supercritical CO_2 demonstrates the highest thermal conductivity and heat capacity, as seen in Figure 19. Also, as observed in Figure 20 (right), at a CO_2 pressure of 85 atm, by comparing different inlet temperatures (0°C to 50°), the inlet temperature of 35°C gives the lowest winding temperature rise, especially at a low flow rate (10 L/min). However, the advantage of setting the inlet temperature at 35°C diminishes for higher flow rates ($\geq$70 L/min), reaching thermal saturation, as explained earlier.

At first, severe electromagnetic saturation limited the model performance. This prompted a redesign of electromagnetics in the baseline models. This was particularly achieved by changing the lamination material from M19 to Hyperco50A and reconfiguring the motor windings with an increase in motor base speed. The increased base speed required the addition of a 2:1 gear box to the electric motor subsystem. The gearbox efficiency was assumed to be 99%. It was realized at this stage that a high-speed geared solution may provide better specific power for the complete electric motor subsystem. A comprehensive parametric study is required to optimize the geared solution, which, however, was not part of this study. Nevertheless, with a 2:1 geared solution, analysis showed significant improvement in specific power, more than 4× from the baseline designs, with a percent improvement in efficiency as well, even though some modeling limitations were encountered, as explained later. At this stage, it was also understood that retrofitting an advanced cooling methodology on an existing design will not provide the optimum solution, as it does not fully benefit from a true co-design approach. A complete co-design of electromagnetics and thermal is needed to avoid saturation effect and the lack of an optimized operation. Also, the built-in thermal nodal model used in this study was developed to investigate traditional cooling systems. This prohibited introducing advanced cooling ideas in this study, such as placing cooling channels near the winding hotspots, using hollow wires for intra-cooling and interlamination cooling. Nevertheless, results from this analysis confirmed that an effective, low resistance, embedded near the heat source, can be highly impactful to improve the specific power while keeping winding temperature sufficiently low. However, that alone may not be enough to meet the program goals set by ASCEND [47], as explained in later sections.

5.4. *Advanced structural materials*

It is common practice in motor design to focus on reducing the weight of torque-producing (active) materials, such as copper and steel. However, structural weight can also become significant when high power densities are desired. This is especially true for air core and superconducting designs, where structural weight may dominate the weight of active materials. The weight of the motor can be significantly reduced when traditionally employed metals are replaced with alloys and composites, such as carbon-fiber-reinforced polymer composites, similar to the advanced structural material used in some aerospace applications due to its lightweight and superior thermal conductivity. The modeling study showed a ∼21% weight reduction for structural components when aluminum alloy-based end caps and

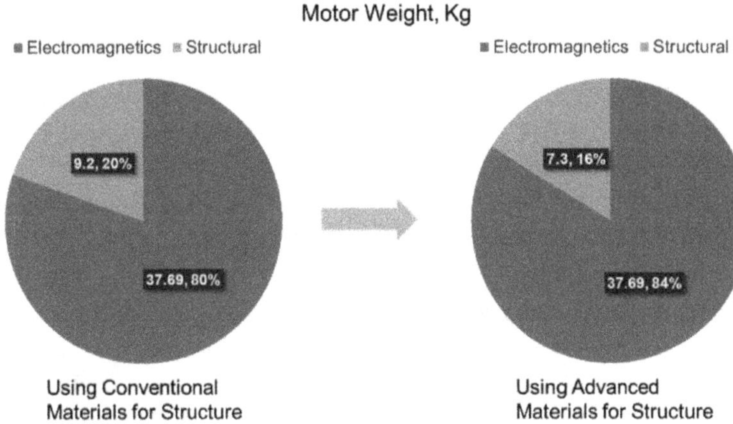

Figure 21. The use of carbon-fiber composite reduced housing and end cap weight by 21%, i.e., 7.3 kg vs. 9.2 kg, reducing the total weight by 4%.

housing were replaced with carbon-fiber-reinforced polymer composites, as seen in Figure 21, reducing the total weight by 4%.

5.5. *Co-design and integration*

A well-planned co-design approach is needed to take advantage of the combined effects of advances such as those referenced above. A simple cooling feature added to an existing baseline motor design did not provide sufficiently compelling results due to electromagnetic saturation, as explained earlier. However, when co-designed with electromagnetics, novel thermal management technologies can be utilized in the electric motors and associated power electronics to enhance the power density and efficiency of the motor and its drive (i.e., power electronics). The co-design of power electronics and electromagnetics can also lead to new topologies with greater torque densities. Similarly, integration of the power electronics with the electric motor will lead to higher power densities and efficiencies, as this will eliminate lengthy interconnections between the subsystems. And with continuous advancements in high-temperature power electronics devices, the full integration of the motor and power electronics is becoming feasible.

A limited co-design and optimization study was done in this study due to the limited scope of the study and software limitations. A detailed finite element analysis (FEA) is needed for model flexibility but was not performed because of time constraints. Hence, the primary focus at this stage of the modeling was to further reduce the electromagnetic saturation using limited FEA and come up with a semi-optimized solution without deviating too much from the selected baseline designs. This was particularly achieved by changing the lamination material from M19 to Hyperco50A. Motor windings were reconfigured (i.e., reduction in copper), along with an increased thickness of the stator segment. The maximum motor drive voltage and current were constrained at 1,000 Vdc and 100 Arms for all studies. This, however, limited further optimization of the complete motor and drive system. Yet, a 10% improvement in specific power was achieved, reaching the 20 kW/kg target with a ~96% efficiency for the electric motor. At this stage, it was well understood that retrofitting the motor with an existing drive will provide a sub-optimal solution. A complete co-design of electromagnetics, thermal, and drive is needed to ensure that saturation of any one of these aspects will not affect the optimization process.

5.6. *Simulation results summary*

Improvements in specific power obtained from the five key technologies detailed in the earlier section are summarized in Figure 22. The results are organized in sequential steps, with cumulative improvements added from previous steps. The "with saturation" results are obtained from modeling direct retrofitting of the five technologies on the baseline design (BMW i3 in this example). The "without saturation" results are obtained when the electromagnetics of baseline models were redesigned to address the magnetic saturation, as explained earlier.

A significant contribution to specific power was achieved when low-resistance cooling was incorporated into the model, followed by enhancements obtained from the advancements in insulation and potting materials. However, a synergistic result from all five

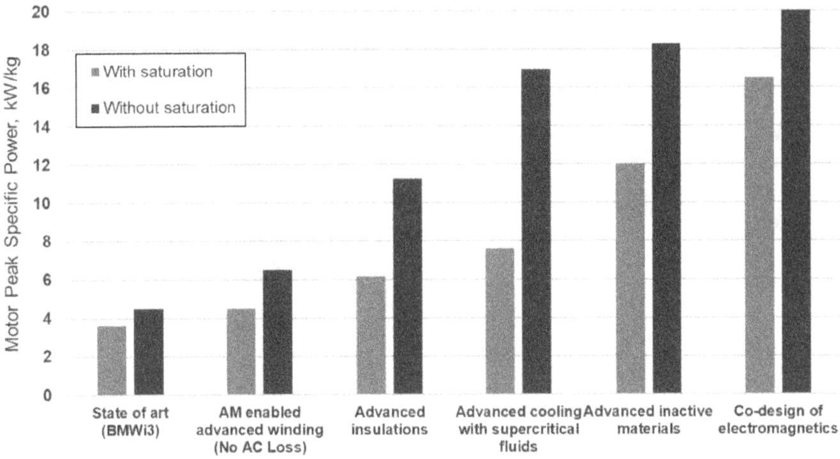

Figure 22. Improvements in specific power obtained from five selected key technologies.

technologies was needed to reach motor power density of 20 kW/kg without including the TMS and gearbox. The study achieved 15 kW/kg with the TMS and gearbox weight included, as seen in Figure 23. The nominal efficiency of the motor in all design cases achieved ≥97% at cruise (continuous) condition and above 95% at takeoff (peak) conditions but not reaching the desired ≥97% target at takeoff. However, such results were obtained from semi-optimized solutions due to modeling restrictions, as stated earlier. Nevertheless, the study shows a viable pathway to meet the ASCEND [47] metrics through the application of the representative technologies with a novel co-design, rather than a retrofitting approach. Further possible solutions that may address the shortcomings but were not included in this study include increased flowrate during takeoff, high-speed geared solutions, and boosting of the drive voltage at takeoff. However, detailed parametric studies are needed for each of these possible solutions to obtain system-level optimization. It was also interesting to observe that the margin between peak and continuous powers was within 10% for the designs discussed in Section 5.4, which is due to a significant reduction

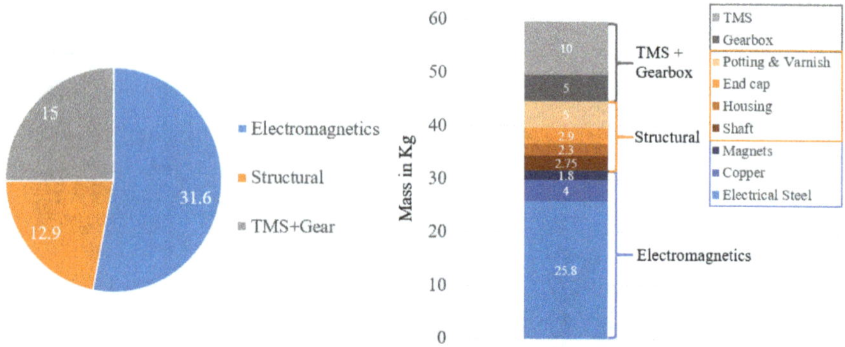

Figure 23. Weight breakdown of the final design.

in the thermal mass. The use of support structures, such as the airframe, to increase the thermal mass without penalizing the motor power density may be another solution to meet such demanding figure of merits.

6. Summary

Light and ultraefficient electric motors are in great demand in many industrial sectors; however, for the electric transportation sector, they are a necessity. Recent advancements made in electric motors may be attributed to the rapid progress of the electric vehicle (EV) industry to meet its high efficiency and volumetric power density (kW/L) targets [53]. However, the constraints for in-land transportation are different from those for future electrified aviation, where ultraefficient and high-specific-power (kW/kg) electric motors are key. EV motors are typically designed to fit existing vehicle architectures; therefore, geared high-speed motors with indirect liquid-cooling solutions are prevalent. Stringent cost targets also play a vital role in materials selection. For the future electrified aviation industry, the platform has yet to be finalized. Nevertheless, it is clear that the efficiency and specific power (kW/kg) requirements for the power train will be much higher (>4×) than those for the EV sector [47]. Among others, the three key differentiators of an aviation motor from an EV motor include the following: (1) the operational speed for the aviation applications is within a much narrower range (e.g., 3,500–4,500 rpm); (2) the operating temperature range of the aviation TMS is much wider (e.g., <-50°C at cruise and >70°C at takeoff); and (3) operating pressures are lower due to operation at high altitudes [54]. Therefore, future aviation-class motors will require aggressive thermal management solutions capable of extracting heat from a comparatively lower-speed motor with thicker insulations and potting materials used for corona discharge prevention.

The role of thermal management and the most recent improvements to the current thermal limitations of electric motors, with emphasis on approaches to increasing the electrification of air transport, were also discussed. The study demonstrated that most of the commercial state-of-the-art electric motors are thermally limited and that a substantial increase in power and torque densities, as well as efficiency of electric motors, can be achieved by mitigating the thermal limitations in electric motors through a co-design approach utilizing electromagnetics and thermo-mechanical design aspects.

Current methods for commercial motor cooling, as discussed in Section 2, impose limitations on power density due to a lack of sufficient absorption of heat. This is particularly important for motors that benefit from operation at high voltages, which in turn require thicker electrical insulation layers that often further limit heat dissipation. Researchers have shown that there exists significant potential for improved cooling. Specifically, the stator windings generate the majority of heat losses in electric motors, and examples show that 43–64% of total motor heat generation occurs due to these resistive losses. As discussed in Section 3, by employing inter-winding and intra-winding thermal management, the power density of electric motors can be increased by a factor of three to five or more. Supercritical fluids that offer the density of a liquid and the viscosity of a gas can also contribute to substantial additional cooling of motors, with a direct impact on enhancing power and torque densities and efficiency. This was discussed in Section 4.

Recent studies have shown that the heat flux levels from semiconductor devices can be increased by more than tenfold through innovative cooling technologies and co-design approaches. For electric motors, the potential for such improvements can be even more significant, realizing the same principles of thermally potent materials, topologies, and fluids. These approaches can be used to increase the power density of electric motors so that their thermal management will no longer lag behind that of semiconductors. An effective thermal management solution will optimally cool the motor itself as well as its integrated drive, the power electronics, and the power conditioning system.

Finally, this chapter also discussed, in Section 5, possible research pathways to meet two critical targets: specific power and efficiency, which are derived for the electric motor subsystem defined in the ARPA-E [48] ASCEND [47] program. The simulation study performed here showed low-resistance cooling with microchannels embedded in the stator slot to be most impactful to improve specific power, even though more advanced cooling approaches were not considered due to the limitations of the existing electric motor modeling and analysis software. These were left as topics for future analysis. Yet, that alone may not be enough, and a synergistic approach is required across the multidisciplinary technical areas (electromagnetic, power electronics, and thermal) in order for the program goals to be met.

Acknowledgment

Some of the motor simulations presented in this study were contracted and performed by Motor Design Limited (https://www.motor-design.com) with the technical guidance and supervision provided by the authors.

References

[1] Z. Wang, Y. Wu, M.H. Mahmud, Z. Zhao, Y. Zhao, H.A. Mantooth, Design and validation of a 250-kW all-silicon carbide high-density three-level t-type inverter, *IEEE J. Emerg. Select. Topics Power Electron.*, 8 (1), November 5, 2019, 578–588.

[2] T.M. Jahns, H. Dai, The past, present, and future of power electronics integration technology in motor drives, *CPSS Trans. Power Electron. Appl.*, 2 (3), November 23, 2017, 197–216.

[3] AVID Technology Limited. EVO axial flux electric motors. Available at: https://avidtp.com/product/evo-motors/ (accessed September 6, 2019).

[4] YASA Limited. YASA 750R E-Motor. Available at: https://www.yasa.com/wp-content/uploads/2018/01/YASA-750-Product–Sheet.pdf: 2018 (accessed September 6, 2019).

[5] D. Moreels, P. Leijnen, High efficiency axial flux machines, MAGNAX White Paper, V 1.9, November 2018.

[6] A.T. De Almeida, F.J. Ferreira, J.A. Fong, Standards for efficiency of electric motors, *IEEE Ind. Appl. Mag.*, 17 (1), November 11, 2010, 12–19.

[7] G.A. McCoy, J.G. Douglass, Premium efficiency motor selection and application guide — A handbook for industry (No. DOE/GO-102014-4107). Washington State University Energy Program, 2014.

[8] A. Dubois, M. van der Geest, J. Bevirt, R. Christie, N.K. Borer, S.C. Clarke, Design of an electric propulsion system for SCEPTOR's outboard nacelle, in: *16th AIAA Aviation Technology, Integration, and Operations Conference*, 2016, p. 3925.

[9] P. Ponomarev, M. Polikarpova, J. Pyrhönen, Thermal modeling of directly-oil-cooled permanent magnet synchronous machine, in: *2012 XXth International Conference on Electrical Machines*, September 2, 2012, pp. 1882–1887. IEEE.

[10] Y. Yang, B. Bilgin, M. Kasprzak, S. Nalakath, H. Sadek, M. Preindl, J. Cotton, N. Schofield, A. Emadi, Thermal management of electric machines, *IET Electron. Syst. Transport.*, 7 (2), June 28, 2016, 104–116.

[11] J. Broughton, V. Smet, R.R. Tummala, Y.K. Joshi, Review of thermal packaging technologies for automotive power electronics for traction purposes, *J. Electron. Packag.*, 140 (4), December 1, 2018, 040801.

[12] S.U. Yuruker, R.K. Mandel, P. McCluskey, M.M. Ohadi, S. Chakraborty, Y. Park, H. Yun, A. Khaligh, L. Boteler, M. Hinojosa, Advanced packaging and thermal management of high-power DC-DC converters, in: *International Electronic Packaging Technical Conference and Exhibition*, October 7, 2019, Vol. 59322, p. V001T06A026. American Society of Mechanical Engineers.

[13] M. Popescu, D. Staton, A. Boglietti, A. Cavagnino, D. Hawkins, J. Goss, Modern heat extraction systems for electrical machines: A review, in: *2015 IEEE Workshop on Electrical Machines Design, Control and Diagnosis (WEMDCD)*, March 26, 2015, pp. 289–296. IEEE.

[14] D.C. Deisenroth, M. Ohadi, Thermal management of high-power density electric motors for electrification of aviation and beyond, *Energies*, 12 (19), January 2019, 3594.

[15] P. Charoensawan, S. Khandekar, M. Groll, P. Terdtoon, Closed loop pulsating heat pipes: Part A: parametric experimental investigations, *Appl. Thermal Eng.* 23 (16), November 1, 2003, 2009–2020.

[16] S. Adera, D. Antao, R. Raj, E.N. Wang, Design of micropillar wicks for thin-film evaporation, *Int. J. Heat Mass Transfer*, 101, October 1, 2016, 280–294.

[17] J. Corman, R. Tompkins, R. Edgar, M. Mclaughlin, Inventors, General Electric Co, assignee. Rotating electrical machine having rotor and stator cooled by means of heat pipes. United States patent US 3,801,843. April 2, 1974.

[18] N. Putra, B. Ariantara, Electric motor thermal management system using L-shaped flat heat pipes, *Appl. Thermal Eng.*, 126, November 5, 2017, 1156–1163.

[19] T. Hassett, M. Hodowanec, inventors; Thermal Motor Innovations LLC, assignee. Electric motor with heat pipes. United States patent US 7,569,955. August 4, 2009.

[20] F.P. McCluskey, Y. Saadon, Z. Yao, A. Camacho, Cooling for electric aircraft motors, in: *2019 18th IEEE Intersociety Conference on Thermal and Thermomechanical Phenomena in Electronic Systems (ITherm)*, May 28, 2019, pp. 1134–1138. IEEE.

[21] F. Song, D. Ewing, C.Y. Ching, Heat transfer in the evaporator section of moderate-speed rotating heat pipes, *Int. J. Heat Mass Transfer*, 51 (7–8), April 1, 2008, 1542–1550.

[22] A. Faghri, S. Gogineni, S. Thomas, Vapor flow analysis of an axially rotating heat pipe, *Int. J. Heat Mass Transfer*, 36 (9), January 1, 1993, 2293–2303.

[23] M.E. Dehshali, M.A. Nazari, M.B. Shafii, Thermal performance of rotating closed-loop pulsating heat pipes: Experimental investigation and semi-empirical correlation, *Int. J. Thermal Sci.*, 123, January 1, 2018, 14–26.

[24] L. Fedoseyev, E.M. Pearce, inventors; Tesla Inc, assignee. Rotor assembly with heat pipe cooling system. United States patent US 9,331,552. May 3, 2016.

[25] J. Pyrhönen, P. Lindh, M. Polikarpova, E. Kurvinen, V. Naumanen, Heat-transfer improvements in an axial-flux permanent-magnet synchronous machine, *Appl. Thermal Eng.*, 76, February 5, 2015, 245–251.

[26] C. Rhebergen, B. Bilgin, A. Emadi, E. Rowan, J. Lo, Enhancement of electric motor thermal management through axial cooling methods: A materials approach, in *2015 IEEE Energy Conversion Congress and Exposition (ECCE)*, September 20, 2015, pp. 5682–5688. IEEE.

[27] W. Sixel, M. Liu, G. Nellis, B. Sarlioglu, Cooling of windings in electric machines via 3-D printed heat exchanger, *IEEE Trans. Ind. Appl*, 56 (5), May 27, 2020, 4718–4726.

[28] D.C. Deisenroth, R. Moradi, A.H. Shooshtari, F. Singer, A. Bar-Cohen, M. Ohadi, Review of heat exchangers enabled by polymer and polymer composite additive manufacturing, *Heat Transfer Eng.*, 39 (19), November 26, 2018, 1648–1664.

[29] S.A. Semidey, J.R. Mayor, Experimentation of an electric machine technology demonstrator incorporating direct winding heat exchangers, *IEEE Trans. Ind. Electron.*, 61 (10), January 30, 2014, 5771–5778.

[30] P. Lindh, I. Petrov, A. Jaatinen-Värri, A. Grönman, M. Martinez-Iturralde, M. Satrústegui, J. Pyrhönen, Direct liquid cooling method verified with an axial-flux permanent-magnet traction machine prototype, *IEEE Trans. Ind. Electron.*, 64 (8), March 15, 2017, 6086–6095.

[31] C. Wohlers, P. Juris, S. Kabelac, B. Ponick, Design and direct liquid cooling of tooth-coil windings, *Electric. Eng.*, 100 (4), December 2018, 2299–2308.

[32] B. Agostini, J.R. Thome, M. Fabbri, B. Michel, High heat flux two-phase cooling in silicon multimicrochannels, *IEEE Trans. Compon. Packag. Technol.*, 31 (3), May 28, 2008, 691–701.

[33] S.V. Garimella, C.B. Sobhan, Transport in microchannels — A critical review, *Annu. Rev. Heat Transfer*, 13, 2003, 1–50. doi: 10.1615/AnnualRevHeatTransfer.v13.30.

[34] J.R. Thome, Boiling in microchannels: A review of experiment and theory, *Int. J. Heat Fluid Flow*, 25 (2), April 1, 2004, 128–139.

[35] M.A. Quaiyum, Experimental investigation of automatic transmission fluid (ATF) in an air cooled minichannel heat exchanger, Master's thesis, University of Windsor, Ottawa, Canada, 2012.

[36] M.H. Kim, J. Pettersen, C.W. Bullard, Fundamental process and system design issues in CO_2 vapor compression systems, *Progress Energy Combust. Sci.*, 30 (2), January 1, 2004, 119–174.

[37] California Air Resources Board. CO_2 (R744) systems technology update, in: *Proceedings of Mobile Air Conditioning Summit*, Sacramento, CA, March 15–16, 2005.

[38] Daimler. From 2017: First vehicles with CO_2 air conditioning. Available at: https://www.daimler.com/sustainability/product/further-environmental-technologies/co2-air-conditioning-system.html (accessed January 4, 2018).

[39] K.J. Geisler, A. Bar-Cohen, *Encyclopedia of Thermal Packaging: Thermal Packaging Techniques, Dielectric Liquid Cooling of Immersed Components*, World Scientific Publishing Company, 2013.

[40] S. Klein, F. Alvardo, EES-Engineering equation solver: User's manual for Microsoft Windows operating systems, F-Chart Software, Madison, WI, 2015.

[41] P. Linstrom, W. Mallard, NIST Chemistry webbook; NIST standard reference database No. 69. NIST: Gaithersburg, MD, 2018.

[42] A. Bar-Cohen, F.L. Robinson, D.C. Deisenroth, Challenges and opportunities in Gen3 embedded cooling with high-quality microgap flow, in: *2018 International Conference on Electronics Packaging and iMAPS All Asia Conference (ICEP-IAAC)*, April 17, 2018, pp. K-1. IEEE.

[43] M. Rosu, P. Zhou, D. Lin, D.M. Ionel, M. Popescu, F. Blaabjerg, V. Rallabandi, D. Staton, *Multiphysics Simulation by Design for Electrical Machines, Power Electronics and Drives*, John Wiley & Sons, December 18, 2017.

[44] M. Popescu, J. Goss, D.A. Staton, D. Hawkins, Y.C. Chong, A. Boglietti, Electrical vehicles: Practical solutions for power traction motor systems, *IEEE Trans. Ind. Appl.*, 54 (3), January 12, 2018, 2751–2762.

[45] T. Burress, Electrical performance, reliability analysis, and characterization. presented at the US Dept. Vehicle Technologies Office Annu. Merit Rev., US Dept. Energy, June 2017.

[46] T. Burress, Benchmarking EV and HEV technologies. Oak Ridge National Laboratory. June 17, 2014.

[47] ARPA-E Program Listing Information. Available at: https://arpa-e.energy.gov/?q=program-listing.

[48] The Advanced Research Projects Agency — Energy (ARPA-E) is an organization within the Department of Energy (DOE) of USA. https://arpa-e.energy.gov.

[49] V. Jagdale, J. Tangudu, Topology optimized end winding for additively manufactured induction motor with distributed winding, SAE Technical Paper, September 20, 2016.

[50] A. Kallaste, T. Vaimann, A. Rassãlkin, Additive design possibilities of electrical machines, in: *2018 IEEE 59th International Scientific Conference on Power and Electrical Engineering of Riga Technical University (RTUCON)*, November 12, 2018, pp. 1–5. IEEE.

[51] W. Yin, A. Yakimov, H. Comanzo, O. Siclovan, P. Smigelski, J. Gao, K. Bodla, A. Duggal, Highly thermally conductive insulation for high power density electric machines, in: *2019 AIAA/IEEE Electric Aircraft Technologies Symposium (EATS)*, August 22, 2019, pp. 1–7. IEEE.

[52] E.W. Lemmon, M.O. McLinden, D.G. Friend, P. Linstrom, W. Mallard, NIST chemistry WebBook, NIST Standard Reference Database number 69. National Institute of Standards and Technology, Gaithersburg, MD, 2011.

[53] Electrical UD. Electronics Technical Team Roadmap. Available at: https://www.energy.gov/eere/vehicles/downloads/us-drive-electrical-and-electronics-technical-team-roadmap (accessed October 20, 2020).

[54] RTCA DO-160 — Section 4.0 — Temperature and altitude. Available at: https://do160.org/temperature-and-altitude/.

https://doi.org/10.1142/9789811279379_bmatter

Index

L

M

N

O

P

R

S

T

Printed in the USA
CPSIA information can be obtained
at www.ICGtesting.com
LVHW020723010224
770406LV00002B/6

9 789811 277931